AF531091

Digital Signal Integrity

Modeling and Simulation with Interconnects and Packages

Brian Young

Prentice Hall PTR

Upper Saddle River, NJ 07458

www.phptr.com

Library of Congress Cataloging-in-Publication Data
Young, Brian
Digital signal integrity: modeling and simulation with interconnects and packages / Brian Young.
p. cm.
Includes index.
ISBN 0-13-028904-3
1. Signal processing--Digital techniques--Computer simulation 2. Telecommunication systems--Reliability--Computer simulation. I. Title.
TK5102.9 .Y68 2000
621.382'2---dc21 00-056676
CIP

Editorial/Production Supervision: *Vincent Janoski*
Acquisitions Editor: *Bernard Goodwin*
Editorial Assistant: *Michelle Vincente*
Marketing Manager: *Julie Tiso*
Manufacturing Manager: *Alexis Heydt*
Cover Design Director: *Jerry Votta*
Composition: *Lori Hughes*

Published by Prentice-Hall PTR
Prentice-Hall, Inc.
Upper Saddle River, NJ 07458

Prentice Hall books are widely used by corporations and government agencies for training, marketing, and resale.

The publisher offers discounts on this book when ordered in bulk quantities.
For more information, contact Corporate Sales Department, phone: 800-382-3419;
fax: 201-236-7141; e-mail: corpsales@prenhall.com
or write: Prentice Hall PTR
Corporate Sales Department
One Lake Street
Upper Saddle River, NJ 07458

Printed in the United States of America.
10 9 8 7 6 5 4 3 2 1

ISBN 0-13-028904-3

Prentice-Hall International (UK) Limited, London
Prentice-Hall of Australia Pty. Limited, Sydney
Prentice-Hall Canada Inc., Toronto
Prentice-Hall Hispanoamericana, S.A., Mexico
Prentice-Hall of India Private Limited, New Delhi
Prentice-Hall of Japan, Inc., Tokyo
Pearson Education Asia Pte. Ltd.
Editora Prentice-Hall do Brasil, Ltda., Rio de Janeiro

CONTENTS

PREFACE

Effects of interconnects on the electrical performance of digital components, such as microprocessors, have historically been small enough to handle with simple rules of thumb. As clock rates, bus widths, and bus speeds have increased, packaging and interconnects have more importance and in some cases actually limit or define the system, where silicon performance is usually found to be the gating factor. This role reversal will become more common, and it may be that packaging and interconnects dominate electrical considerations at some point in the future as networks become more prominent.

The relatively recent growth of packaging and interconnects as signficant issues in electrical performance means that relatively few resources exist for learning and training. Much of it exists as scattered applications notes, many of which are quite useful but are sometimes somewhat dated (i.e., notes on ECL rather than CMOS) or are from the less accessible technical literature. Since many organizations are newly finding the need for expertise in the field, in-house experts may not be available to act as mentors.

This book represents my efforts at collecting and deriving the necessary material to support a career in digital signal integrity modeling and simulation. A huge part of such a job is package and interconnect modeling from electromagnetic simulation and/or measurements. By necessity, the book spans a broad spectrum of techniques, including electromagnetic simulation, transmission line theory, frequency-domain modeling, time-domain modeling, analog circuit simulation, digital signaling, and some architectural issues, to put it all in perspective. Such a broad technological reach makes for a very interesting and challenging job. Since I believe that the number of engineers working this area will need to increase dramatically to support the technological trends, I hope that this book will provide a sufficient set of tools to help engineers succeed in this field.

The goal of the book is to provide detailed introductory material that is self-consistent and self-contained. As such, there are very few references. There is nothing in the book that is new to the field, so technical credit must go to the innumerable contributors to the technical literature, application notes, and standards.

The book is organized to move gradually from broad, general topics to specific modeling techniques. Particular emphasis is placed on rigorous derivation and on multiconductor interconnects. Chapter 1 discusses the role of signal integrity in digital systems. Chapters 2 and 3 then cover issues in signaling and signal integrity. Chapters 4 through 7 cover detailed concepts in basic passive circuit components, with particular emphasis on multiconductor interconnects. One of the more difficult aspects of detailed simulations in signal integrity is the need to model multiconductor interconnects such as sockets, packages, edge connectors, and buses. Experimental characterization of interconnects is covered in chapter 8, where emphasis is on measurements of very small parasitics for high leadcount interconnects. Interconnect modeling is covered in chapters 9 and 10, where distinction is drawn between low-frequency lumped modeling and high-frequency wideband modeling. Because interconnects are often physically small, lumped modeling is often the optimal choice. Finally, chapter 11 provides extended coverage of signal integrity topics and represents advanced application of material and concepts from prior chapters.

The manuscript was typeset using LaTeX 2_ε running under Linux on a PC clone based on a Tyan motherboard with a Cyrix processor. The text was prepared using a custom text editor written in Tcl/Tk. Circuit simulations used Berkeley SPICE 3f4. The figures were prepared using Xfig. Symbolic manipulation used *Mathematica*™.[1]

1. Mathematica is a trademark of Wolfram Research, Inc.

I came to the digital packaging field from analog microwave packaging and numerical methods in electromagnetics. As such, I needed a kick start to get me going. In particular, thanks go to Aubrey Sparkman for many interesting discussions when I first entered the field. Later on I had the pleasure of working with Greg Pitner, who challenged me with many tough questions, so thanks to Greg. Thanks go to Joel Dietz for encouraging me to start writing the book. I would like to thank Francis Bostick and Robert Flake for letting me use an early version of the book to teach a graduate course on packaging at the University of Texas. Thanks go to Bruce Wagner for providing the line card from which a few of the photographs are taken. My early technical mentors deserve much credit, so thanks to my dad, Charles Young, and to Joe Owens. Writing a book is an enormous undertaking, so special thanks to my lovely wife, Monica Carlson, for sacrificing a big chunk of time over the many years required.

Brian Young
Austin, Texas

Chapter 1

DIGITAL SYSTEMS AND SIGNALING

Computing performance can be enhanced with a wide variety of techniques, including dense integration, fast circuit styles such as dynamic logic, advanced compilers, high-throughput architectures, wide instructions and data, parallel processing, and faster clock rates, to name a few. For maximum performance, all high-speed components of a system should be integrated onto a single die; however, technological difficulties (especially yields for large die) prevent the implementation of system-on-a-chip except for simple cases. A continuing problem then is interconnection of high-speed parts, and as component clock rates increase, the bandwidth of the interconnections must increase too.

The principal ways to increase interconnect bandwidth are to clock it faster or to increase the width, or both. The factors that complicate the design are the signaling standard, compatibility, power dissipation, and costs associated with packaging, printed circuit board (PCB) area, and silicon area. A comprehensive system design effort will trade these considerations off against architectural issues and related costs. To help support system design tradeoff optimization, modeling and simulation can be performed to estimate performance. The critical issues in accurate estimates are model availability, model accuracy, and appropriate model use.

1.1 Tradeoffs for Performance Enhancement

To increase the performance of a digital system, the basic approach is to make changes that enable faster clocking, more processing, and/or more data transfer per clock. A few of the factors at the disposal of the system designer are architectural tradeoffs, bus widths, bus speed, signaling standard, logic family, topology, and loading. These issues must be balanced to achieve the required performance at a suitable cost within the allowed time frame and within regulatory limits on radiated electromagnetic emissions.

1.1.1 Architecture

Systems are best partitioned so that high performance can be achieved using high-speed interconnects only where necessary. Because system architectures cannot be addressed in general terms, consider as an example a microprocessor system. Since high-speed memory is expensive, a memory hierarchy is required where small amounts of expensive, high-speed memory are used in conjunction with large amounts of inexpensive, low-speed memory. How the memory is partitioned and implemented strongly affects the interconnect speeds required and the overall system performance.

The fastest memory, called the *level 1*, or *L1*, cache, is available on the microprocessor itself, but in limited quantity, to both maximize the speed (smaller memories are faster) and to minimize the die size for least cost. Additional memory is supplied by the *level 2*, or *L2*, cache, which is constructed of static RAM (SRAM, essentially the same type as used for the microprocessor L1 cache but constructed using a process optimized for memories). However, the cost of SRAM is many times that of the cheapest memory available, capacitance-based dynamic RAM (DRAM), so the L2 cache is not sized to completely fulfill the memory needs of the system. The bulk of the memory requirements of the system are satisfied by the main memory composed of slow (relative to SRAM) DRAM. Even the main memory is insufficient, so virtual operating systems swap data between memory and disk storage, but a severe performance penalty results if too much swapping is performed.

The memory hierarchy can be organized in many ways, with four options depicted in Figure 1.1. A basic layout is shown in Figure 1.1a, where a microprocessor with its L1 cache and the main memory form the memory hierarchy. High-speed operation on the system bus is inhibited by the complex topologies of the nets needed to connect many memory devices and by their heavy loading. For the usual case where the main memory is slower than the system bus, performance can be improved by placing a small quantity of high-speed L2 cache memory on the system bus as shown in Figure 1.1b. For data stored in the cache, access can be achieved at the top speed of the system bus. However, the system bus is still complex with many devices connected to it, so the speed of the system bus is limited.

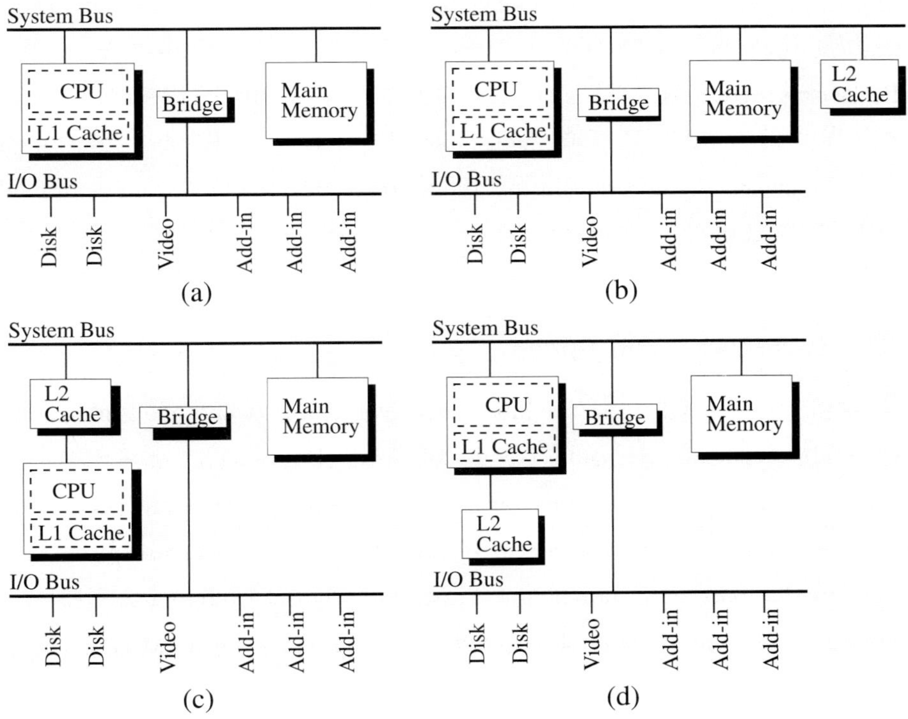

Figure 1.1. Four memory hierarchies for microprocessor-based systems: (a) no L2 cache, (b) L2 cache on system bus, (c) inline L2 cache, and (d) back-door cache.

Considerable performance can be gained by moving the L2 cache off of the slow system bus. One alternative, called an *inline cache*, is to place the cache between the microprocessor and the system bus, as shown in Figure 1.1d. The interconnect between the microprocessor and cache is very simple, just point-to-point connections, so very high clock speeds are possible. For system bus transactions not involving the cache, the cache controller simply passes data through to the slower system bus. Performance can be further enhanced by moving the interface to the L2 onto the microprocessor itself, as shown in Figure 1.1c. The so-called *backdoor cache* enables the microprocessor to access L2 data on a fast dedicated interface simultaneous to executing transactions on the system bus.

The microprocessor example illustrates one of the principal methods for enhancing system performance: system partitioning to enable faster interconnects between critical components. For microprocessors, the critical off-chip interface is to the L2 cache. Multiprocessor systems may also find a need for such dedicated interfaces between processors. Added performance due to dedicated buses has a price: higher pin counts, more complex routing on the system card, and more noise from the additional switching by the extra output buffers.

1.1.2 Bus Width and Speed

Another direct way to increase the information carrying capacity of an interconnect is to simply make the interconnect wider or faster. Doubling the bus width gives an immediate doubling of the bus bandwidth; however, wider buses require more package pins and more PCB real estate, and the extra drivers use more silicon real estate, dissipate more power, and create more noise. Doubling the bus speed (if possible) also doubles the bus bandwidth while using no additional physical resources, but designs are much more difficult, noise and radiation can be worse, and power dissipation increases. Wider and faster buses create switching noise that can only be minimally damped with the addition of bypass capacitance, so buses must be carefully designed and executed to ensure reliable operation.

The evolution of the personal computer expansion bus illustrates the use of higher clocks and wider buses for higher bus bandwidth. The original ISA bus was 16 bits wide and clocked at 8.3MHz for a peak bus bandwidth of 16.6MB/s. To enable higher throughput, the PCI bus was introduced at 32 bits wide and 33.3MHz for a peak bandwidth of 133MB/s. Further throughput enhancements included increases in bus width to 64 bits and in speed to 66.6MHz speed for 533MB/s peak bandwidth. Each jump in bus width and speed requires ever-tighter control of bus topology, layout, driver and receiver design, connector and package selection, and timing.

1.1.3 Power Distribution

Higher levels of integration enable integrated circuits to process more data per clock, but more transistors also means more power is required in less time. Package parasitics, while very small, can produce large voltage swings relative to the power supply levels. Reliable operation requires low levels of noise, so packaging parasitics must be minimized. However, high-speed circuits generate considerable levels of high-frequency noise due to the sharp edges of the digital waveforms, and package parasitics help block their access to the system level where electromagnetic interference (EMI) is more easily launched due to the larger circuit dimensions. High-speed circuits create a tradeoff between power supply ripple and EMI suppression. Bypass capacitance can help smooth the ripple and suppress EMI.

1.1.4 Topology and Loading

Topology plays a critical role in determining the maximum speed at which an interconnect can be clocked. The fastest interconnect is a unidirectional point-to-point connection between two devices over impedance-controlled transmission lines with a matching scheme to absorb any reflections that may be generated by the source or load. Any deviation from this setup will lower the maximum attainable clock speed. In particular, when three or more devices must be connected (called a *multidrop net*), impedance control is compromised due to reflections generated at the impedance discontinuity where the extra components must connect.

Multidrop nets can be configured in a wide variety of ways. Two major methods are the *daisy chain route* and the *star cluster*. For the daisy chain, one long route passes near each component, which is then connected with a short stub. The principal needs are to keep the route and the stubs as short as possible. For the star cluster, each device is connected to a central hub with equal-length transmission lines, and the key design criteria is that the lengths and loadings should be very similar.

After topology, loading is the next most critical aspect regulating clock speeds on interconnects. Each input of a device adds capacitive loading to the net, so large numbers of devices (as in memory subsytems) result in large loadings and slow speeds due to RC charging times. Light loading is required for high speed. Capacitive loading is mostly due to packaging, electrostatic discharge (ESD) protection diodes, and output buffers in their high-impedance state (for I/O cells). Since ESD structures and packaging are hard to avoid, highest speed is attainable for unidirectional nets (such as address buses).

Terminations can be used to dramatically enhance signal quality for higher speeds by dissipating unwanted reflections. Series termination can slow the signaling by increasing RC charging times, while parallel termination increases power dissipation. Both types of termination increase parts count, cost, and PCB board complexity and area.

Architectural needs for high-speed signaling can eventually dictate the use of point-to-point interconnections. When several parts must be interconnected with point-to-point interconnects, one or more crossbars are required to establish the connections. Crossbars can require large numbers of pins, leading to expensive packaging.

1.1.5 Logic Levels and Signaling

There is a well known hierarchy of techniques for improving signal integrity on interconnects to enable higher clock rates. In rough order of performance-enhancing effect, these techniques are

1. Topology
2. Termination
3. Reduced voltage swing
4. Slew rate control
5. Differential receiver with a reference voltage
6. Fully differential drivers and receivers

These techniques can be used singly or in combination to balance performance, power dissipation, and cost.

At the low end of performance is standard CMOS or TTL signaling utilizing large voltage swings, no termination, and generally no slew rate control. Used on relatively slow and heavily loaded buses, CMOS and TTL drivers are usually optimized to switch large capacitive loads. However, power dissipation is relatively small.

At the high end of performance is emitter-coupled logic (ECL) utilizing reduced voltage swing, controlled edge rates, termination, and fully differential signaling. The price paid for using ECL is high power dissipation and larger pin counts. These features also appear in the Low-Voltage Differential Signaling (LVDS) standard, which provides for more convenient voltages and lower power dissipation.

A wide variety of intermediate signaling standards are possible, a few of which are discussed in section 1.2. In addition, mixed standards are also utilized. For example, in many systems the clock is transmitted fully differential to minimize clock skew, but the signals are not differential (termed *single-ended*).

1.1.6 Power Dissipation

The level of power dissipation can become a limiting factor in interconnect design. Consider an unterminated clock, which must charge and then discharge its capacitive load every cycle with period T. The average power dissipation per clock period is then

$$\begin{aligned} P_{\text{avg}} &= \frac{\varepsilon}{T} \\ &= \frac{2\left(\frac{1}{2}CV^2\right)}{T} \\ &= CV^2 f, \end{aligned} \tag{1.1}$$

where $f = 1/T$ is the clock frequency, and V is the power supply voltage. For a data line with transitions only on the rising or falling edge of the clock, the average power dissipation is $\frac{1}{2}CV^2 f$. For example, a 72-bit bus operating at 33MHz with a 3.3V interface and a maximum load of 50pF per bit results in a maximum average power dissipation of 0.65W.

Because of its square dependence, power is most easily reduced through lower interface voltages, which also helps reduce noise. After that, a linear reduction with lower capacitance is possible, although a lower capacitance means less loading and a consequent reduction in fan out. Frequency is generally increasing to provide higher performance, and this is usually the cause of efforts to reduce power through voltage and capacitance reduction.

The power dissipation for parallel-terminated nets is

$$P_{\text{avg}} = \frac{V^2}{R + Z_o},$$

where R is the driver impedance and Z_o is the characteristic impedance of the transmission line. Power dissipation can also reach this level on long unterminated nets when the round trip delay equals half the clock period. With the square dependence on voltage, again reduction in voltage is the most effective means of reducing power dissipation.

Differential signaling with terminated current-mode drivers can minimize the dependence of power dissipation on frequency. In contrast, power dissipation for unterminated voltage-mode signaling increases linearly with frequency. For minimization of power dissipation, unterminated voltage-mode signaling is preferred at low frequencies, while terminated current-mode signaling is preferred at high frequencies. This power relationship is sketched in Figure 1.2.

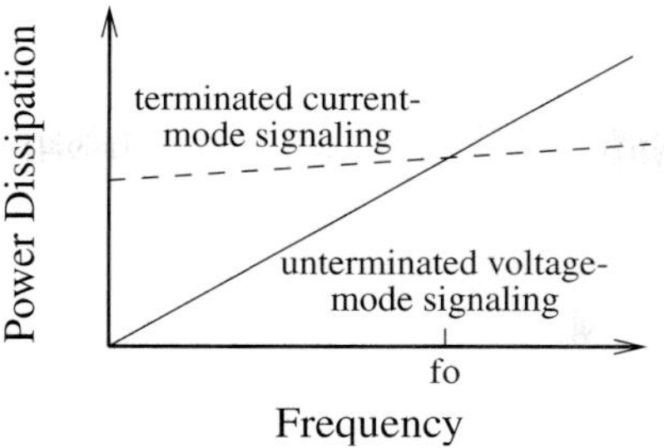

Figure 1.2. Frequency dependence of power dissipation for unterminated voltage-mode signaling and terminated current-mode signaling.

1.2 Signaling Standards and Logic Families

Successful communication requires that the driver and receiver agree on what voltage levels constitute logic low and logic high. For voltage-mode signaling such as TTL, the driver directly sets the output voltage. For current-mode signaling, the driver sets the current level, and the voltage drop across a termination resistor at the receiver is detected. In either case, it is sufficient to discuss signaling in terms of the voltages set by the driver and the voltages detected by the receiver.

A signaling standard specifies the voltages for logic high and logic low for both the driver and receiver. A noise margin is built into the specification. In addition, termination requirements are specified. A wide variety of signaling standards exist to address different performance and cost design points. A few of the common standards are listed in Tables 1.1 and 1.2, where they appear roughly in order of the maximum speed attainable. Higher speeds are obtained through termination, reduced voltage swing, slew rate control, and differential receivers. Parallel termination and differential receivers consume static power, so as high-speed options are added, power consumption increases.

A logic family is a specific implementation of a signaling standard with additional requirements on timing-related performance, including slew rate (rise/fall times), setup and hold times, and fan out. Faster circuits have greater power dissipation

Table 1.1. Comparison of driver specifications for several common signaling standards.

Standard	V_{DDQ}, V	V_{OL}, V		V_{OH}, V		$V_{differential}$, V		V_{offset}, V		Termination	Driver[a]
		Min	Max	Min	Max	Min	Max	Min	Max		
TTL	$5 \pm 10\%$		0.4	2.4						None	PP
LVTTL[b]	$3.3 \pm 10\%$		0.4	2.4						None	PP
GTL[c]	$1.2 \pm 5\%$		0.4	V_{DDQ}-0.4	V_{DDQ}					None	PP
GTL			0.4							R to $1.2 \pm 5\%$ V[d]	OD
HSTL[e]	1.5 ± 0.1		0.4	$V_{DDQ} - 0.4$						[f]	PP
ECL[g]	$-5.2 \mp 5\%$	-1.810	-1.620	-1.025	-0.880					50Ω to -2V	CM
PECL[h]	$5.0 \pm 5\%$	3.190	3.380	3.975	4.120					50Ω	CM
LVPECL[i]	$3.3 \pm 5\%$	1.490	1.680	2.275	2.420					50Ω	CM
LVDS[j]		0.925			1.474	0.250	0.400	1.125	1.275	50Ω	CM

a. PP - push-pull; OD - open-drain (active pull-down); CM - current-mode

b. JEDEC Standard, "Interface standard for nominal 3V/3.3V supply digital integrated circuits," Electronic Industries Association, JESD8-A, June 1994.

c. JEDEC Standard, "Gunning transceiver logic (GTL) low-level, high-speed interface standard for digital integrated circuits," Electronic Industries Association, JESD8-3. The standard defines unterminated and terminated implementations. GTL is patented.

d. The resistor value is chosen to match the impedance of the system.

e. EIA/JEDEC Standard, "High-speed transceiver logic (HSTL) a 1.5V output buffer supply voltage-based interface standard for digital integrated circuits," Electronic Industries Association, EIA/JEDEC8-6, August 1995.

f. HSTL defines driver types I–IV for four cases of unterminated, source terminated, far-end terminated, and far-end plus near-end parallel terminated, respectively. All must satisfy the same output specifications.

g. JEDEC Standard, "Standard for operating voltages and interface levels for low voltage emitter-coupled logic (ECL) integrated circuits," Electronic Industries Association, JESD8-2, March 1993.

h. PECL (Positive ECL) is ECL run with a +5V supply. See Cleon Petty and Todd Pearson, "Designing with PECL (ECL at +5.0V)," Motorola Application Note AN1406, Feb. 1998.

i. LVPECL (Low-Voltage PECL) is ECL run with a +3.3V supply.

j. IEEE Standard, "Standard for low voltage differential signals for scalable coherent interface (SCI)," IEEE P1596.3/D1.3, Nov. 1995.

Table 1.2. Comparison of receiver specifications for several common signaling standards.

Standard	V_{DDQ}, V	V_{IL}, V		V_{IH}, V		V_{range}, V		$V_{\mathrm{differential}}$, V		$V_{\mathrm{common\ mode}}$, V		Receiver[a]
		Min	Max	Min	Max	Min	Max	Min	Max	Min	Max	
TTL	$5 \pm 10\%$		0.8	2.0								SE
LVTTL	$3.3 \pm 10\%$	-0.3	0.8	2.0	$V_{\mathrm{DD}} - 0.3$							SE
GTL	$1.2 \pm 5\%$[b]		V_{REF}[c]-0.05	V_{REF}+0.05								PD
HSTL[d]	1.5 ± 0.1	-0.3	V_{REF}[e] $- 0.1$	$V_{\mathrm{REF}} + 0.1$	$V_{\mathrm{DDQ}} + 0.3$							PD
HSTL	1.5 ± 0.1					-0.3	V_{DDQ}+ 0.3	0.2	V_{DDQ}+ 0.6	0.68	0.9	D
ECL	$-5.2 \mp 5\%$	-1.810	-1.475	-1.165	-0.880							D
PECL	$5.0 \pm 5\%$	3.190	3.525	3.835	4.120							D
LVPECL	$3.3 \pm 5\%$	1.490	1.825	2.135	2.420							D
LVDS		0.9	1.5	1.1	1.7			0.200	0.800			D

a. SE - single-ended; D - differential; PD - pseudo-differential/comparitor

b. Specified for unterminated case only.

c. $V_{\mathrm{REF}} = 2/3 V_{\mathrm{DDQ}}$

d. HSTL specifies both pseudo-differential and differential receivers.

e. $V_{\mathrm{REF}} = V_{\mathrm{DDQ}}/2$

and generate more noise, so generally the slowest family that meets timing should be used.

For example, the original TTL standard has been implemented several times to achieve higher speed or lower power dissipation, where the goal is to improve the speed-power product. The 74XX family founded the standard. In the Schottky TTL (74SXX) family, the transistors are augmented with integral Schottky diodes across the base-collector junction to prevent them from saturating. Faster switching speeds result since the junctions do not have to be cleared of excess charge before the transistors can turn off. In addition, resistor values are reduced to achieve smaller RC time constants at the expense of higher power dissipation. Other TTL families include low-power Schottky (74LSXX), advanced Schottky (75ASXX), Fast (74FXX), and advanced low-power Schottky (74ALSXX).

Two circuits using different signaling standards require interface circuits to enable communication. Two circuits using different families of the same signaling standard can communicate, but the interface speed is limited to that of the slower family.

Reliable signaling is achieved when the signaling standard specifications are met under full, adverse noise conditions as well as device variations due both to process deviations in manufacturing and to changes over the operating temperature range.

1.2.1 Noise Margins

Noise margins are defined by the signaling standard. For a given standard, assume that the desired voltage levels for logic high and low are V_H and V_L, respectively. Also assume, for convenience of discussion, that $V_H > V_L$, although signaling could easily be worked the other way around.

For logic high, the driver is specified to reach and exceed a minimum output voltage, $V_{OH} \leq V_H$. To successfully communicate, then the receiver must accept as logic high any voltage above $V_{IH} < V_{OH}$. The noise margin is then $\text{NM}_H = V_{OH} - V_{IH}$.

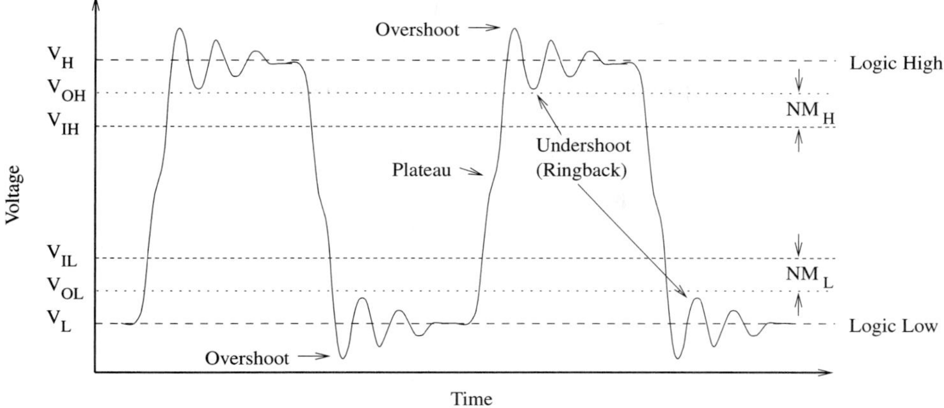

Figure 1.3. Voltage levels and specifications for digital signaling.

For logic low, the driver is specified to fall and hold below a maximum output voltage, $V_{OL} \geq V_L$. The receiver must accept as logic low any voltage below $V_{IL} > V_{OL}$, so the noise margin is $\text{NM}_L = V_{IL} - V_{OL}$.

For example, the levels for TTL are $V_{IL} \leq 0.8V$, $V_{IH} \geq 2V$, $V_{OL} \leq 0.4V$, and $V_{OH} \geq 2.4V$. Therefore, the noise margins for TTL are at least $0.4V$.

The logic family specifies how quickly the voltages must transition between valid logic levels. Fundamentally, all digital signals are analog in nature. The logic voltage level specifications are shown in Figure 1.3 along with a representative noisy waveform showing some common defects and their names. If the ringback violates the signaling specification, then additional time must be allowed so that the waveform can settle within the specification. Settling time reduces the maximum clock rate of the system.

1.2.2 Setup and Hold Times

Receivers sample the waveform, and this process takes time. Within a logic family, receiver specifications include limits on the timing and duration of input waveforms in addition to V_{IH} and V_{IL}. For clocked systems, the two primary specification are the *setup* and *hold* times.

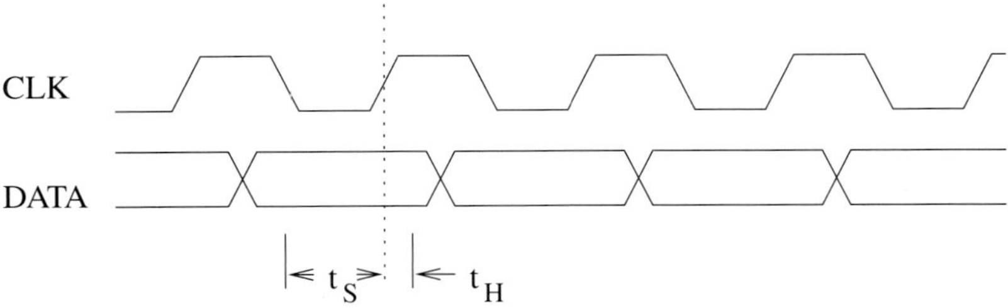

Figure 1.4. Setup and hold times define when the waveform must meet input voltage specifications at the receiver.

The setup time is the time the waveform must meet input voltage specifications (in other words, the waveform is settled) before the clock edge signals the receiver to sample. The hold time is the time after the clock edge during which the waveform must still meet the input voltage specifications. The setup and hold times are graphically represented in Figure 1.4, where idealized time-dependent voltage waveforms represent the digital signals. The speed of an interface is strongly affected by setup and hold times and driver slew rate (see section 2.8.1).

1.2.3 Drivers

From a signal integrity point of view, driver designs fall into two basic categories: *push-pull* and *current-steering*. The push-pull driver uses transistors to connect the output to either of two voltage levels to indicate the logic level. The current-steering driver uses two different levels of current to indicate logic levels, where the current is converted to voltage at a termination resistor to enable detection. From the receiver point of view, the driver type is immaterial since voltage levels are what need to be detected. Push-pull drivers can be very simple and generally dissipate low power. Current-steering drivers typically dissipate higher power but offer the best signal integrity.

Drivers are analog circuits that must provide shaped outputs under varying load conditions. Because transistor I-V curves are nonlinear, the output impedance is also nonlinear. High-accuracy signal integrity simulations require either a transistor-level driver model or I-V curves, such as those used in IBIS (Input/Output Buffer Interface Specification), to capture the nonlinear effects. The most important nonlinear effect is the increase in driver impedance and edge rate as the power supply droops, and an accurate simulation of this relationship is required to obtain accurate estimates of simultaneous switching noise (SSN).

Push-Pull Drivers

The basic configuration for a CMOS push-pull (or totem-pole) driver is shown in Figure 1.5, where a PMOS enhancement MOSFET (PFET) pulls the output up to logic high V_{DD} while an NMOS enhancement MOSFET (NFET) pulls down to logic low V_{SS}. The gate signals A and $\overline{A}$ may be tied together (in which case the driver is a basic CMOS inverter) or driven separately to manage crowbar current. When both A and $\overline{A}$ are high, then the pull-up transistor is off and the pull-down transistor is on, so the output is pulled towards V_{SS}. Conversely, if A and $\overline{A}$ are low, then the output is pulled towards V_{DD}. The CMOS totem-pole driver as discussed here is a full-swing driver from V_{SS} to V_{DD}, and while such a large voltage range provides for plenty of noise margin at the receiver, it also creates considerable noise.

Ideally, the pull-up and pull-down transistors are operated out of phase so that

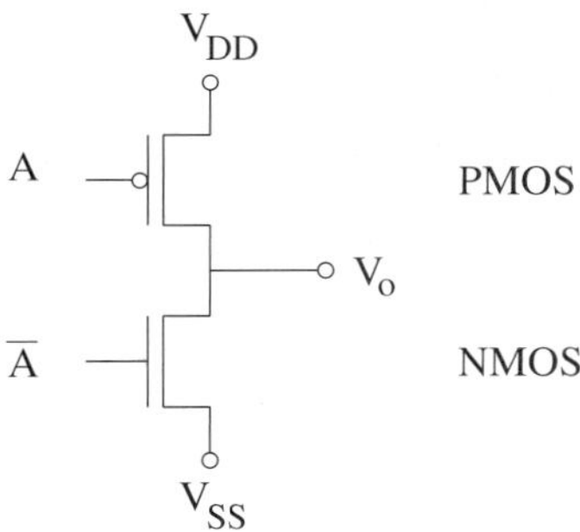

Figure 1.5. Schematic of a push-pull driver implemented in CMOS.

one is off while the other is on; however, to obtain maximum speed, the pull-up and pull-down transistors are operated simultaneously, so inevitably there is a small interval when both transistors are on so that a low-impedance path exists directly from the power supply to ground. The current through this path, called *crowbar current*, *feedthrough current*, or *overlap current*, performs no logic function and represents wasted power and extra noise.

Since the driver itself consists of two transistors in series where one is always off (except during transition), the driver dissipates very little power (set by the current leakage of the off transistor) while holding a logic level. For capacitive loads, then the totem-pole driver dissipates very little power once the capacitor is charged. In other words, the power dissipation is dominated by that required to charge and discharge the load capacitance. In this configuration, totem-pole drivers are said to have no static power dissipation.

If additional circuitry is added to the basic gate, both pull-up and pull-down transistors can be turned off so that the impedance looking into the output is very high. The high-impedance state is useful for implementing buses where multiple drivers are connected simultaneously to the same circuit. Multiple connections to a bus create the potential for a low-impedance path across the power supply. Just as crowbar current can exist in a single totem-pole driver, buses can create the same problem if one driver tries to pull the line high while another tries to pull the line low, as illustrated in Figure 1.6. This situation is called *bus contention*, and it dissipates excess power with the threat of component burnout while causing a logical fault. Bus contention is avoided by careful control of timing.

The output impedance of the totem-pole driver varies depending on which transistor is active. For low-to-high transitions, the pull-up transistor does the work, and its I-V characteristics plus any series resistance sets the output impedance. Similarly, for high-to-low transitions, the pull-down transistor plus any series resistance sets the output impedance.

Depending on the logic state, the push-pull driver utilizes different paths to

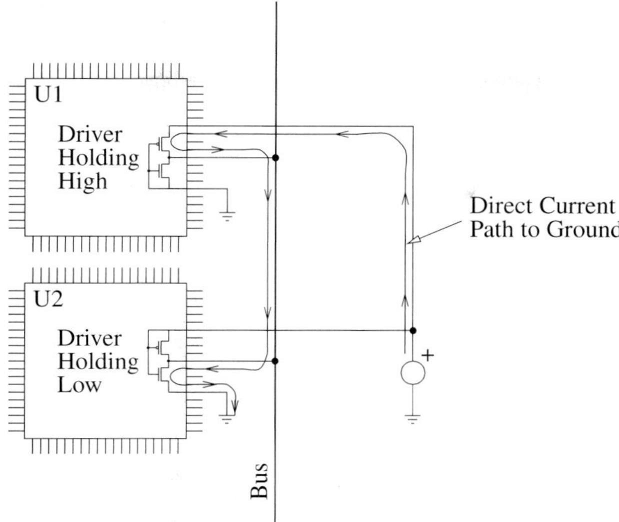

Figure 1.6. Bus contention creates a high-current path that dissipates excess power and can cause component burnout.

provide current to the load. As illustrated in Figure 1.7, the logic high state uses the signal line and V_{DD}. The pull-up transistor discharges any capacitance to V_{DD} while providing a path to charge capacitance to V_{SS}. Similarly, the logic low state uses the signal line and V_{SS}, where the pull-down transistor charges capacitance to V_{DD} while discharging capacitance to V_{SS}. The crowbar current uses only V_{DD} and V_{SS} and performs no logic function since it does not help to charge or discharge any load capacitance.

To model the switching characteristics of the push-pull driver, all three legs of the path must be present in the model: V_{DD}, signal, and V_{SS}. Assuming that V_{SS} is taken as the reference, then a four-port model is required to provide V_{DD} and signal connections at the driver and at the load. When modeling the interconnect with lumped elements or general multiport networks, the port requirements are easily met. For transmission line modeling, many simulators provide only a two-port

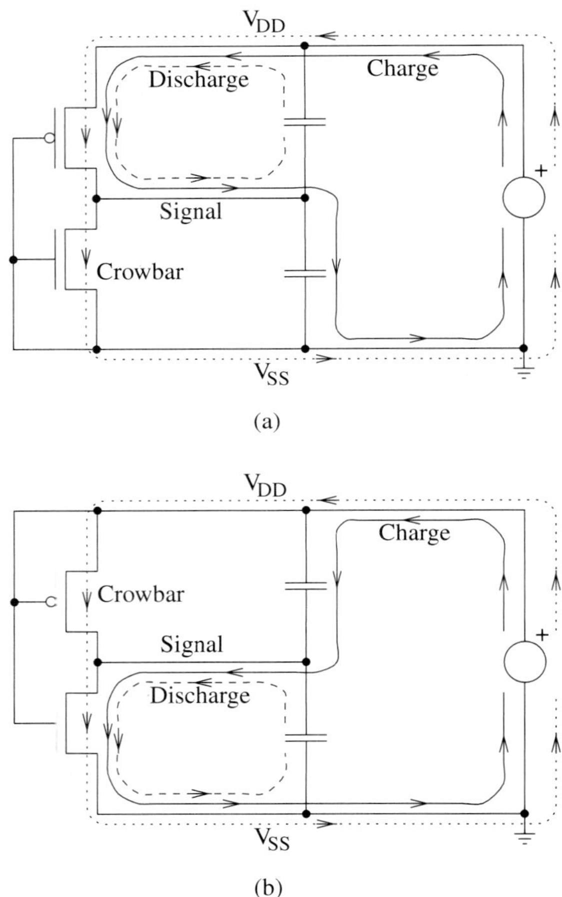

Figure 1.7. Current paths during signaling with totem-pole drivers in CMOS with rail-to-rail voltage swings: (a) logic high, (b) logic low.

transmission line model, which is insufficient without several simplifying assumptions.

To use a two-port transmission line model with a push-pull driver, two conditions must be met. First, the pull-up and pull-down output impedances must be equal. Second, infinite bypass capacitance must exist both at the driver and at the load so that V_{DD} can be treated as a broadband AC ground. In other words, power distribution is ideal with no power supply disturbances. While stringent, these conditions are often sufficiently met when using high-speed drivers, which must match pull-up and pull-down output impedances for best performance, and when just a few signals are switching so that power distribution perturbations are minimal. When a significant fraction of a device's drivers are switching, simultaneous switching noise will significantly disturb the power supply levels, so two-port transmission line models are not appropriate for this case.

Open-Drain and Open-Collector Drivers

One transistor of a push-pull driver can be replaced with a resistor so that the power supply itself pulls the output to a logic state. Due to the much lower mobility of holes compared to electrons, the PFET transistor in a CMOS push-pull driver is approximately three times larger than the NFET transistor, so it is advantageous to replace the pull-up PMOS rather than the pull-down NFET transistor to save the additional area and capacitance.

A typical open-drain CMOS driver setup is shown in Figure 1.8a, where the power supply pulls the line to V_{tt} through the resistor when the NFET is off. The resistor also acts as a termination to improve signal integrity. A good location for the resistor is at the receiver, but bus topology may dictate placement at another location, or even splitting the resistor into multiple locations.

Open-drain and open-collector drivers are convenient for creating wired-OR buses as shown in Figure 1.8b. When all drivers are off (all NFETs off), the bus is pulled high through the pull-up resistors. Any driver can pull the bus low by

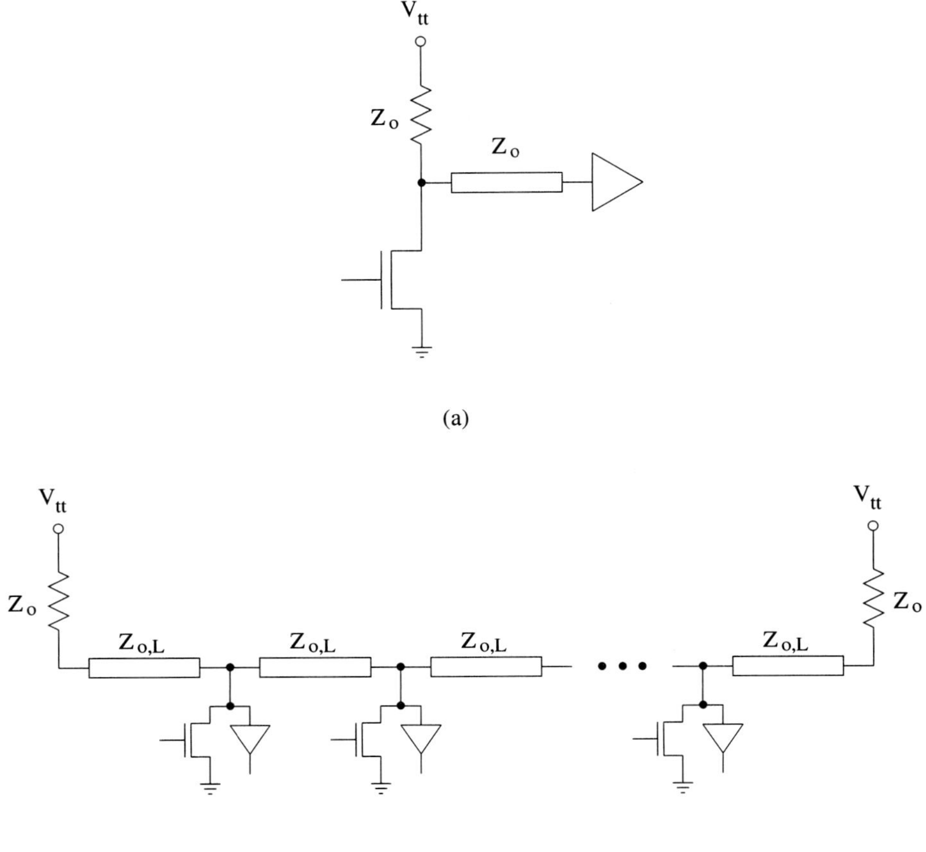

Figure 1.8. Open-drain drivers used to (a) drive a single line, or (b) construct a wired-OR bus.

turning on its driver (one NFET on), giving the bus the logical OR function. The driver will launch a wave in both directions, due to symmetry, and each wave will propagate to the ends of the bus where the waves are terminated by the pull-up resistors. The effective impedance seen by the driver is $Z_o/2$. The transmission lines forming the bus are loaded by the capacitance of the drivers and should take this loading into account (see chapter 2, section 2.6).

The advantages of open-drain/open-collector design include lower output capacitance, smaller driver area on silicon, adjustable termination impedance, and the ability to create a wired-OR bus. The main disadvantage is the static dissipation of DC power.

Current-Steering Drivers

A couple of configurations for the current-steering (or current-mode) driver are shown in Figure 1.9. There are two hallmark features of a current-steering driver: a constant-current source and parallel routing paths for the current. The driver operates by routing the current through one leg or the other so that the current drawn from the power supply is almost constant, generating little noise due to small dI/dt fluctuations. There are two complementary outputs, so the current-steering driver is inherently differential. Circuit board routing is by differential pair, with characteristic impedance $Z_{oo} = R_L$, to a fully differential receiver.

The driver in Figure 1.9a uses two identical NFETs in an open-drain configuration. The output voltage is developed across the load resistors, R_L, and is adjustable by the termination voltage, V_{tt}. Symmetry is critical to balance the current drawn down each leg. When V_{in} is high, Q_1 is on and Q_2 is off, and the current routes through the left-hand leg to produce a voltage drop of IR_L across the termination resistor. While Q_1 pulls $\overline{V_{\text{out}}}$ down to $V_{tt} - IR_L$, the power supply pulls V_{out} up to V_{tt}. A differential receiver detects the difference in voltage across the outputs, IR_L. Note that this configuration develops a sizable common-mode voltage of $V_{tt} - IR_L/2$ at the input of the receiver, leaving less of the receiver's common-mode range for noise cancellation.

The driver in Figure 1.9b uses four FETs to route the current in opposite directions through the single termination resistor. When V_{in} is low, then Q_3 and Q_2 are on while Q_1 and Q_4 are off, so the current routes through the load from top

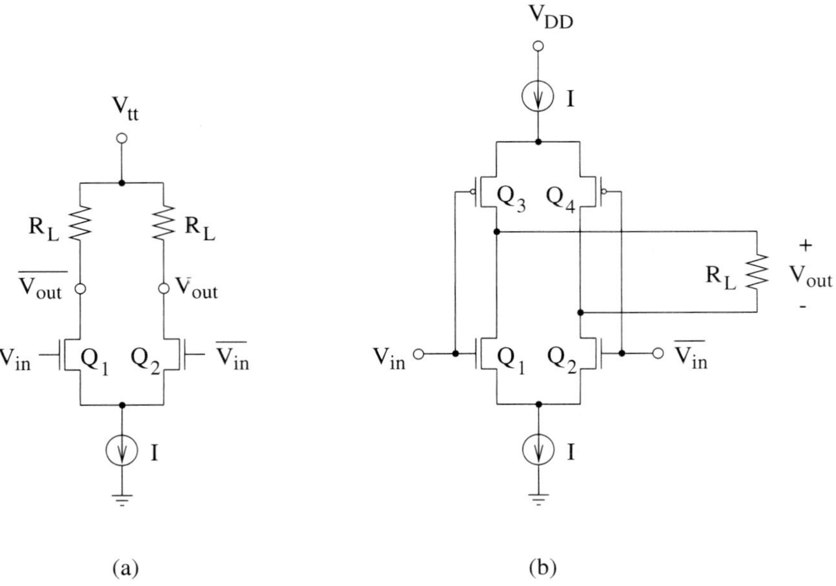

Figure 1.9. Two configurations for current-steering drivers: (a) open-drain, (b) fully differential.

to bottom, developing a voltage $+IR_L$. When V_{in} is high, then Q_1 and Q_4 are on while Q_3 and Q_2 are off, so the current routes through the load from bottom to top, developing a voltage $-IR_L$. A differential receiver detects the voltage developed across the load resistor.

In both designs, logic high is a differential voltage of $+IR_L$, while logic low is $-IR_L$. The differential voltage swing is then $\pm IR_L$.

Power Dissipation Revisited

The average power dissipation for a clock-like signal is shown in (1.1) to be $P_{\text{avg}} = CV^2 f$ using energy arguments. For a simple switch model of the totem-pole driver with a capacitive load, the same result can be found by explicitly integrating the instantaneous power

The power supply charges the capacitor through the driver impedance holding high for half the clock cycle. In the second half of the cycle, the power supply provides no power since the capacitor discharges through the driver impedance holding low. While charging, the voltage across the capacitor is

$$v_o(t) = V\left(1 - e^{-t/RC}\right),$$

so the instantaneous power drawn from the supply is

$$p(t) = Vi(t) = VC\frac{dv_o}{dt} = \frac{V^2}{R}e^{-t/RC}.$$

Integrating the power over a half a cycle and dividing by the full period finds that

$$\begin{aligned} P_{\text{avg}} &= \frac{1}{T}\int_0^{T/2} \frac{V^2}{R}e^{-t/RC}dt \\ &= CV^2 f \end{aligned}$$

for $T \gg RC$ and $f = 1/T$.

1.2.4 Linear Driver Modeling

When the effects of power supply perturbations are negligible, linear driver models often provide sufficient accuracy for significant work in signal integrity simulations. Any nonlinear circuit can be linearized and represented as a Thevenin equivalent circuit. To model a driver with equal pull-up and pull-down output impedances as a Thevenin equivalent, the Thevenin source voltage is assigned the driver's output voltage swing and slew rate, while the Thevenin impedance is assigned the driver's output impedance. For example, if a driver can swing a given load from 0V to 5V in 2ns, then the Thevenin source voltage is assigned a linear independent voltage source swinging from 0V to 5V in 2ns. The Thevenin impedance is a resistance equal to the output impedance of the driver at a strategic point, such as halfway through transition. The resulting driver model is shown in Figure 1.10.

The voltage swing is generally well known, especially for CMOS, which swings from rail to rail. The appropriate edge rate depends on the loading and may need

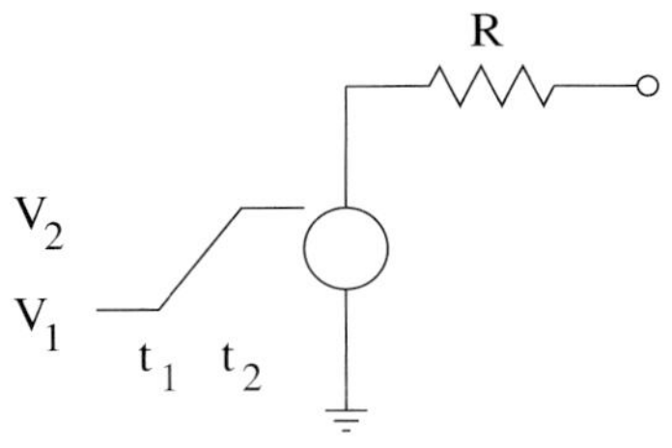

Figure 1.10. Thevenin equivalent circuit of a nonlinear driver having equal pull-up and pull-down output impedance under the condition of no power supply variation.

modification as the load changes. A given driver may have substantially different edge rates when driving a long transmission line versus a large capacitive load.

1.2.5 Receivers

The primary function of the receiver is to detect the logic level of an analog waveform in the presence of noise. The receiver may also need to adjust voltage levels from the external interface to the internal core logic. A common problem here is adapting full-swing 5V or 3.3V CMOS levels to core voltages of less than 2V, a situation that creates reliability issues for gate oxides. Receivers are usually integrated with ESD structures and may be integrated with drivers to form I/O buffers. Receivers may need to include JTAG boundary scan circuits[1] to facilitate board-level testing. Finally, receivers must perform all of these duties while meeting specifications for delay, power consumption, and logic threshold variation, which contributes to skew.

From a signal integrity point of view, receivers can be classified as either *single-ended*, *differential*, or *pseudo-differential*. For single-ended receivers, the input voltage is compared to local ground to determine the output state. Differential receivers have two inputs that are compared using a differential amplifier to determine output

1. Boundary scan circuits added to components enable board-level testing of complex systems at low cost. Work by the Joint European Test Action Group (JETAG) and the Joint Test Action Group (JTAG) in North America led to a procedure standardized by IEEE Std 1149.1-1990, IEEE Standard Access Port and Boundary-Scan Architecture.

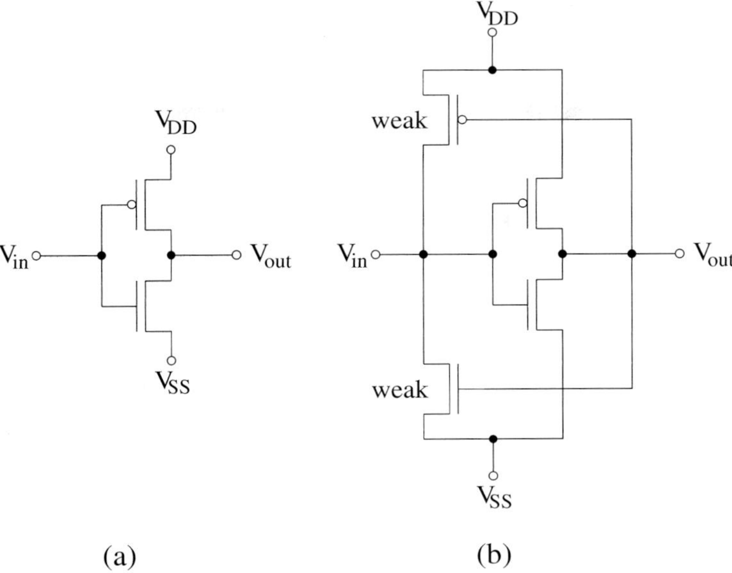

Figure 1.11. CMOS inverter-based single-ended receivers: (a) simple inverter, (b) inverter with feedback.

state. A pseudo-differential receiver has a fully differential receiver with one input tied to a reference voltage.

Single-ended receivers are the simplest to implement, requiring fewest transistors and least silicon area, with low power consumption. However, these receivers are sensitive to common-mode noise such as EMI and SSN. Inverters are a common method of implementation, and Figure 1.11 shows a CMOS inverter plus an implementation with feedback, which consists of a weak inverter that helps pull the input to the power rails to limit power consumption and to avoid metastability. Feedback slows the receiver by fighting the change in logic transition. Inverters introduce considerable skew because the logic trip point is difficult to control and is dependent on process and temperature variations.

Differential receivers accept two out-of-phase inputs and subtracts them to obtain a clean waveform for level detection, and a couple of topologies are shown in

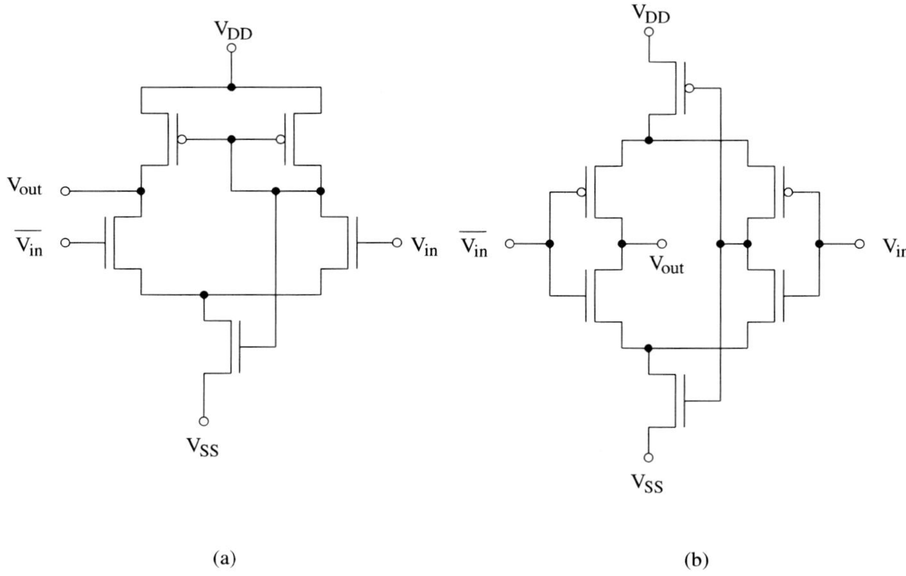

Figure 1.12. Two circuit topologies for implementing differential receivers.

Figure 1.12. Many sources of noise inject a common noise waveform onto all signal lines, so the ability of differential receivers to reject common-mode noise allows differential systems to operate to much higher frequencies than single-ended systems. In addition, differential current-mode drivers minimize the amount of noise injected into the system in the first place. Differential receivers also have more stable trip points for less skew. Differential receivers require biasing currents that are always on, so power dissipation is relatively high. Combined with the fact that differential signaling doubles the pin count compared to single-ended signaling, the costs of differential can be high. For this reason, differential signaling is often used on clocks to reduce jitter while remaining signals use single-ended drivers and receivers to minimize power dissipation and costs. However, the number of power and ground connections can be reduced with differential because SSN is far less of a concern.

Pseudo-differential signaling is a cross between single-ended and differential sig-

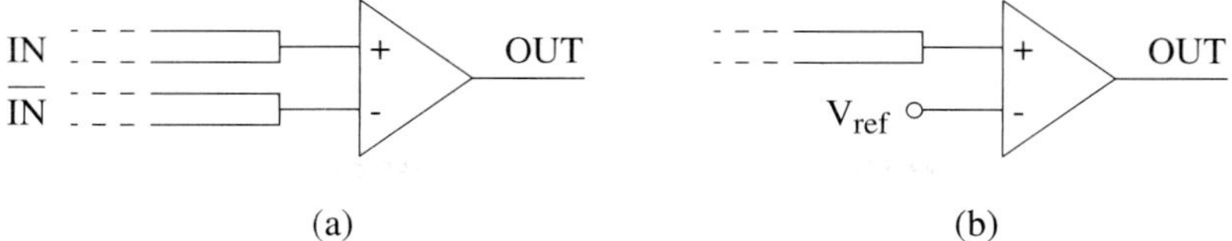

Figure 1.13. Differential amplifier used as a receiver: (a) fully differential, (b) pseudo-differential with reference voltage.

naling. As shown in Figure 1.13, one input of a fully differential receiver is tied to an externally supplied reference voltage. The reference voltage allows the receiver trip point to be adjusted as needed to enhance signal integrity. However, it can be difficult to supply a clean reference voltage in a noisy package environment.

Noise Immunity

Any signal applied to a receiver must have sufficient width and amplitude for the receiver to recognize the input. Narrow voltage spikes, or glitches, arise from several sources: slight timing variations through combinational logic, switching noise, and constructive interference. The ability of a receiver to ignore noise spikes is called *receiver noise immunity.* In general, narrower pulses require more amplitude to trip the receiver, and faster logic families are more sensitive to a given spike than are slower families.

Receiver noise immunity can be computed or measured once a suitable definition has been defined. For example, the noise passed through a receiver could be required to not exceed some value, such as 50mV, or the requirement could be that the noise does not set a latch. Tests should be performed over corners and receiver loading. A simple example is shown in Figure 1.14 for a 3.3V inverter-based receiver with feedback, using 50mV as the maximum allowed feedthrough noise. For wide pulses, the maximum pulse height is near the trip point of 3.3/2=1.65V, but as the pulse narrows below 1ns, more amplitude is needed to fail the 50mV criteria. For pulse widths below 0.1ns, the pulse amplitude must exceed the 3.3V supply voltage.

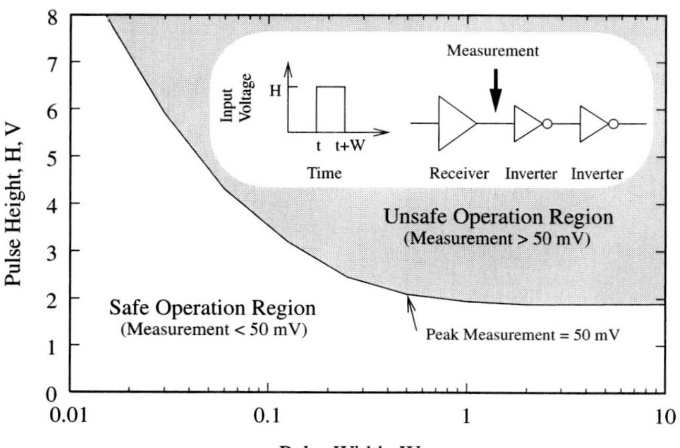

Figure 1.14. Example noise immunity curve showing the dependence of transmitted noise on pulse width and height.

1.2.6 Receiver Modeling

While detailed models of drivers are required to obtain accurate simulations of signal integrity, the same is not true for receivers. Signal integrity analyses are typically concerned with computing measures of signal quality, such as eye diagrams, up to the input of the receiver. Receivers clean up signals, so their outputs are not good measures of signal quality. Coming into a receiver, a signal encounters the ESD protection circuits, gate capacitance of the input FET, and any integral termination resistance. Therefore, an adequate receiver model consists of ESD diodes, capacitance, and resistance (if any). In particular, the ESD diodes are important because they can clip large excursions above and below the power supply rails.

1.3 Interconnects

Digital systems utilize baseband signaling, so the interconnect must be able to carry a DC component. Therefore, two conductors are required, and interconnects provide the electrical connection between devices by providing a current path to carry signals

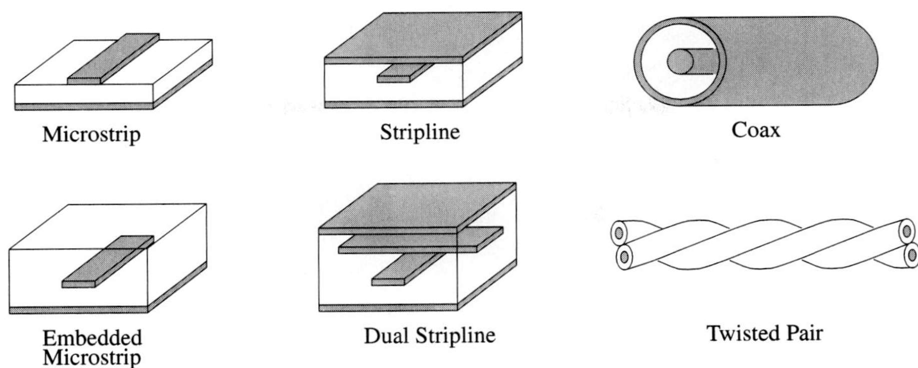

Figure 1.15. Several common types of transmission line.

and a return path to close the loop.[2] In high-speed signaling, the configuration of these two wires becomes very important, and the return path must be treated as carefully as the signal wire. Interconnects are usually referred to by function, such as packages, sockets, connectors, PCBs and many others.

If the cross section of an interconnect does not change along its length, then the interconnect forms a transmission line, which has special properties allowing more detailed modeling with less effort. Although the terminology is redundant, transmission lines are often referred to as having *controlled impedance*. Interconnects that are not transmission lines are often referred to as *nonuniform transmission lines*. Several common transmission line types are sketched in Figure 1.15. Microstrip and stripline are most used on PCBs, where microstrips are naturally formed by traces routed over planes and striplines naturally occur for traces routed between planes in a multilayer stackup. If two routing layers are allocated between planes, then dual stripline results, where the two layers are often routed orthogonally to minimize crosstalk. Connections between subassemblies are often made with coax and twisted pair. Twisted pair is particularly convenient for differential signaling.

2. Clocks and encoded signals (such as with 8B/10B encoding) have constant DC components that can be filtered out with capacitive or inductive coupling. Even without the DC component, two-wire interconnects are still used for their wide bandwidth and low cost.

Figure 1.16. High-performance digital systems, such as this microcomputer motherboard, use many different interconnects. Photograph courtesy of Tyan Computer Corporation.

Typical high-performance digital systems use many types of interconnects. The microcomputer motherboard shown in Figure 1.16 uses, among others, two high-speed microprocessor edge connectors (with plenty of bulk charge storage nearby) at the top, four high-speed memory edge connectors in the center, and multiple package styles including ball-grid array (BGA) (three are visible) and quad flat pack (QFP). The motherboard itself is a high-performance multilayer PCB with controlled-impedance transmission lines. A similar example is the microprocessor daughtercard in Figure 1.17. Cache memory is available close by to ensure high-speed access. Note the use of extensive meandering of the connections, visible for the left-hand memory, for improved timing. A few high-speed edge connectors are shown close up in Figure 1.18.

Packaging for individual dice comes in a variety of styles. A high-end BGA is shown in Figure 1.19. The die is visible and mounted face down with solder balls

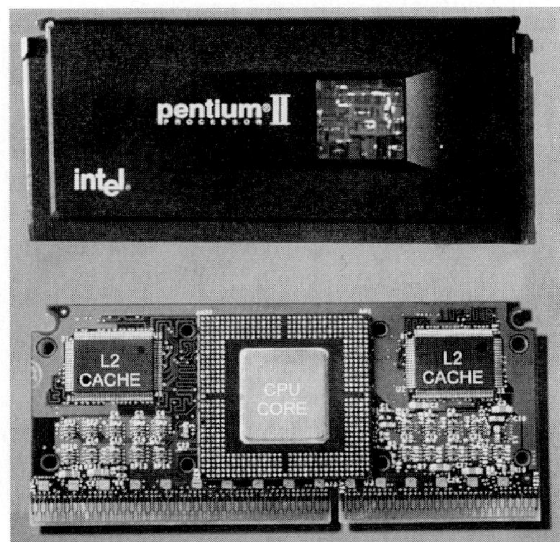

Figure 1.17. A high-performance daughtercard featuring a microprocessor with high-speed cache memory nearby. Photograph courtesy of Intel Corporation.

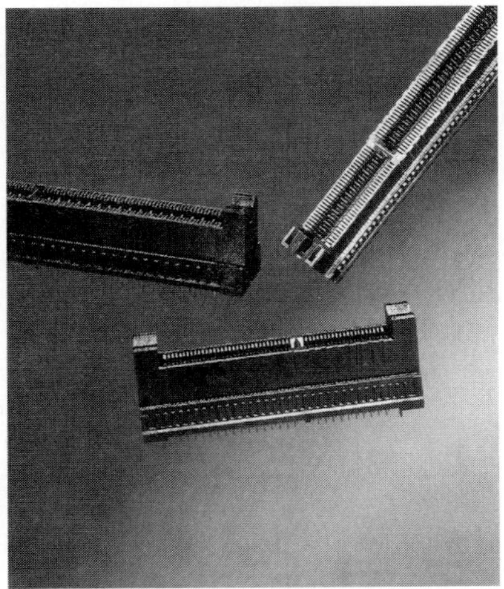

Figure 1.18. High-speed edge connectors provide hundreds of compact contacts with impedance control and low insertion force. Photograph courtesy of Molex Incorporated.

Figure 1.19. A microprocessor packaged in a C4-CBGA. Photograph courtesy of Motorola, Inc.

(C4). The package substrate is multilayer ceramic, hence the acronym CBGA. A regular array of solder balls on the back side provides for attachment to the PCB. A less expensive, single-layer BGA is shown in Figure 1.20. The die is encapsulated in epoxy in the center, and the fine-pitch routing from the die to the solder balls is visible. The workhorse QFP is shown in Figure 1.21.

A few close-up views of interconnects from a single high-performance line card are shown in Figures 1.22 through 1.30. A large variety of package, connector, and socket styles are present on this one card. Note that not all of the components require modeling as high-speed interconnects.

1.4 Modeling of Digital Systems

The focus of this text is the modeling of digital signals as analog waveforms. At high-performance levels, digital signals require simulation as analog waveforms to capture the effects of imperfections in the system. All systems require nonideal components to facilitate construction, including the use of packages, sockets, connectors, and

Figure 1.20. A single-layer BGA. Photograph courtesy of Motorola, Inc.

Figure 1.21. A QFP with bumpers. Photograph courtesy of Motorola, Inc.

Figure 1.22. The plastic BGA, or PBGA, on the left uses a fine-pitch substrate with wiring density much higher than the PCB. A small leadcount J-lead QFP is mounted just to the right of the PBGA.

Figure 1.23. An angled view of the same packages from the prior photo shows the height difference between the packages. The solder balls on the BGA are just visible below the substrate.

Figure 1.24. Packages can vary substantially in lead pitch and body thickness.

Figure 1.25. High-performance systems require many packaging styles. This photograph shows a gull-wing QFP (lower right), a J-lead QFP (lower left), and a BGA (top).

Figure 1.26. A gull-wing QFP with bumpers to protect the leads is mounted in the upper left. A high-speed connector is mounted on the right.

Figure 1.27. On the left side of the QFP, two of the traces are meandered to equalize the line lengths for improved timing.

Figure 1.28. High-speed parts require many bypass capacitors to minimize rail collapse. This photograph shows the capacitors mounted on the back side of the PCB opposite a BGA.

Figure 1.29. The part in the lower right corner is socketed. If the I/O on the part are sufficiently fast, then package and socket models are required in system simulations.

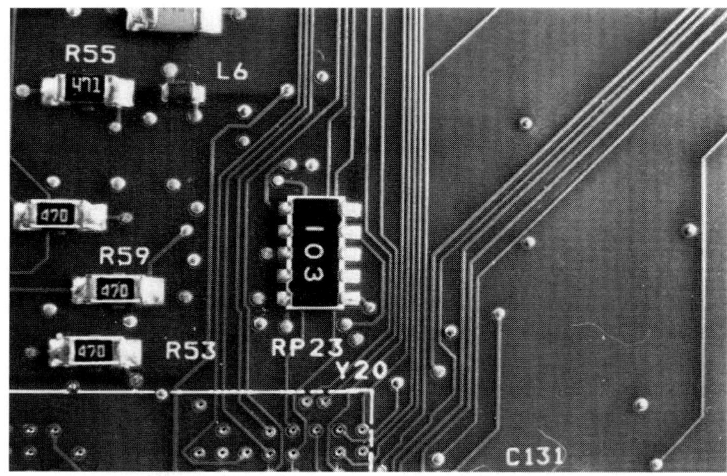

Figure 1.30. Trace routing is usually constrained to 45° increments. Notice the large size of the vias and how they limit routing density.

tight packing. A prototypical system is shown in Figure 1.31, which demonstrates the major points of interest.

A full simulation of the prototypical system requires models of the significant components: packages, sockets, connectors, PCBs, capacitors, drivers, and receivers. Of course, each model must be a standalone implementation of the proper physics (bandwidth, frequency dependence, coupling, and so on), but in addition, all of the models must work together properly. A variety of component styles must be accommodated. Figure 1.31 also represents a most basic, but often overlooked, feature of digital system simulation: signals are interpreted only at the silicon dice with respect to their local power supply.

Most of the effort in assembling a detailed simulation of a complex digital system is in constructing and/or obtaining suitable models. It is especially difficult to obtain adequate package and socket models because of the difficulty in characterizing components with large leadcounts. Commonly, simple models involving one or a few leads are widely available, but these are useful only under limited circumstances. It can also be difficult to obtain good driver and receiver models. The IBIS standard

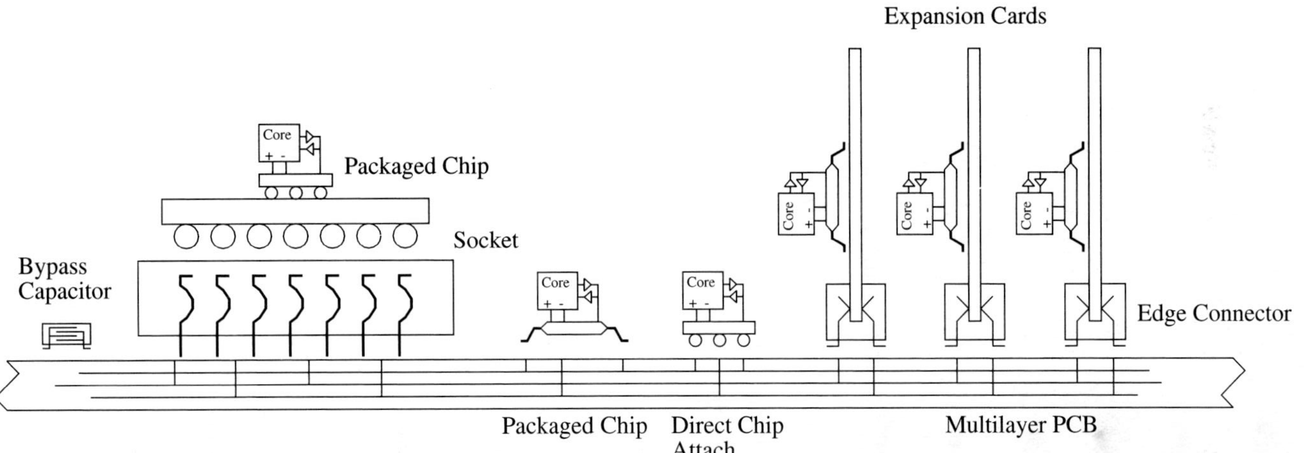

Figure 1.31. Prototypical digital system with components needing modeling.

provides for a good set of voltage-current relationships to describe drivers and receivers, and most manufacturers provide data in IBIS format. The catch is that the data must be well-implemented in the circuit simulator being used for the digital simulation.

Once the system model is complete, detailed simulation can be performed to investigate the performance of signals from silicon to silicon. Performance tradeoffs can be examined to verify operation and/or to converge the design towards an optimal solution.

1.4.1 Analog Quality of Digital Waveforms

Interconnects degrade the quality of digital signals through several effects. Interconnect capacitance must be charged or discharged at every logic transition, causing slower edge rates, higher current draw, and increased power consumption. Interconnect inductance introduces voltage offsets due to the time-rate of change of the current, causing slower edge rates, non-monotonic level transitions, delay, and potential logic level violations. Interconnect capacitance and inductance together form transmission lines that introduce delay due to finite propagation speed and to voltage plateaus. Interconnect resistance can become important for short rise/fall times and/or for long interconnects by rounding off edges, damping noise, and reducing crosstalk.

Waveforms in a digital system are time-varying voltages and currents just as in any other electronic system, but with the special properties that they are principally intended to switch from one logic level to another. The effects of interconnects and parasitics are determined by the rate of change of either voltage for capacitive effects or current for inductive effects. High-speed digital systems imply fast edge rates (or slew) on logic transitions, so significant parasitic effects can arise from remarkably small capacitances and inductances. Ever-faster digital systems imply disruptive effects for ever-smaller parasitic components.

Accurate prediction of high-speed digital systems requires good modeling of the power distribution system and the signaling interconnects. Since all waveforms are

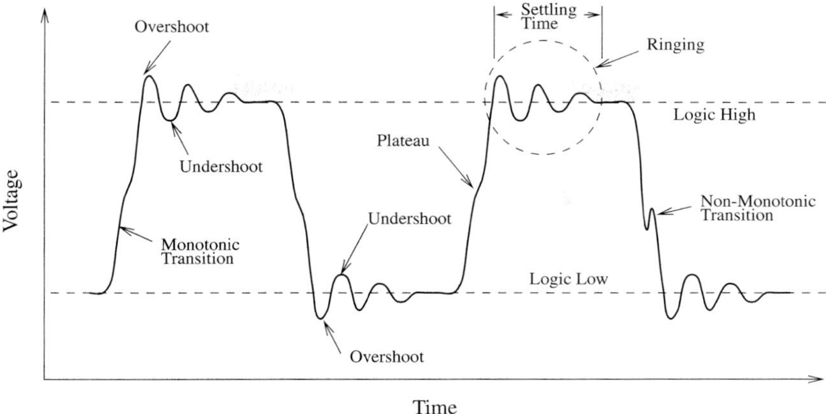

Figure 1.32. Detailed appearance of a digital waveform with several common analog defects.

analog in nature and can be of poor quality, analog simulations are required and SPICE and SPICE-like simulators are very commonly used. The analog appearance of a digital waveform is sketched in Figure 1.32. Overshoot and undershoot occur due to impedance mismatch between the driver and transmission line, crosstalk between signal lines, ground and power bounce due to device packaging, and multiple reflections. Plateaus are common for longer interconnects and can introduce added delay depending on where the signal is sampled. After simulation, the analog waveform is examined for violations of the logic signaling specification and for excessive delay. In addition, some control circuits require that the signal switch between logic levels in a monotonic fashion to prevent double clocking.

The deviation of a waveform from its ideal shape can be thought of as superimposed noise. Noise, in this context, describes unwanted voltage or current deviations. In contrast, the usual use of the term involves random processes, such as $1/F$ noise and thermal noise in semiconductor devices. In digital systems, imperfect waveforms are repeatable and hence deterministic. Good circuit models can accurately predict the deterministic noise in the system.

1.4.2 Modeling, Frequency Content, and Bandwidth

One of the more critical aspects of modeling is ensuring that the model is accurate over a sufficient bandwidth. Lumped models are simplest and can be used when the physical dimensions of the component are small compared to the wavelength at the highest frequency of interest. When this is the case, the *lumped approximation* is said to hold, or that the component is *electrically small* or has *short electrical length*. When the lumped approximation is not applicable, wideband modeling is required and the component is said to be *electrically large* or has *long electrical length*. For example, surface mount resistors can be modeled as lumped elements well into the GHz range. On the other hand, a long interconnect on a PCB cannot be treated as lumped. Lumped models are preferred when applicable, since wideband modeling is considerably more complex.

Bandwidth requirements depend on the shape of the time-domain waveform, where sharp edges and corners require more bandwidth than smooth waveforms. Since it is impossible to know beforehand the precise waveform that will be encountered, rules of thumb must be used to estimate bandwidth needs. Digital waveforms tend to fall into two categories. For unterminated TTL and CMOS signaling, charging of parasitic and load capacitances tend to result in waveforms with exponential edges due to RC charging. For terminated signaling, such as ECL and LVDS, the waveforms tend towards a trapezoidal shape. Bandwidth requirements of both cases are covered below. In addition, bandwidth requirements are estimated from a simple RC interconnect. All three derivations lead to similar requirements.

Once the required bandwidth is identified, a modeling strategy can be developed and implemented. In many situations, such as for packages, sockets, and connectors, lumped models are completely appropriate. In others, wideband models are required. The job of the analyst is to define the important physical effects that must be captured by the model and then to construct a model that satisfies those requirements. In addition to bandwidth, frequency dependence and coupling play a major role in modeling. The level of complexity rises rapidly as additional physical effects are added.

Trapezoidal Edges

For circuits with little capacitance, such as fast terminated nets, digital signals can have edges that are nearly trapezoidal. Consider the pulse train shown in Figure 1.33, where the falling edge is given by

$$v(t) = \left(\frac{\frac{T_W + \tau}{2} - t}{\tau} \right) V_o.$$

Because the pulse train is periodic and even, it can be written as the Fourier series

$$v(t) = a_o + \sum_{n=1}^{\infty} a_n \cos\left(\frac{2\pi n}{T} t\right),$$

where

$$a_o = \frac{2}{T} \int_0^{T/2} v(t) dt$$

and

$$a_n = \frac{4}{T} \int_0^{T/2} v(t) \cos\left(\frac{2\pi n}{T} t\right) dt.$$

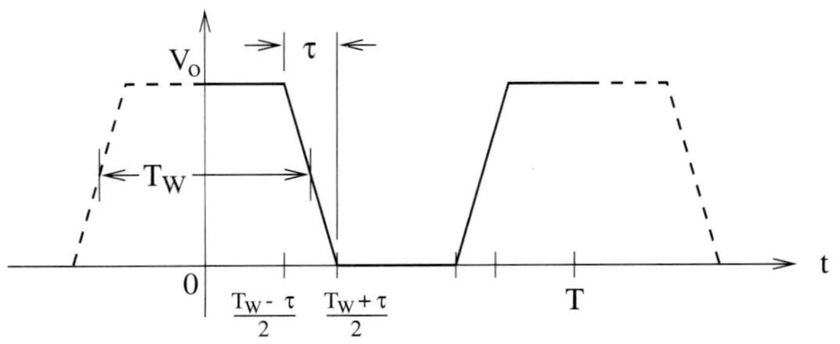

Figure 1.33. Pulse train with trapezoidal edges set up for even symmetry.

Executing the integrals yields the complete result as

$$v(t) = V_o \frac{T_W}{T}\left(1 + 2\sum_{n=1}^{\infty} \frac{\sin(n\pi T_W/T)}{n\pi T_W/T}\frac{\sin(n\pi\tau/T)}{n\pi\tau/T}\cos\left(\frac{2\pi n}{T}t\right)\right). \tag{1.2}$$

The power spectrum is a plot of the squared magnitude of each frequency component of the signal. Since the amplitude of each frequency component $\cos\left(\frac{2\pi n}{T}t\right)$ at the radian frequency $\omega_n = \frac{2\pi n}{T}$ is set by two sinc functions ($\text{sinc}(x) = \sin(x)/x$), the sinc function defines the spectrum of the waveform. While ω_n is actually discrete, it is convenient for discussion to let it assume continuous values.

The generic sinc function and its spectrum are plotted in Figure 1.34. The spectrum is 3dB down when $x = 0.4425\pi$ and rolls off at 20dB/decade (6dB/octave). The first sinc in (1.2) rolls off the spectrum at $2.78/T_W$ with 20dB/decade, while the second rolls it off at $2.78/\tau$ with an additional 20dB/decade for a total of 40dB/decade. The envelope of the spectrum for (1.2) is shown in Figure 1.35. Since $\tau < T_W$, the rolloff in the spectrum associated with the edge rate is higher in frequency than that for the pulse width. Therefore, bandwidth considerations must be based on the edge rate rather than on the pulse width.

At the corner frequency, the spectrum is only 3dB down. To accurately model the pulse edge, more bandwidth is required. With a rolloff of 40dB/decade, an additional 12dB can be gained in only one octave. The radian bandwidth requirement for modeling a trapezoidal pulse train is then

$$\omega_{n,\min} = 5.56/\tau, \tag{1.3}$$

or in terms of frequency, $f_{n,\min} = 0.885/\tau$. Given the approximate nature of the needed additional bandwidth, a good rule of thumb for the required bandwidth is $f_{n,\min} = 1/\tau$.

The number of harmonics can be computed by replacing ω_n with its original definition as $2\pi n/T$ in (1.3). Solving for n yields

$$n = 0.885\frac{T}{\tau}.$$

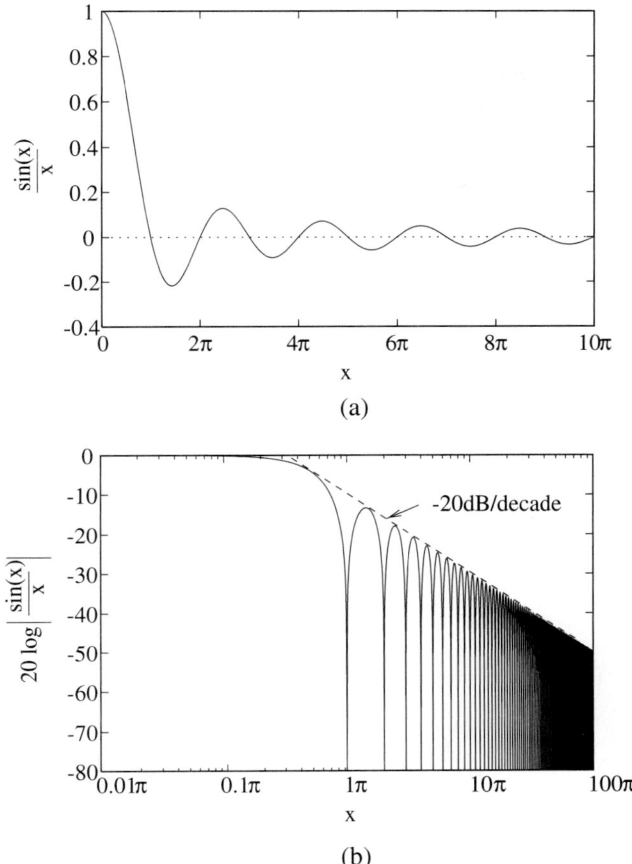

Figure 1.34. The generic sinc function: (a) plot, (b) power spectrum.

For a typical edge rate of $T/\tau \approx 10$, about nine harmonics are required. For this edge rate, the plots in Figure 1.36 show that 10 harmonics is reaching the point of diminishing returns, while the lower numbers of included harmonics create significant overshoot and edge rate error.

As a simple example, consider a system operating with a 100MHz clock and with data transitions on the rising edge of the clock. The clock period is 10ns. If the

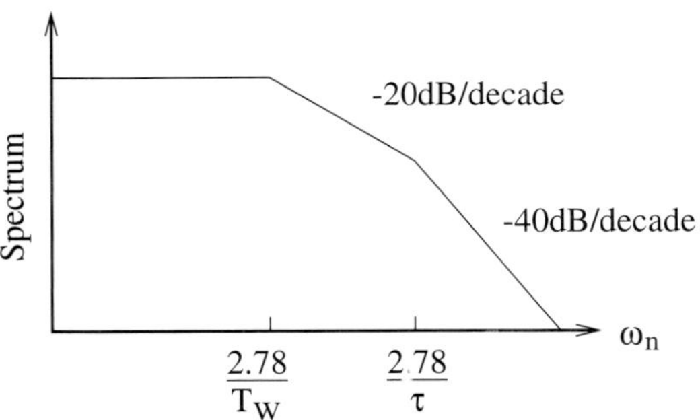

Figure 1.35. Envelope of the spectrum for the trapezoidal pulse train in Figure 1.33.

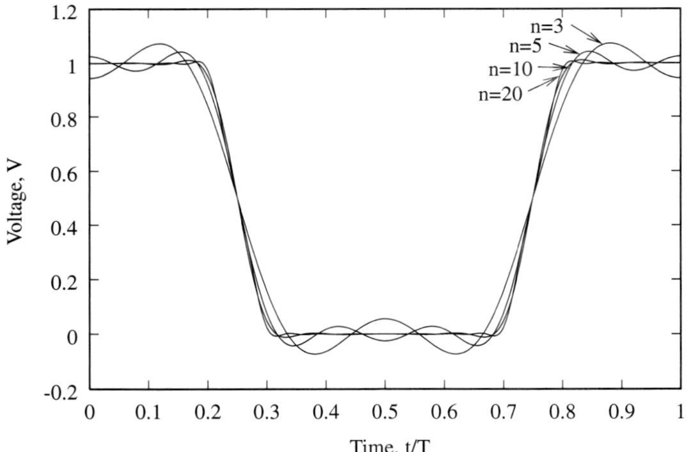

Figure 1.36. Trapezoidal pulse train with $T_W = 0.5T$ and $T/\tau = 10$ reconstructed from its Fourier series with the number of harmonics varying from 3 to 20.

rising and falling times of the waveforms are designed at 20% of the period, then $t_r = 2\text{ns}$. The required bandwidth for any circuit model is $1/2 \times 10^{-9} = 500\text{MHz}$.

Exponential Edges

Due to the RC time constant of charging capacitive loads, digital signals often have exponential edges. Consider the isolated pulse defined by

$$v(t) = \begin{cases} 0, & t \le 0 \\ v_o \frac{1-e^{-\alpha t}}{1-e^{-\alpha T/2}}, & 0 \le t \le T/2 \\ v_o \left(1 - \frac{1-e^{-\alpha(t-T/2)}}{1-e^{-\alpha T/2}}\right), & T/2 \le t \le T \\ 0, & t \ge T, \end{cases}$$

where the parameter α sets the sharpness of the rising and falling edges, and T is the total pulse width. The rise/fall times are directly related to α but not through a simple formula. Several pulses with $v_o = T = 1$ are superimposed in Figure 1.37 for a few 10% to 90% rise times.

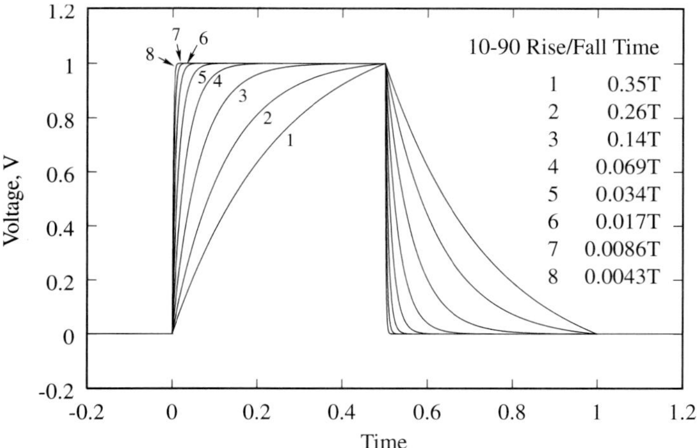

Figure 1.37. Exponential pulse with a 50% duty cycle for several 10% to 90% rise/fall times.

A Fourier transform pair can be defined by

$$F(\jmath\omega) = \int_{-\infty}^{\infty} f(t)e^{-\jmath\omega t}dt \tag{1.4}$$

and

$$f(t) = \frac{1}{2\pi}\int_{-\infty}^{\infty} F(\jmath\omega)e^{\jmath\omega t}d\omega. \tag{1.5}$$

The Fourier transform pair is denoted symbolically as $F(\jmath\omega) = \mathcal{F}[f(t)]$ and $f(t) = \mathcal{F}^{-1}[F(\jmath\omega)]$. The power spectrum of a time-domain waveform is defined as the magnitude squared of the Fourier transform. Using (1.4), the spectrum of each waveform in Figure 1.37 is plotted in dB in Figure 1.38, where it is easily observed in Figure 1.38a that faster rise/fall times lead directly to stronger high-frequency components. In Figure 1.38b the envelopes of several of the spectra are plotted to demonstrate that to capture frequency components down to a given level (30dB in the figure), more bandwidth is required for faster edges.

To estimate the bandwidth required to accurately represent the exponential pulses, the spectra are truncated at several bandwidths, then inverse transformed using (1.5). For a rise/fall time of $0.14T$, these various representations of the original waveform are plotted in Figure 1.39. The plot shows that excellent shape is achieved for bandwidths exceeding $4/t_r$, where t_r is the rise/fall time. Good waveform shape is achieved for bandwidths as low as $1/t_r$. The rectangular window used in this analysis is conservative since interconnects act as low-order, low-pass filters. For the bulk of interconnect work, bandwidths of $1/t_r$ are adequate.

RC Filter

A common signaling scenario is a short, unterminated transmission line, which can be modeled with the simple RC circuit in Figure 1.40. The driver is modeled as a linear circuit with resistance R, while the load is purely capacitive and modeled with capacitance C. The voltage across the capacitor for a unit step input is

$$v_o(t) = u(t)\left(1 - e^{-t/RC}\right). \tag{1.6}$$

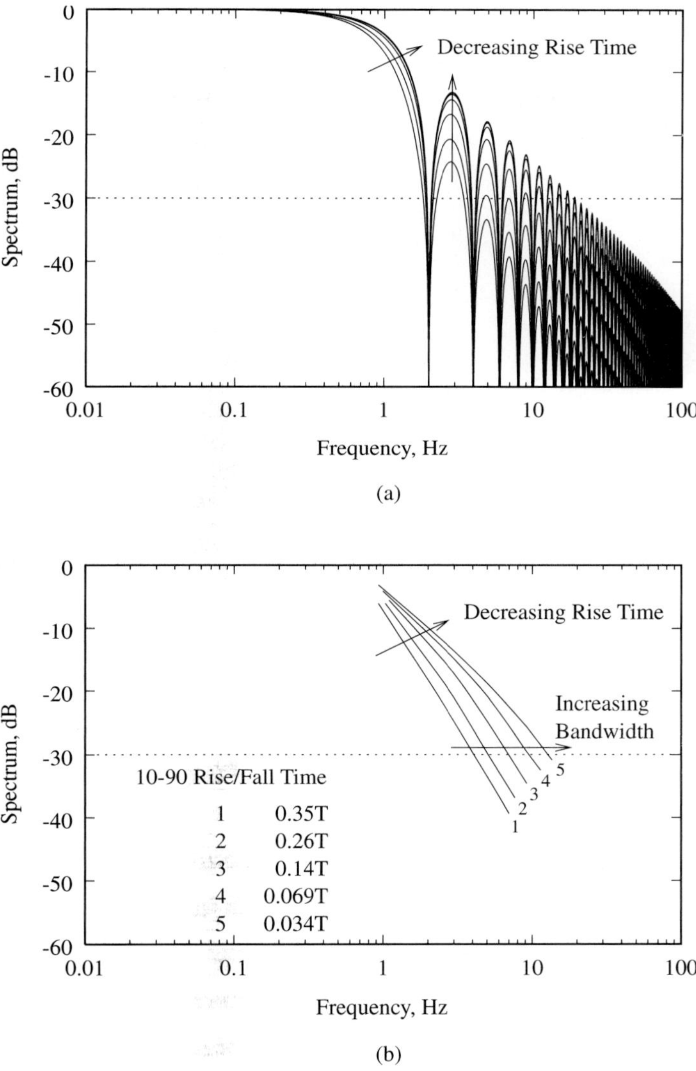

Figure 1.38. (a) Power spectra of the exponential pulses in Figure 1.37. Note that the waveforms with shorter rise times (sharper edges) have stronger high-frequency components. (b) Envelopes for the spectra in (a) show that more bandwidth is required to capture frequency components with values above a given level (-30dB here) as the edges become sharper.

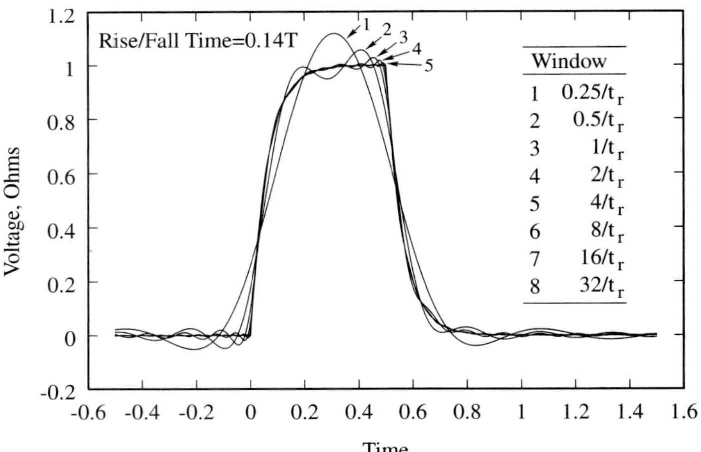

Figure 1.39. Reconstruction of a pulse after rectangular windowing of its spectrum shows that frequency components above $4/t_r$ are not needed for high accuracy. Acceptable accuracy can be achieved with as little bandwidth as $1/t_r$.

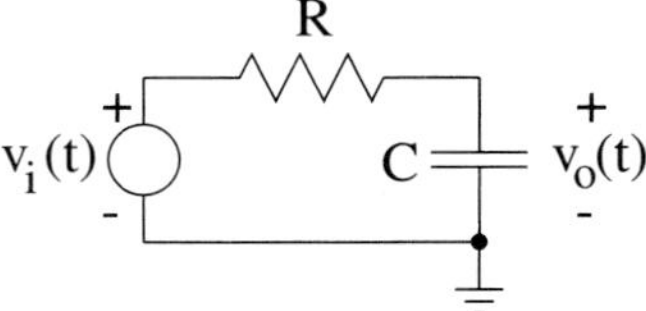

Figure 1.40. Simple RC circuit for estimation of bandwidth in interconnects.

For $t > 0$, the time for a given voltage is

$$t(v_o) = -RC\ln(1 - v_o).$$

The 10% to 90% rise time is then

$$t_r = t(0.9) - t(0.1) = RC\ln\left(\frac{1-0.1}{1-0.9}\right) = 2.197RC.$$

The delay to the 50% point is

$$t_d = t(0.5) = -RC\ln(1 - 0.5) = 0.69RC.$$

Note that $t_d \neq t_r/2$ since the waveform is much steeper (and so more prone to cause noise) at the start of the transition than it is at the end.

The differential equation for an arbitrary input waveform for the circuit in Figure 1.40 is

$$v_i(t) = RC\frac{dv_o(t)}{dt} + v_o.$$

Taking the Fourier transform and solving for the transfer function yields

$$\begin{aligned} H(\omega) &= \frac{\mathcal{F}[v_o(t)]}{\mathcal{F}[v_i(t)]} \\ &= \frac{1}{1 + \jmath\omega RC}, \end{aligned} \tag{1.7}$$

so the power spectrum is

$$|H(\omega)|^2 = \frac{1}{1 + (\omega RC)^2}.$$

$|H(\omega)|^2 = 1$ while $(\omega RC)^2 \ll 1$ but drops to 1/2, the -3dB half-power point, at the corner frequency defined by $(\omega_{3\mathrm{dB}}RC)^2 = 1$. RC can be eliminated using the 10% to 90% rise time to obtain

$$\omega_{3\mathrm{dB}} = \frac{2.197}{t_r},$$

which in terms of frequency is

$$f_{3\mathrm{dB}} = \frac{0.35}{t_r}.$$

The 3dB bandwidth is not nearly adequate to model the RC charging waveform. The transfer function (1.7) is a one-pole response that decays at 10dB/decade or 6dB/octave. An additional two octaves of bandwidth gains 12dB of spectrum, implying a bandwidth requirement of

$$BW = \frac{1.4}{t_r},$$

which is similar to the bandwidth requirements derived in the prior two sections. In general, bandwidths of $1/t_r$ or more are required for interconnect simulations.

1.4.3 Process Variations

To ensure that performance targets are met over a range of operating conditions and component tolerances, simulations should be repeated with adjusted models and power supply values. Semiconductor device performance varies with processing, and performance is usually modeled as best-case (BCS), worst-case (WCS) and typical (TYP). In addition, performance varies with temperature, which ranges from low to nominal to high. Power supplies always specify a variation about the nominal output voltage. Printed circuit boards and lumped passive components have tolerances and sometimes temperature-dependent resistance. Combining and simulating all possible combinations is practical only for small-scale simulations.

One method to reduce the number of required simulations is to define process corners to bound the performance of the system. For example, a two-corner simulation could be based with one corner on low-temperature BCS silicon, high-impedance transmission lines, high-power supply voltage, and small capacitor and resistor values. This corner covers the fast system using fast silicon with least capacitive loading and shortest RC charging times. The other corner would be the opposite: high-temperature WCS silicon at low voltage with low-impedance transmission lines and large resistances. Corners based on process, voltage, and temperature variations are often referred to as *PVT corners*.

Process corners should not be invariant rules. Different circuits can require different corners. In practice, best-case and worst-case corners can be too pessimistic, causing overly conservative designs. Product performance in the field can be used to adjust the corners accordingly.

1.4.4 Challenges for Modeling High-Speed Systems

For sufficiently slow clock and slew rates, the effects of interconnects can be incorporated through their capacitive loading on drivers. As clock rates increase, transmission line effects begin and significant voltages appear on small parasitics. For simulation, high speed means that that transmission line and parasitic models are required to achieve good accuracy.

Simulations of high-speed systems require models of all components, and these models must capture all of the significant physical effects: frequency dependence, losses, coupling, delay, and parasitic circuit effects. Models can be constructed through analytical analysis, measurement, and/or simulation.

Analytical models can be quite suitable in many instances; however, they carry the significant risk of poor application, and mistakes can lead to serious simulation errors. In the modeling of interconnects, packaging, and sockets, the formulas for partial inductance are particularly susceptible to misapplication. The primary limitation of analytical modeling is the lack of formulas because most problems are too complex.

Measurements of interconnects are difficult for a couple of reasons. First, the parasitic values are often very small. For example, in packages and sockets the connections to be characterized are intentionally designed to be good short circuits. Small parasitic values can still lead to significant voltage levels. For instance, 80pH of power supply inductance in a microprocessor package can lead to high noise levels because microprocessors can draw current at the rate of several A/ns. Second, interconnects often involve large numbers of ports that must be characterized. For example, a 100-lead QFP nominally requires 20,100 two-port swept-frequency measurements for a full characterization. Shortcuts and approximations are often required.

Simulations of interconnects can be difficult if the problem size exceeds the capacity of the simulator, the operating system, or the computer. Electromagnetic simulation is often required in some form, and these are usually resource intensive. Just as for measurements, shortcuts and approximations are often required. For example, 2D simulations require far less resources than 3D simulations, so long as interconnects can be approximated as transmission lines and analyzed in 2D to obtain per-unit-length resistance, inductance, and capacitance values. Unfortunately, it can require considerable skill and experience to deploy a suitable range of electromagnetic techniques (and can be expensive, too).

Simulations over corners can be difficult due to the need to manage models, schematics, and simulation results. Often, models over corners are unavailable.

Finally, even if a full interconnect model is produced, a circuit simulation must successfully run with the model. Interconnect models can easily overwhelm a circuit simulator. For example, the partial inductance matrix alone of a 100-lead QFP requires 5,050 elements in a SPICE simulation. Specialized simulations can be performed that enable the interconnect models to be reduced to small and manageable levels.

In summary, the job of the signal integrity analyst is very challenging because the specialization cuts across so much of the field of electrical engineering. Significant expertise is required in digital systems concepts; analog circuit behavior; analog circuit simulation; driver and receiver design; electromagnetic simulation; lumped modeling; transmission line theory; multiconductor transmission lines and interconnects; the physics of resistance, inductance, and capacitance; frequency-domain analysis; time-domain analysis; and potentially, laboratory techniques and microwave theory for characterizing very small parasitics and large leadcount interconnects.

Chapter 2

SIGNAL INTEGRITY

Digital systems rely on signaling from drivers to receivers to pass information between their components. Reliable signaling is achieved when the signaling specifications are met under full adverse noise conditions as well as device behavior variations due to both process deviations in device manufacturing and normal changes over the operating temperature. A breakdown in communication leads to glitches where unintended or incorrect data is transferred, a situation called *false signaling*.

Critical data paths within systems often contain safeguards against false signaling, and the primary system for this is data redundancy through parity. Parity can enable the system to detect small-scale signaling failures, while an error correction scheme can be optionally included to correct data faults detected using parity information. However, parity and error correction cannot be relied upon to make a noisy system stable and reliable. In addition, it is often not practical in terms of cost or performance to include parity and error correction on every circuit.

Superimposed on the desired signals are unwanted waveforms (i.e., noise) generated from many sources. The principal sources are crosstalk, impedance mismatch, simultaneous switching noise, and multiple reflections. Each can be independently characterized to facilitate an understanding of the mechanisms that degrade signal quality and to help guide design decisions. In real systems, all act simultaneously and require detailed circuit simulation to obtain good estimates of the total waveform on each signal line.

2.1 Transmission Lines

A transmission line is a two-conductor interconnect (so that it can carry signal frequency components down to DC) that is long compared to the conductor cross section and uniform along its length. Because many interconnects are dominated by long runs over unbroken ground planes (to minimize radiation and EMI susceptibility), they can be accurately modeled as transmission lines, and much of signal integrity analysis is based on them.

If a short length of a transmission line is considered, then the lumped approximation applies and the transmission line can be modeled, as shown in Figure 2.1, with series resistance and inductance and with shunt capacitance and conductance. Applying Kirchhoff's voltage law around the loop, then

$$v(z + \Delta z, t) - v(z, t) = -Ri(z, t) - L\frac{di(z, t)}{dt}. \tag{2.1}$$

Similarly, Kirchhoff's current law applied at $z + \Delta z$ yields

$$i(z + \Delta z, t) - i(z, t) = -Gv(z + \Delta z, t) - C\frac{dv(z + \Delta z, t)}{dt}. \tag{2.2}$$

Divide through by Δz and let $\Delta z \rightarrow 0$, then (2.1) and (2.2) transform from difference equations to the differential equations

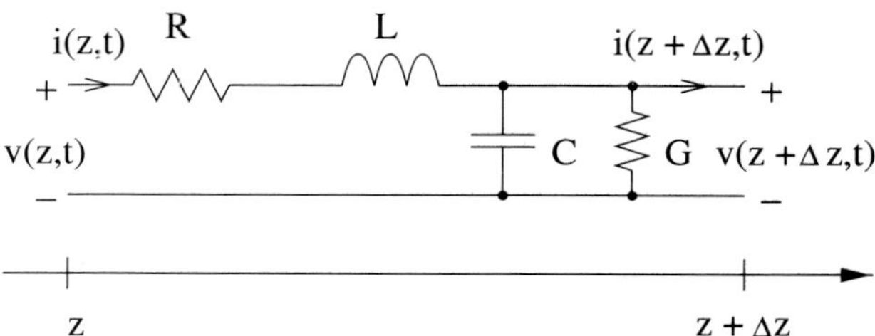

Figure 2.1. Lumped model of a short length of a transmission line.

$$\frac{\partial v(z,t)}{\partial z} = -ri(z,t) - \ell\frac{\partial i(z,t)}{\partial t} \tag{2.3}$$

and

$$\frac{\partial i(z,t)}{\partial z} = -gv(z,t) - c\frac{\partial v(z,t)}{\partial t}, \tag{2.4}$$

where the lumped component values transition to the per-unit-length quantities r, ℓ, c, and g due to normalization by Δz. Simultaneous solution of the transmission line equations (2.3) and (2.4) yields the voltage and current at any point on the transmission line.

2.1.1 Time-Domain Solution

Transmission lines for digital signaling often have low losses. For example, the effect of half an Ohm of loss on a 50Ω transmission line driven with a 50Ω driver is negligible for most applications. To facilitate investigations of the effects of various system imperfections on signal integrity, losses will be neglected. The lossless transmission line equations are recovered from (2.3) and (2.4) by setting $r = g = 0$ yielding

$$\frac{\partial v(z,t)}{\partial z} = -\ell\frac{\partial i(z,t)}{\partial t} \tag{2.5}$$

and

$$\frac{\partial i(z,t)}{\partial z} = -c\frac{\partial v(z,t)}{\partial t}. \tag{2.6}$$

An important property of the lossless transmission line is that pulses propagate undistorted along the length of the line. Consider an arbitrary waveform such as the one in Figure 2.2, where the wave shape is described by the function $f(\tau)$ with τ an independent variable. Since there are no losses and no frequency dependence to ℓ or c, the waveform must move down the line unchanged in shape (see section 2.1.3 for a proof) and can be described mathematically by $f(z,t) = f(\tau) = f(z - \nu t)$, where $\tau = z - \nu t$. A maximum or minimum of the waveform occurs when $\partial\tau/\partial t = 0$, so

$$\frac{\partial \tau}{\partial t} = 0 = \frac{dz}{dt} - \nu,$$

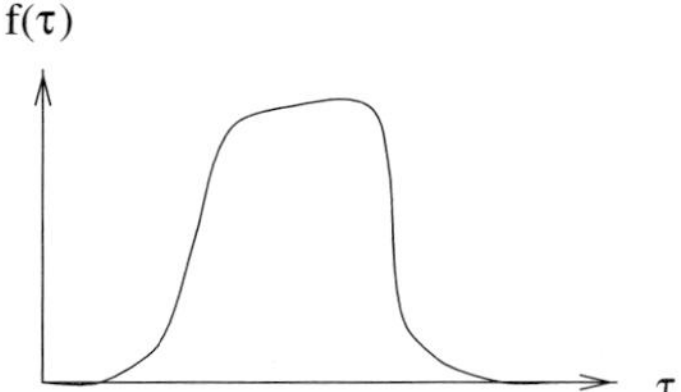

Figure 2.2. Arbitrary waveform for propagation down a transmission line.

or

$$\nu = \frac{dz}{dt},$$

indicating that the maximum or minimum point is moving in the $+z$ direction with velocity ν.

The partial derivatives in (2.5) and (2.6) can be rewritten in terms of τ by noting that

$$\frac{\partial}{\partial z} = \frac{\partial}{\partial \tau}\frac{\partial \tau}{\partial z} = \frac{\partial}{\partial \tau}\frac{\partial (z - \nu t)}{\partial z} = \frac{\partial}{\partial \tau}$$

and

$$\frac{\partial}{\partial t} = \frac{\partial}{\partial \tau}\frac{\partial \tau}{\partial t} = -\nu\frac{\partial}{\partial \tau},$$

then

$$\frac{\partial v(\tau)}{\partial \tau} = \ell\nu\frac{\partial i(\tau)}{\partial \tau} \tag{2.7}$$

and

$$\frac{\partial i(\tau)}{\partial \tau} = c\nu\frac{\partial v(\tau)}{\partial \tau}. \tag{2.8}$$

Eliminating $\frac{\partial i(\tau)}{\partial \tau}$ between (2.7) and (2.8) and cancelling $\frac{\partial v(\tau)}{\partial \tau}$ yields

$$\nu = \frac{1}{\sqrt{\ell c}}, \tag{2.9}$$

so the velocity of an arbitrary pulse can be directly computed from the per-unit-length inductance and capacitance of the transmission line. Integrating (2.7) with respect to τ while assuming that no static charge is on the transmission line (so that the integration constant vanishes), then

$$v(\tau) = \sqrt{\frac{\ell}{c}}\, i(\tau),$$

where (2.9) is used. Therefore, the voltages and currents of an arbitrary waveform on a lossless transmission line are in phase and related by the impedance

$$Z_o = \sqrt{\frac{\ell}{c}},$$

called the *characteristic impedance* of the transmission line.

For a transmission line of length d, the time for a wave to travel the length of the line is called the *delay* or the *time of flight* (TOF) and can be computed as

$$\text{TOF} = d\sqrt{\ell c}. \tag{2.10}$$

The lossless transmission line is completely specified by its characteristic impedance and delay. Note that

$$\ell = Z_o \frac{\text{TOF}}{d} \tag{2.11}$$

and

$$c = \frac{1}{Z_o} \frac{\text{TOF}}{d}. \tag{2.12}$$

The analyses can be repeated with $\tau = z + \nu t$ with identical results, except that the waveform travels in the $-z$ direction with velocity $\nu = 1/\sqrt{\ell c}$ and the voltage and current are related by

$$v(\tau) = -Z_o i(\tau).$$

Effective Dielectric Constant

Since the TOF can be found given the length of a transmission line and the velocity of a wave on it, the velocity is often the unknown parameter that must be found.

In a transmission line where the electric and magnetic fields are completely encased in a dielectric with dielectric constant ϵ_r, then the velocity is

$$\nu = \frac{c_o}{\sqrt{\epsilon_r}}, \tag{2.13}$$

where $c_o = 3 \times 10^8$m/s is the velocity of a wave in a vacuum ($\epsilon_r = 1$) [which is also called the *free-space speed of light*]. For transmission lines such as stripline, dual stripline, embedded microstrip, and coax, the velocity is easily computed once the dielectric constant of the filler material is known.

When the electric and magnetic fields run through two dielectrics, the wave still propagates with some velocity. Generalizing (2.13) yields

$$\nu = \frac{c_o}{\sqrt{\epsilon_{\text{eff}}}},$$

where ϵ_{eff} is an effective dielectric constant. For a transmission line like microstrip, the two dielectrics are air with $\epsilon_r = 1$ and the substrate. The effective dielectric constant must lie between these two, and since most of the field is below the strip in the substrate, the effective dielectric constant must be closer to the dielectric constant of the substrate than to that of air.

Effective dielectric constants are convenient because they offer a handy shortcut in many situations and can be easily estimated for approximate calculations. For example, if $\epsilon_r = 4$ for the substrate in microstrip, then $\epsilon_{\text{eff}} \approx 3$.

For the lossless case, the formulas in (2.11) and (2.12) are easily modified for the effective dielectric constant to be

$$\ell = \frac{Z_o\sqrt{\epsilon_{\text{eff}}}}{c_o} \tag{2.14}$$

and

$$c = \frac{\sqrt{\epsilon_{\text{eff}}}}{Z_o c_o}. \tag{2.15}$$

2.1.2 Directional Independence

The analysis of lossless transmission lines can be carried further to show that two waveforms traveling in opposite directions do not interact. Let v^+ denote a voltage waveform launched in the $+z$ direction, while v^- indicates one traveling in the $-z$ direction. Due to the linearity of Maxwell's equations, and assuming linear materials, the total voltage must be the superposition of these two, so

$$v(z,t) = v^+(z - \nu t) + v^-(z + \nu t). \tag{2.16}$$

The total current is then

$$i(z,t) = \frac{1}{Z_o} v^+(z - \nu t) - \frac{1}{Z_o} v^-(z + \nu t). \tag{2.17}$$

Substituting these expressions into (2.5) and (2.6) yields

$$\frac{\partial v^+}{\partial z} + \frac{\partial v^-}{\partial z} = -\frac{\ell}{Z_o}\frac{\partial v^+}{\partial t} + \frac{\ell}{Z_o}\frac{\partial v^-}{\partial t}$$

and

$$\frac{\partial v^+}{\partial z} - \frac{\partial v^-}{\partial z} = -cZ_o\frac{\partial v^+}{\partial t} - cZ_o\frac{\partial v^-}{\partial t}.$$

Adding these two results in

$$\frac{\partial v^+}{\partial z} = -\frac{1}{2}\left(\frac{\ell}{Z_o} + cZ_o\right)\frac{\partial v^+}{\partial t} + \frac{1}{2}\left(\frac{\ell}{Z_o} - cZ_o\right)\frac{\partial v^-}{\partial t},$$

while subtracting provides

$$\frac{\partial v^-}{\partial z} = -\frac{1}{2}\left(\frac{\ell}{Z_o} - cZ_o\right)\frac{\partial v^+}{\partial t} + \frac{1}{2}\left(\frac{\ell}{Z_o} + cZ_o\right)\frac{\partial v^-}{\partial t}.$$

These can be further simplified by noting that $\left(\frac{\ell}{Z_o} + cZ_o\right) = 2/\nu$ while $\left(\frac{\ell}{Z_o} - cZ_o\right)$ $= 0$, then

$$\frac{\partial v^+}{\partial z} = -\frac{1}{\nu}\frac{\partial v^+}{\partial t}$$

and

$$\frac{\partial v^-}{\partial z} = +\frac{1}{\nu}\frac{\partial v^-}{\partial t}.$$

Note that these two equations are decoupled, with each equation a function only of v^+ or v^-; therefore, the two waves cannot interact and travel along the transmission line without influencing each other. The waves interact only at boundaries between transmission lines with different impedances or between a transmission line and a component. This property is exploited to facilitate analyses and to construct bounce diagrams.

2.1.3 Frequency-Domain Solution

When losses are significant on transmission lines, the frequency domain is convenient for analytic solutions. After Fourier [4] transformation with (1.4), the lossy transmission line equations in (2.3) and (2.4) become

$$\frac{\partial V(z,\omega)}{\partial z} = -(r + \jmath\omega\ell)I(z,\omega) \tag{2.18}$$

and

$$\frac{\partial I(z,\omega)}{\partial z} = -(g + \jmath\omega c)V(z,\omega), \tag{2.19}$$

where V and I are the Fourier transforms of v and i. Eliminating I from (2.18) and (2.19) provides the second-order differential equation

$$\frac{\partial^2 V}{\partial z^2} - (r + \jmath\omega\ell)(g + \jmath\omega c)V = 0. \tag{2.20}$$

Define $\gamma^2 = (r + \jmath\omega\ell)(g + \jmath\omega c)$, then

$$\gamma(\omega) = \sqrt{(r + \jmath\omega\ell)(g + \jmath\omega c)}, \tag{2.21}$$

and the solution to (2.20) is

$$V(z,\omega) = \mathrm{A}(\omega)e^{-\gamma(\omega)z} + \mathrm{B}(\omega)e^{\gamma(\omega)z}. \tag{2.22}$$

Note in particular that A, B, and γ are functions of ω. γ is called the *complex propagation constant.*

The complex propagation constant can always be written in terms of its real and imaginary parts as

$$\gamma(\omega) = \alpha(\omega) + \jmath\beta(\omega);$$

then (2.22) becomes

$$V(z,\omega) = \mathrm{A}(\omega)e^{-\alpha(\omega)z}e^{-\jmath\beta(\omega)z} + \mathrm{B}(\omega)e^{\alpha(\omega)z}e^{\jmath\beta(\omega)z}.$$

The time-domain voltage can then be found using the inverse Fourier transform given in chapter 1 by (1.5) to obtain

$$\begin{aligned} f(t) &= \frac{1}{2\pi}\int_{-\infty}^{\infty}\left(\mathrm{A}(\omega)e^{-\alpha(\omega)z}e^{-\jmath\beta(\omega)z} + \mathrm{B}(\omega)e^{\alpha(\omega)z}e^{\jmath\beta(\omega)z}\right)e^{\jmath\omega t}d\omega \\ &= \frac{1}{2\pi}\int_{-\infty}^{\infty}\left(\mathrm{A}(\omega)e^{-\alpha(\omega)z}e^{-\jmath(\beta(\omega)z-\omega t)}\right. \\ &\quad \left. + \mathrm{B}(\omega)e^{\alpha(\omega)z}e^{\jmath(\beta(\omega)z+\omega t)}\right)d\omega, \end{aligned} \tag{2.23}$$

where it can be seen in the second form that the total voltage consists of the superposition of many complex exponentials. Each $\mathrm{A}(\omega)e^{-\alpha(\omega)z}e^{-\jmath(\beta(\omega)z-\omega t)}$ represents a weighted sinusoidal wave traveling in the $+z$ direction that is attenuated exponentially with distance. A fixed point on the exponential can be identified when $\beta(\omega)z - \omega t = \text{constant}$. Solving for z and taking the derivative with respect to time yields the phase velocity

$$\nu_p = \frac{dz}{dt} = \frac{\omega}{\beta(\omega)}. \tag{2.24}$$

Since β is a function of frequency, then the phase velocity is a function of frequency. In a similar fashion, each $\mathrm{B}(\omega)e^{\alpha(\omega)z}e^{\jmath(\beta(\omega)z+\omega t)}$ represents a weighted and attenuated wave traveling in the $-z$ direction. Due to their roles in wave propagation, α is called the *attenuation constant* and β is called the *propagation constant*.

The phase velocity and attenuation of the sinusoidal waves are different at every frequency, so the shape of the time-domain waveform must change as it moves down the line. Attenuation is stronger for higher-frequency components, so the waveform tends to spread, or *disperse*, with distance as low-frequency components take over. For this reason, the effects of losses on wave shape is called *dispersion*. Sometimes dispersion is attributed to the source of the losses, such as conductor loss dispersion or dielectric loss dispersion.

Because each frequency component has a different phase velocity, the wave velocity is not equal to the phase velocity of any given component. For lossy lines, the wave velocity is found by finding the TOF of a pulse between two points and dividing into the distance traveled.[1]

The characteristic impedance can be found by solving for I in (2.18) with the general voltage solution in (2.22) to obtain

$$I(z,\omega) = \sqrt{\frac{g + \jmath\omega c}{r + \jmath\omega\ell}}\left(\mathrm{A}(\omega)e^{-\gamma(\omega)z} - \mathrm{B}(\omega)e^{\gamma(\omega)z}\right). \tag{2.25}$$

The characteristic impedance for a lossy transmission line is then

$$Z_o(\omega) = \sqrt{\frac{r + \jmath\omega\ell}{g + \jmath\omega c}}, \tag{2.26}$$

and it is, in general, frequency-dependent.

The directional components of voltage and current are clearly apparent with a change of notation in (2.22) to

$$V(z,\omega) = V^{+}(\omega)e^{-\gamma(\omega)z} + V^{-}(\omega)e^{+\gamma(\omega)z}, \tag{2.27}$$

then the current from (2.25) is

$$I(z,\omega) = \frac{1}{Z_o(\omega)}\left[V^{+}(\omega)e^{-\gamma(\omega)z} - V^{-}(\omega)e^{+\gamma(\omega)z}\right]. \tag{2.28}$$

These two expressions are often a good starting point in solving circuit problems involving transmission lines.

Low-Loss Transmission Lines

When losses on a transmission line are small, additional mathematical manipulation can yield good insight into waveform propagation on transmission lines. The complex propagation constant can be rearranged as

$$\begin{aligned}
\gamma &= \sqrt{(r + \jmath\omega\ell)(g + \jmath\omega c)} \\
&= \sqrt{\jmath\omega\ell(1 + \frac{r}{\jmath\omega\ell})\jmath\omega c(1 + \frac{g}{\jmath\omega c})} \\
&= \jmath\omega\sqrt{\ell c}\sqrt{1 + \frac{r}{\jmath\omega\ell}}\sqrt{1 + \frac{g}{\jmath\omega c}}.
\end{aligned} \tag{2.29}$$

1. For narrow-band signals, the group velocity can be computed from the frequency dependence of the phase velocity to find the velocity of a wave packet. Digital signals are baseband, so the group velocity is not applicable since the narrow-band assumption does not apply.

For low losses, the approximations

$$\begin{aligned}\sqrt{1+\tfrac{r}{\jmath\omega\ell}} &\approx 1+\tfrac{1}{2}\tfrac{r}{\jmath\omega\ell}, \quad \tfrac{r}{\omega\ell}\ll 1\\ \sqrt{1+\tfrac{g}{\jmath\omega c}} &\approx 1+\tfrac{1}{2}\tfrac{g}{\jmath\omega c}, \quad \tfrac{g}{\omega c}\ll 1\end{aligned} \tag{2.30}$$

can be applied to (2.29) to obtain

$$\begin{aligned}\gamma &\approx \jmath\omega\sqrt{\ell c}\left(1+\frac{1}{2}\frac{r}{\jmath\omega\ell}+\frac{1}{2}\frac{g}{\jmath\omega c}-\frac{1}{4}\frac{rg}{\omega^2\ell c}\right)\\ &\approx \jmath\omega\sqrt{\ell c}+\frac{1}{2}\left(\sqrt{\frac{c}{\ell}}r+\sqrt{\frac{\ell}{c}}g\right),\end{aligned} \tag{2.31}$$

where the term $\frac{1}{4}\frac{rg}{\omega^2\ell c}$ is dropped as a negligible second-order small term. The attenuation and propagation constants for low losses are then given by

$$\alpha=\frac{1}{2}\left(\sqrt{\frac{c}{\ell}}r+\sqrt{\frac{\ell}{c}}g\right) \tag{2.32}$$

and

$$\beta=\omega\sqrt{\ell c}, \tag{2.33}$$

and the phase velocity is

$$\nu_p=\frac{1}{\sqrt{\ell c}}.$$

Note that the attenuation constant and the phase velocity are frequency-independent, so all frequency components of the waveform travel together with uniform attenuation. Therefore, the waveform propagates down the transmission line with no change in wave shape except for reduction in amplitude. In this case, the wave velocity does equal the phase velocity. This propagation is dispersionless.

Substituting the attenuation and propagation constants in (2.32) and (2.33) for the low-loss line into (2.23) yields

$$\begin{aligned}f(t) &= \frac{1}{2\pi}e^{-\alpha z}\int_{-\infty}^{\infty}A(\omega)e^{-\jmath\omega\sqrt{\ell c}z}e^{\jmath\omega t}d\omega\\ &\quad+\frac{1}{2\pi}e^{+\alpha z}\int_{-\infty}^{\infty}B(\omega)e^{+\jmath\omega\sqrt{\ell c}z}e^{\jmath\omega t}d\omega\\ &= e^{-\alpha z}a(t-\sqrt{\ell c}z)+e^{+\alpha z}b(t+\sqrt{\ell c}z),\end{aligned} \tag{2.34}$$

where $a(t) = \mathcal{F}^{-1}[A(\omega)]$ and $b(t) = \mathcal{F}^{-1}[B(\omega)]$. The time-shifting theorem for Fourier transforms,

$$\mathcal{F}[f(t-\xi)] = \mathcal{F}[f(t)]e^{-\jmath\omega\xi},$$

is also utilized. The result in (2.34) generalizes to low-loss lines the directional independence established in section 2.1.2 for lossless lines. In addition, the result shows that waveforms travel unchanged except for amplitude on low-loss lines.

Conductor Loss-Dominated Transmission Lines

Losses are often dominated by conductors, so while r is often significant, g is often not. Setting $g = 0$ in the lossy expressions for Z_o and γ yields

$$Z_o = \sqrt{\frac{r + \jmath\omega\ell}{\jmath\omega c}} \tag{2.35}$$

and

$$\gamma = \sqrt{(r + \jmath\omega\ell)\jmath\omega c}. \tag{2.36}$$

At low frequencies, $r \gg \omega\ell$, so

$$Z_o \approx \sqrt{\frac{r}{\omega c}}\, e^{-\jmath\pi/4}$$

and

$$\gamma \approx \sqrt{\omega r c}\, e^{\jmath\pi/4}.$$

For both Z_o and γ, the real and imaginary parts are equal (ignoring the sign) with a square root dependency on frequency. These results are very different from those obtained from the lossless transmission line model, yet the lossless model is often used in signal integrity work where baseband digital signals include significant low-frequency content.

An approximate metric is available for determining if lossless transmission line modeling is appropriate. At low frequencies, the lumped approximation applies, and the lossy transmission line is simply a loop of wire that can be modeled as a series RLC circuit. For losses to be negligible, the RLC circuit must be strongly underdamped, a condition that occurs when

$$R \ll 2\sqrt{\frac{L}{C}}.$$

Using per-unit-length quantities for a transmission line with length d, then

$$d \ll \frac{2}{r}\sqrt{\frac{\ell}{c}}. \quad (2.37)$$

If the line length is sufficiently short, then lossless transmission line modeling can be appropriate.

To demonstrate the line length dependence, consider a 50Ω transmission line with $\epsilon_{\text{eff}} = 3$ and 10Ω/m of loss. The per-unit-length quantities from (2.14) and (2.15) are 289nH/m and 115pF/m, respectively. The line length limit from (2.37) is $d \ll 10$m. Simulation results for lossy and lossless versions of this transmission line are shown in Figure 2.3. The results show that the lossless model is accurate for line lengths less than 1m, so the prediction from (2.37) works very well for this case. Board-level signal integrity work deals with lines less than a meter long, so lossless transmission line modeling can be appropriate if the edge rates are sufficiently slow. Waveforms with fast edge rates experience significant frequency-dependent losses due to the skin effect, so frequency-dependent lossy simulations can be required (see chapter 10, section 10.3).

At high frequencies, $r \ll \omega\ell$, then

$$Z_o \approx \sqrt{\frac{\ell}{c}}\left(1 - \jmath\frac{1}{2}\frac{r}{\omega\ell}\right)$$

and

$$\gamma \approx \jmath\omega\sqrt{\ell c}\left(1 - \jmath\frac{1}{2}\frac{r}{\omega\ell}\right).$$

In this case, the real part of Z_o is constant and dominates the imaginary part, while for γ the imaginary part dominates with a simple linear dependence on frequency. When high-frequency losses are negligible, the characteristic impedance and propagation constant simplify to that of the lossless transmission line.

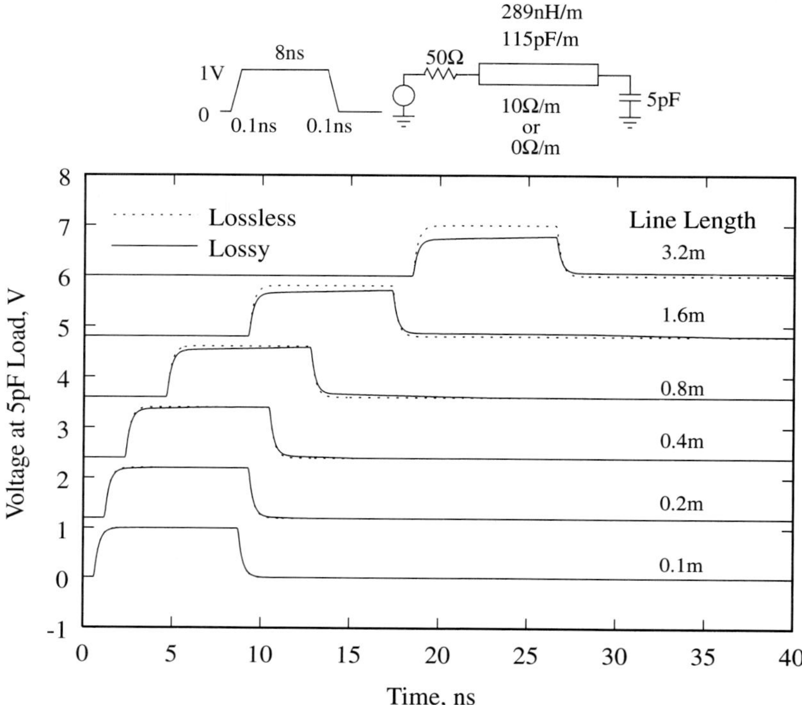

Figure 2.3. The accuracy of the lossless transmission line model depends on the line length.

General Formulas Using Real Math

Without assumptions on r and $\omega\ell$, complex math is required for calculating Z_o and γ, so calculator-based calculations using (2.35) and (2.36) are somewhat difficult. Convenient formulas based only on real math are possible after some manipulation. First,

$$Z_o = \sqrt{\frac{\ell}{c}}\sqrt{1 - \jmath\frac{r}{\omega\ell}},$$

then the real part is

$$\mathrm{Re}[Z_o] = \frac{1}{2}\sqrt{\frac{\ell}{c}}\left(\sqrt{1 - \jmath\frac{r}{\omega\ell}} + \sqrt{1 + \jmath\frac{r}{\omega\ell}}\right).$$

The term in parenthesis can be rearranged as

$$\begin{aligned}\sqrt{1 - \jmath\frac{r}{\omega\ell}} + \sqrt{1 + \jmath\frac{r}{\omega\ell}} &= \pm\sqrt{\left(\sqrt{1 - \jmath\frac{r}{\omega\ell}} + \sqrt{1 + \jmath\frac{r}{\omega\ell}}\right)^2} \\ &= \pm\sqrt{2\left(1 + \sqrt{1 + \left(\frac{r}{\omega\ell}\right)^2}\right)}.\end{aligned}$$

The real part of the characteristic impedance is then

$$\mathrm{Re}[Z_o] = \sqrt{\frac{\ell}{c}}\sqrt{\frac{1}{2}\left(1 + \sqrt{1 + \left(\frac{r}{\omega\ell}\right)^2}\right)},$$

where the positive root is taken to represent a passive structure. For the imaginary part of the characteristic impedance,

$$\mathrm{Im}[Z_o] = \frac{1}{2\jmath}\sqrt{\frac{\ell}{c}}\left(\sqrt{1 - \jmath\frac{r}{\omega\ell}} - \sqrt{1 + \jmath\frac{r}{\omega\ell}}\right).$$

The term in parenthesis can be rearranged as

$$\begin{aligned}\sqrt{1 - \jmath\frac{r}{\omega\ell}} - \sqrt{1 + \jmath\frac{r}{\omega\ell}} &= \pm\sqrt{\left(\sqrt{1 - \jmath\frac{r}{\omega\ell}} - \sqrt{1 + \jmath\frac{r}{\omega\ell}}\right)^2} \\ &= \pm\sqrt{2\left(1 - \sqrt{1 + \left(\frac{r}{\omega\ell}\right)^2}\right)},\end{aligned}$$

then

$$\mathrm{Im}[Z_o] = -\sqrt{\frac{\ell}{c}}\sqrt{\frac{1}{2}\left(-1 + \sqrt{1 + \left(\frac{r}{\omega\ell}\right)^2}\right)},$$

where the negative sign is taken to match up to the low-loss approximation above. The complex propagation constant can be similarly addressed to find that

$$\mathrm{Re}[\gamma] = \alpha = \omega\sqrt{\ell c}\sqrt{\frac{1}{2}\left(-1 + \sqrt{1 + \left(\frac{r}{\omega\ell}\right)^2}\right)},$$

and

$$\mathrm{Im}[\gamma] = \beta = \omega\sqrt{\ell c}\sqrt{\frac{1}{2}\left(1+\sqrt{1+\left(\frac{r}{\omega\ell}\right)^2}\right)}.$$

2.1.4 Impedance Boundaries

Waves traveling in opposite directions couple at the interfaces between transmission lines with different characteristic impedances. Suppose multiple lossless or lossy transmission lines are cascaded. At the interface between the nth and $(n+1)$th line, the voltage and the current must be continuous. For the voltage,

$$v_n^+ + v_n^- = v_{n+1}^+ + v_{n+1}^-,$$

while for the current,

$$\frac{v_n^+ - v_n^-}{Z_{o,n}} = \frac{v_{n+1}^+ - v_{n+1}^-}{Z_{o,n+1}}.$$

Eliminating v_{n+1}^- yields

$$v_{n+1}^+ = \frac{1}{2}\left(1+\frac{Z_{o,n+1}}{Z_{o,n}}\right)v_n^+ + \frac{1}{2}\left(1-\frac{Z_{o,n+1}}{Z_{o,n}}\right)v_n^-, \tag{2.38}$$

while eliminating v_{n+1}^+ finds

$$v_{n+1}^- = \frac{1}{2}\left(1-\frac{Z_{o,n+1}}{Z_{o,n}}\right)v_n^+ + \frac{1}{2}\left(1+\frac{Z_{o,n+1}}{Z_{o,n}}\right)v_n^-. \tag{2.39}$$

The waveforms in both directions on each line are mixed at the boundary, and waveforms are generated as necessary to fulfill the boundary condition.

For example, if the only wave that exists is v_n^+, after hitting the boundary a reflected wave

$$v_n^- = -v_n^+\frac{1-\frac{Z_{o,n+1}}{Z_{o,n}}}{1+\frac{Z_{o,n+1}}{Z_{o,n}}} = v_n^+\frac{Z_{o,n+1}-Z_{o,n}}{Z_{o,n+1}+Z_{o,n}}$$

is formed according to (2.39), yielding the classic reflection coefficient formula

$$\Gamma = \frac{v_n^-}{v_n^+} = \frac{Z_{o,n+1}-Z_{o,n}}{Z_{o,n+1}+Z_{o,n}}. \tag{2.40}$$

Similarly, a transmitted wave is formed according to (2.38), leading to the transmission coefficient

$$T = \frac{v_{n+1}^{+}}{v_{n}^{+}} = \frac{2Z_{o,n+1}}{Z_{o,n+1} + Z_{o,n}}.$$

Note that $v_{n+1}^{-} = 0$ throughout as it is assumed that the only wave introduced is v_n^+.

One of the primary techniques for maximizing the quality of signals on an interconnect is to eliminate or minimize the magnitude of impedance discontinuities to minimize the generation of reflections. When two components of an interconnect, such as a driver and a transmission line, have the same impedance, then they are said to be *matched* and will produce no reflection.

2.2 Ideal Point-to-Point Signaling

Point-to-point signaling can achieve the highest signal quality to support the fastest clock rates. The basic setup for CMOS signaling is shown in Figure 2.4a, where a driver signals a receiver through a three-conductor transmission line. The receiver is represented by capacitive parasitics to both power and ground.

To facilitate discussions on signaling, the setup can be simplified with a few strong but reasonable assumptions. If the power (V_{DD}) and ground (V_{SS}) connections have sufficient bypass capacitance, they can be wired in parallel. In this case, V_{DD} is said to be an AC ground. If the driver pulls up and down with equal strength, it can be linearized and modeled as a Thevenin equivalent circuit. With these assumptions, the schematic in Figure 2.4a can be replaced with that in Figure 2.4b.

Energy injected into the system must be dissipated to prevent it from accumulating as noise. Resistance is required at the source, at the load, or at both. The driver has some resistance that varies depending on the application. Often, drivers with very low impedance are used to drive large capacitive loads on short interconnects, such as with memory buses. Higher impedances can be used for lighter loads or when timing is more relaxed. Reflections at the driver can be minimized

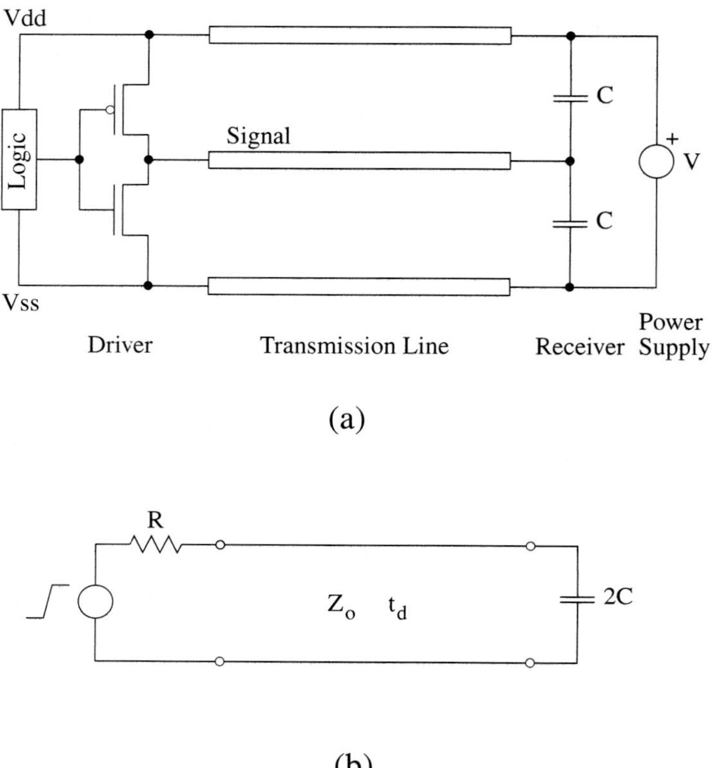

Figure 2.4. CMOS point-to-point signaling: (a) full model, (b) model with infinite bypass, symmetric and linear driver.

by designing its impedance, R, to be equal to the transmission line impedance, Z_o. This configuration is called *source matched*, or *source termination*.

At the load, the small receiver capacitance acts nearly as an open circuit that reflects the entire signal with a reflection coefficient of almost +1. These reflections can be virtually eliminated by adding a resistor in parallel to the load with a value of Z_o. This configuration is called *parallel termination*.

For a point-to-point interconnect, either source or parallel termination is sufficient, although some demanding applications use both. After a short discussion on edge rates, the cases of source plus parallel termination and of source termination alone are considered.

2.2.1 Fast and Slow Edges

Generally, there are two situations to consider. For slow rise times, the transmitted wave has sufficient time to reflect off the load and return to the driver before the driver has completed its transition. The load then modifies the impedance seen by the driver and affects its switching characteristics. In other words, the driver "feels" the load while it switches. For this situation to occur, the edge rate must be greater than twice the TOF.

For fast rise times with edge rates less than twice the TOF, the driver completes its transition before any of the transmitted wave can reflect from the load and return. During the logic transition, the driver sees only the transmission line's characteristic impedance. In this situation, the loading does not affect the driver's switching behavior.

The waveforms on an interconnect are quite different depending on whether the edges are fast or slow. The switching characteristics of the driver can also vary substantially depending on the TOF.

2.2.2 Source and Parallel Termination

With both ends of the interconnect damping reflections, the signal quality on the transmission line is excellent. Since there are no reflections on the line, the driver always sees a constant load of Z_o, and the behavior of the interconnect is mostly independent of the edge rate.[2]

A voltage divider is formed between the driver impedance and the transmission line's characteristic impedance. With a good source match, the voltage launched onto the transmission line is just half the source voltage. The driver must continuously supply current to hold this voltage. In other words, the static logic condition dissipates power, and this is the main tradeoff for parallel termination. As a rule,

2. The predominant exception is for heavy capacitive loading, but parallel termination would not generally be applied in this case.

enhanced signal quality can always be purchased with higher power dissipation. For example, this signaling approach is taken with ECL, a logic family well known for clean signals and high power dissipation.

SPICE simulations of the topology in Figure 2.4b are shown in Figures 2.5 and 2.6 for both fast and slow rise times. The 50Ω lossless transmission line has a delay of 1ns. The source ramps from 0V to 1V through 50Ω of source resistance in either 1ns or 4ns for the fast and slow cases, respectively. The capacitive load is 5pF with a 50Ω resistor in parallel for the parallel termination case.

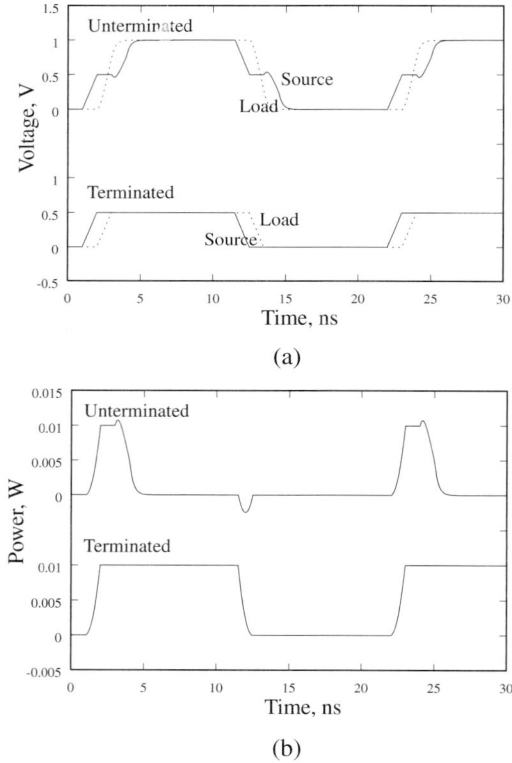

Figure 2.5. Signaling with fast edges on source terminated and on source plus parallel terminated transmission lines; 1ns rise and fall times with 1ns delay: (a) voltage, (b) power.

With both source and parallel termination, the simulations show no difference between the fast and slow cases except for the edge rates. Power is constantly expended while holding logic high. Since all resistances equal the characteristic impedance of the transmission line, the peak voltage on the line is just 0.5V. Signal quality is excellent.

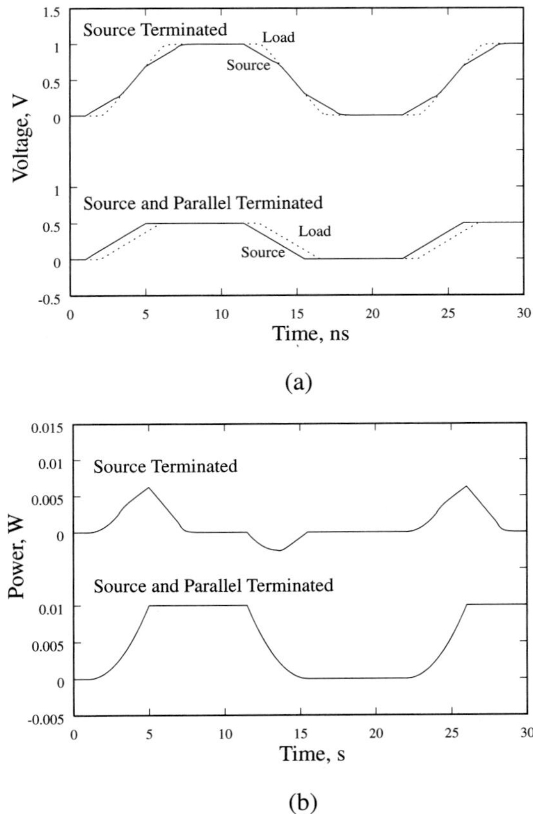

Figure 2.6. Signaling with slow edges on source terminated and on source plus parallel terminated transmission lines; 4ns rise and fall times with 1ns delay: (a) voltage, (b) power.

2.2.3 Source Termination Only

Without parallel termination at the load, the +1 reflection coefficient at the load causes the voltage to essentially double. The large reflected wave travels back to the driver where it is absorbed by the matched source impedance. Due to the presence of the reflected waveform, the signal integrity along the transmission line is not good, but it can be quite good at the load, which is where it matters for a point-to-point net. Once the load capacitance is charged, then the driver no longer needs to supply current to hold the receiver in the logic high state; therefore, the static logic condition does not dissipate power.

Unterminated nets are common with both TTL and CMOS to minimize power dissipation. With good source match, the signal integrity at the load can be quite good. However, the reflected wave is signficant and can cause difficulties on more complex topologies (such as multidrop nets). Also, a bidirectional bus using source termination can be slower because it must wait for the signal to return and terminate in the source impedance before the bus can be turned around into receive mode.

For the fast edges in Figure 2.5, the voltage at the output of the driver is the same for both the unterminated and terminated loads for a time span of 2ns (2×TOF). At that time, the reflection from the unterminated capacitive load arrives at the driver and terminates. Power dissipation is the same for the two cases for the first 2ns, then it drops off for the unterminated case because the load capacitor becomes charged and blocks DC current flow.

For the slow edges in Figure 2.6, the reflected wave returns to the driver while it is still transitioning. The voltage at the source is the sum of the source voltage and the reflected voltage, so the shape is more complex. The waveform shape at the load is still a simple ramp. Power dissipation is reduced because the driver does not have to launch a complete waveform onto the transmission line before the capacitive load begins charging.

2.3 Nonideal Signaling

Ideal signaling for CMOS is represented by point-to-point controlled-impedance nets with matched source impedance. For very high-speed signaling situations, point-to-point nets are required to achieve the required performance. In most cases, point-to-point nets are a luxury that cannot be afforded due to pin counts, board space, power dissipation, cost, and so on. In these cases, many factors arise that disrupt the signal quality, and these factors must be controlled to guarantee system performance. In nonideal situations, the basic signaling requirements that must be met are the switching incidence, which is strongly related to the receiver threshold, and the avoidance of false signaling.

2.3.1 Synchronous vs. Asynchronous

Inputs to a device are either asynchronous or synchronous. Asynchronous inputs are sensitive to voltage at all times, and they are commonly used for clocks, resets, and interrupts. Synchronous inputs sample voltage only at times specified by a clock, and they are commonly used for data, address, mode, and configuration signals. For asynchronous receivers, any time the voltage falls below V_{IL} or rises above V_{IH} for a sufficient time, a logic transition will be registered. If noise causes the voltage level to inadvertently approach valid logic levels, then a system fault can occur. For synchronous receivers, voltage levels only matter for times around clock transitions, so synchronous circuits are much more tolerant of poor signal integrity than are asynchronous circuits. A conservative design approach is to treat all signals as asynchronous.

2.3.2 Switching Incidence

Driver impedances are often not matched to the characteristic impedance of the interconnect, and higher driver impedances are often used to save power and silicon area. Due to voltage division, higher driver impedances launch lower amplitude voltage waves towards the receiver. If the voltage at the receiver is sufficient when the voltage wave arrives, then the receiver switches on the first incidence of the wave, and signaling is said to be *first incidence.*

If the first incidence is insufficient, then the voltage must build to a sufficient level to trigger the receiver. The voltage wave reflects off of the receiver, propagates back to the source, reflects there, and propagates back to the receiver. If the voltage is sufficient at this second arrival, then the receiver switches, and the signaling is said to be *second incidence.* If the signal is again insufficient, then switching may occur at the third incidence, fourth incidence, and so on. Anything other than first incidence switching involves a time penalty of two TOFs per incidence. Note that a matched source impedance requires first incidence switching because there is no reflected wave from the source.

Second incidence switching is demonstrated with the SPICE simulation in Figure 2.7. On the first incidence, the voltage at the receiver is just

$$V_{1st} = 5\frac{50}{50+250} \times 2 = 1.67V,$$

where the factor of 2 accounts for voltage doubling at the small capacitive load. A second incidence is required to take the voltage above V_{IH} (2V). First incidence switching requires either a drop in V_{IH} below 1.67V (plus some noise margin) or a

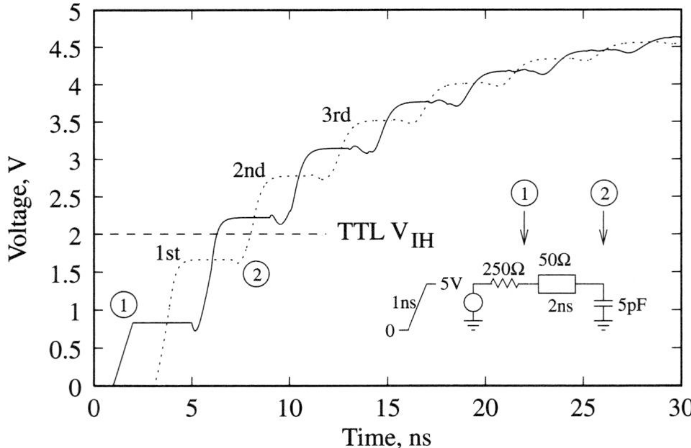

Figure 2.7. Example SPICE simulation showing second incidence switching.

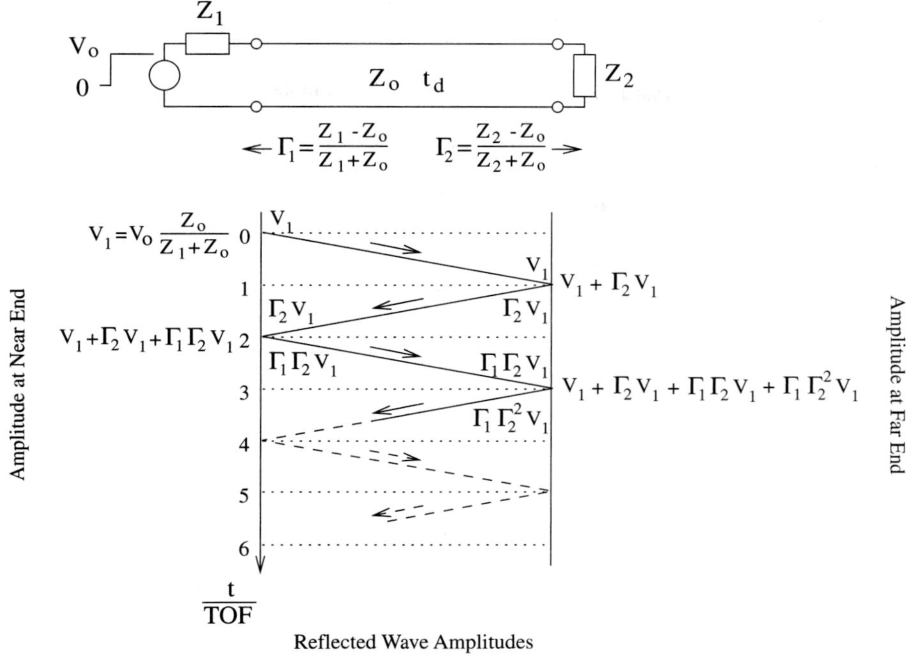

Figure 2.8. Bounce diagram for arbitrary reflection coefficients.

decrease in the driver impedance. For example, a driver impedance of 158.3Ω would produce first incidence switching with a noise margin of 0.4V.

Multiple reflections on a lossless transmission line can be visualized with the help of a bounce diagram, which tracks the amplitude of a reflected wave as it bounces off of imperfect terminations. A bounce diagram for arbitrary terminations is shown in Figure 2.8. The pulse from the voltage source is divided across the source impedance and the transmission line's characteristic impedance to launch a pulse onto the transmission line. One TOF later, the wave arrives at the far end termination. The reflected amplitude is given by the reflection coefficient Γ_2 and the total voltage is the sum of the incident and reflected waves. This total is tallied to the right. After another TOF, the wave arrives back at the source, where a reflected wave is generated with amplitude determined by Γ_1. The total voltage is the incident plus reflected waves plus the voltage that was already there. This total

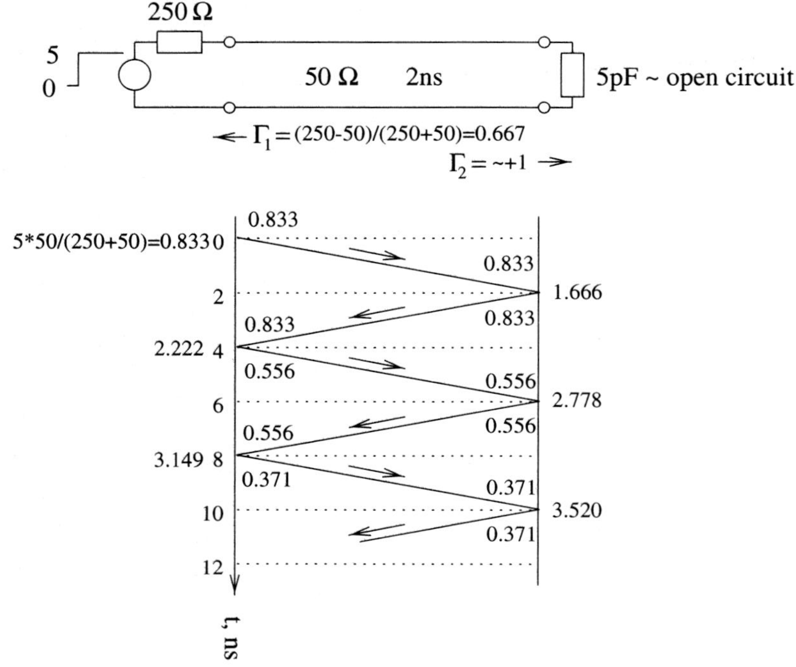

Figure 2.9. Bounce diagram for the circuit in Figure 2.7.

is tallied to the left. The process continues at each interface. Running totals down the sides of the bounce diagram show the voltage amplitudes as a function of time in increments of the TOF.

As a numerical example, a bounce diagram for the circuit in Figure 2.7 is constructed in Figure 2.9. To simplify the calculations, the load capacitance is assumed to be vanishingly small so that the reflection coefficient at the load is simply +1. The amplitudes and timing closely match the SPICE simulation in Figure 2.7.

2.4 Discontinuities

Except for very high-performance systems, high-quality point-to-point nets are rare, and in practice, signal lines experience real-world discontinuities where the impedance changes. As shown in section 2.1.4, any change in impedance alters the signal while generating a reflected wave. A few of the basic discontinuities are series inductance, shunt capacitance, capacitive loads, and impedance steps.

Many interconnects introduce series inductance, such as wirebonds, connectors, and vias, in PCBs and packages. Note that the complete current loop must be considered. For example, a signal wirebond located far from a power or ground wirebond creates a far more inductive loop than one located next to a power or ground wirebond. For very high-speed edges, when the lumped approximation fails, series inductive discontinuities can be modeled as high-impedance transmission lines.

Shunt capacitance results whenever a signal line passes near other metal. For example, a via through a hole in a ground plane results in series inductance for the via plus shunt capacitance from the via to plane. For very high-speed edges, shunt capacitive discontinuities can be modeled as low-impedance transmission lines.

Unterminated signaling results in a capacitive load at the far end due to the loading of the receiver. The capacitive load creates a reflection while affecting timing due to RC charging times.

Impedance steps occur when the characteristic impedance of the transmission line changes. For example, a 70Ω microstrip PCB line connected to a 50Ω coaxial cable introduces a significant impedance step.

These basic discontinuities are examined in detail. A general result is that discontinuities have characteristic time constants that can be used to estimate the impact of a discontinuity on an interconnect.

2.4.1 Laplace Transform

Analytical time-domain analysis requires the use of the Laplace transform. The Laplace transform pair is defined as

$$F(s) = \int_{0^-}^{\infty} f(t)e^{-st}\, dt$$

and

$$f(t) = \frac{1}{2\pi\jmath} \int_{\sigma-\jmath\infty}^{\sigma+\jmath\infty} F(s)e^{st} ds.$$

The Laplace transform pair is denoted symbolically as $F(s) = \mathcal{L}[f(t)]$ and $f(t) = \mathcal{L}^{-1}[F(s)]$. An important fact from Laplace transform theory is that

$$\mathcal{L}\left[\frac{df(t)}{dt}\right] = sF[s] - f[0^-].$$

For interconnect studies, the Laplace transforms of step functions are needed. For the unit step function

$$u(t) = \begin{cases} 0, & t < 0 \\ 1, & t \geq 0 \end{cases},$$

$$\mathcal{L}[u(t)] = \frac{1}{s}. \tag{2.41}$$

The unit ramp function is defined as

$$r(t) = \begin{cases} 0, & t < 0 \\ t/t_o, & 0 \leq t \leq t_o \\ 1, & t > t_o \end{cases},$$

for which

$$\mathcal{L}[r(t)] = \frac{1 - e^{-st_o}}{t_o s^2}.$$

The unit exponential ramp is

$$e(t) = (1 - e^{-t/t_o})u(t),$$

which is highly convenient because it is expressible without limits and because it captures the exponential rolloff typically seen in circuits. Its Laplace transform is

$$\mathcal{L}[e(t)] = \frac{1}{s} - \frac{1}{s + 1/t_o}. \tag{2.42}$$

2.4.2 Capacitive Load

The voltage-current relationship for a capacitor is $i = C\,dv/dt$. In terms of traveling waves, using (2.27) and (2.28) with $z = 0$, the relationship is

$$\frac{v^+}{Z_o} - \frac{v^-}{Z_o} = C\left(\frac{dv^+}{dt} + \frac{dv^-}{dt}\right).$$

Taking the Laplace transform and collecting terms leads to

$$V^- = \frac{1 - cZ_o s}{1 + cZ_o s} V^+. \tag{2.43}$$

For the unit step input, $v^+(t) = u(t)$, so substituting (2.41) into (2.43) and computing the inverse Laplace transform for the circuit initially at rest yields

$$v^- = (1 - 2\,e^{-t/\tau})u(t),$$

where $\tau = CZ_o$. The total voltage is then

$$v = v^+ + v^- = 2(1 - e^{-t/\tau})u(t),$$

which shows the familiar exponential charging of a capacitive load with voltage doubling for large t/τ. Because of the capacitive loading, the waveform at the load does not rise to one-half its full value ($v = 1/2 \times 2 = 1$) until $t = \tau \ln 2$, so the capacitive load introduces a delay adder of $\tau \ln 2$.

For the exponential step, $v^+(t) = e(t)$, and repeating the above process yields

$$v^- = \left(1 + \frac{2e^{-t/\tau} - (\frac{t_o}{\tau} + 1)e^{-t/t_o}}{\frac{t_o}{\tau} - 1}\right) u(t).$$

For very fast rise times, $t_o \ll \tau$, and

$$v^- \approx (1 - 2\,e^{-t/\tau})u(t),$$

which is the same result as for the unit step input. For very slow rise times, $t_o \gg \tau$, and

$$v^- \approx (1 - \,e^{-t/t_o})u(t),$$

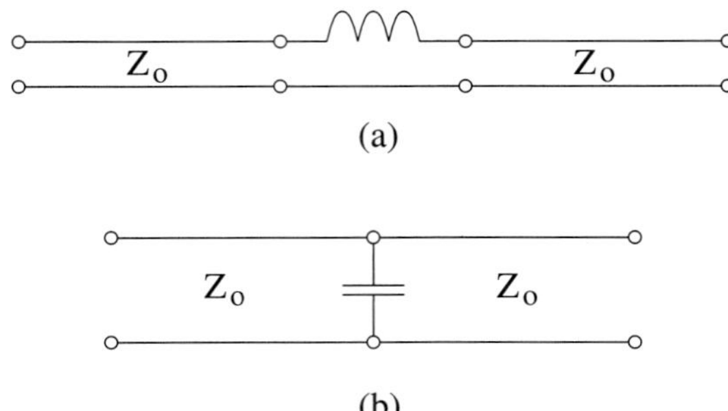

Figure 2.10. Common discontinuities on transmission lines: (a) series inductance, (b) shunt capacitance.

which is the same as the incident wave, so the capacitive load acts like a perfect open circuit with a reflection coefficient of +1.

In general, the effect of a discontinuity can be determined by comparing the rise time to the discontinuity's time constant, τ. If the rise time is short compared to τ, then the behavior is like that for the unit step. On the other hand, if the rise time is long compared to τ, then the behavior is that for the idealized version of the discontinuity.

2.4.3 Series Inductance

For a wave incident from the left of the series inductance discontinuity in Figure 2.10a, the voltage and current are

$$v_1 = v_1^+ + v_1^-$$

and

$$i_1 = \frac{1}{Z_o}v_1^+ - \frac{1}{Z_o}v_1^-.$$

On the right of the inductor, assume that the line is long so there is no reflected wave, then the voltage and current there are

$$v_2 = v_2^+$$

and

$$i_2 = \frac{1}{Z_o} v_2^+.$$

Applying Kirchhoff's voltage law yields

$$v_1^+ + v_1^- = L\frac{di}{dt} + v_2^+, \tag{2.44}$$

while continuity of current requires

$$\frac{1}{Z_o} v_1^+ - \frac{1}{Z_o} v_1^- = \frac{1}{Z_o} v_2^+. \tag{2.45}$$

Eliminating v_1^+ and substituting $i = \frac{1}{Z_o} v_2^+$ yields

$$v_1^- = \tau \frac{dv_2^+}{dt}, \tag{2.46}$$

where $\tau = \frac{1}{2}\frac{L}{Z_o}$ is the time constant of the discontinuity.

Approximate Solution for Weak Reflections

For small inductances causing weak reflections, the incident wave passes through almost unchanged, then $\frac{dv_2^+}{dt} \approx \frac{dv_1^+}{dt}$ and

$$v_1^- \approx \tau \frac{dv_1^+}{dt}. \tag{2.47}$$

The reflected wave is a pulse with width approximately equal to the edge rate of the incident wave. The pulse amplitude is positive for low-to-high transitions and negative for high-to-low.

Exact Solution for Step Input

Eliminating v_1^- between (2.44) and (2.45) with $i = \frac{1}{Z_o} v_2^+$ yields

$$v_1^+ = v_2^+ + \tau \frac{dv_2^+}{dt}, \tag{2.48}$$

Assuming that the circuit is initially at rest, then after Laplace transformation,

$$V_2^+ = V_1^+ \frac{1/\tau}{s + 1/\tau}. \tag{2.49}$$

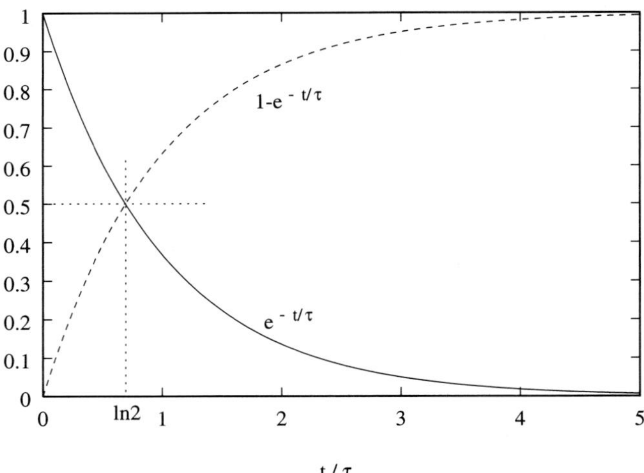

Figure 2.11. Exponential functions describe the reflected and transmitted waves for a step function incident onto series inductance and shunt capacitance.

If v_1^+ is a unit step, then

$$V_2^+ = \frac{1}{s} - \frac{1}{s + 1/\tau}.$$

Inverse Laplace transformation then provides the voltage propagating past the inductor as

$$v_2^+ = (1 - e^{-t/\tau})u(t),$$

and from (2.45), the reflected wave is

$$v_1^- = e^{-t/\tau}u(t).$$

These two functions are plotted in Figure 2.11. The reflected pulse is an exponential spike falling to half its original value by $t = \tau \ln 2$. The transmitted pulse is smoothed, rising to half its final value by time $t = \tau \ln 2$. A SPICE simulation showing step reflection from a series inductor is shown in Figure 2.12.

Considering the frequency domain, at low frequencies the inductor is a short

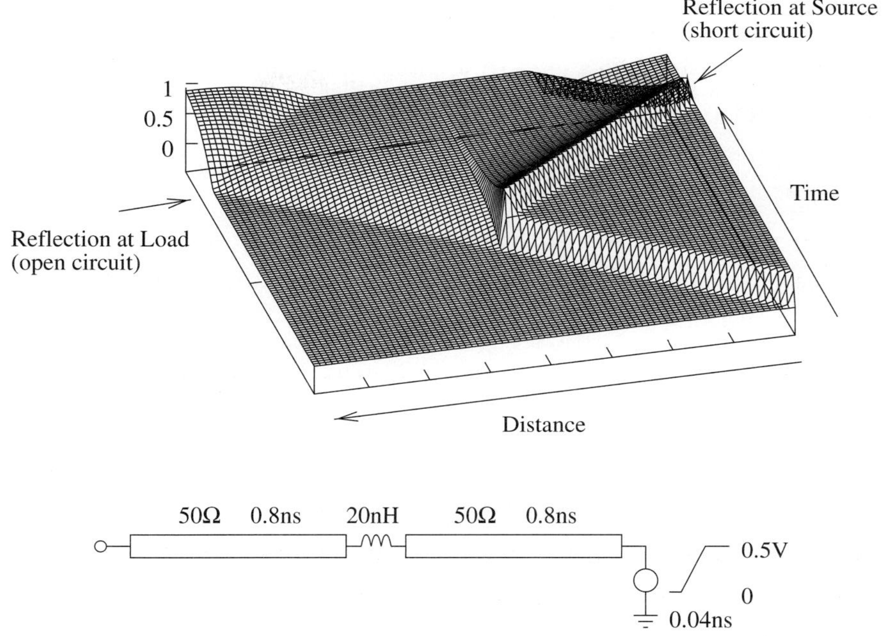

Figure 2.12. A series inductor reflects an exponential pulse back towards the source and low-pass filters the wave that continues down the line.

circuit, so these components readily pass through. At high frequencies, the inductor is an open circuit and reflects these components with a reflection coefficient of +1.

With the high-frequency components reflected back towards the source, the edge rate of the transmitted pulse is rounded off. At the receiver, the slower edge rate causes the wave to be detected later than it would be with a faster edge. If the receiver trip point is $1/2V_{\text{DD}}$, then the delay adder due to the inductor is $\tau \ln 2$.

In total, the series inductor causes noise by creating a reflected pulse and adds delay to the signal due to edge rate degradation. For the step input, the reflected pulse voltage peak equals the amplitude of the step, while the delay adder is $\tau \ln 2$.

Exact Solution for Exponential Input

Let the incident wave v_1^+ be a unit exponential ramp, substitute (2.42) into (2.49), and inverse Laplace transform to get

$$v_2^+ = 1 - \frac{e^{-t/\tau}}{1 - t_o/\tau} - \frac{e^{-t/t_o}}{1 - \tau/t_o}, \tag{2.50}$$

assuming that the circuit is initially at rest.

The delay of the transmitted wave from time $t = 0$ can be found by setting $v_2^+ = 1/2$, then

$$\frac{1}{2} = \frac{e^{-t/\tau}}{1 - t_o/\tau} + \frac{e^{-t/t_o}}{1 - \tau/t_o}, \tag{2.51}$$

but a simple solution for t is not possible. The delay adder is the delay of the transmitted wave minus the delay of the incident wave. For very fast edges with $t_o \ll \tau$, the delay is $t = \tau \ln 2$, so the delay adder is $t_d = \tau \ln 2 - t_o \ln 2 \approx \tau \ln 2$, the same as for the step input. For very slow edges with $t_o \gg \tau$, the delay is $t = t_o \ln 2$ for a delay adder of zero. Just as is seen with the capacitive load, the behavior of the discontinuity is bounded between the fast edge step response and the slow edge ideal case (short circuit, for the series inductor).

An approximate solution to (2.51) is

$$t = \tau \ln 2 + t_o \ln 2,$$

which is asymptotically exact at the extremes of t_o and τ, but reaches -17.4% error near $t_o = \tau$. Using this formula, the delay adder is $t_d = \tau \ln 2$, so to a good approximation, the delay adder for any series inductive discontinuity for any edge rate is just the delay adder for the step function. The approximate solution can be fitted to improve the accuracy near $t_o = \tau$. One solution with maximum error of 1.45% is

$$t = (\tau \ln 2 + t_o \ln 2)\left(1 + 0.194\, e^{-0.174(t_o/\tau + \tau/t_o - 2)}\right).$$

The reflected pulse is found from (2.45), using (2.50), to be

$$v_1^- = \frac{1}{1 - t_o/\tau}\left(e^{-t/\tau} - e^{-t/t_o}\right).$$

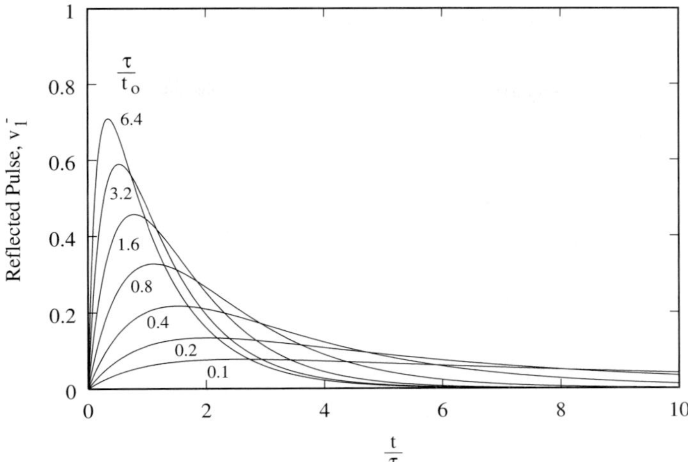

Figure 2.13. Pulses reflected from a series inductive discontinuity for an incident exponential step depend on the rise time of the step.

Several pulses are plotted in Figure 2.13, where it can be seen that larger discontinuities (increasing τ) and/or faster edges (decreasing t_o) lead to sharper pulses with higher peak amplitudes but with faster decay rates. The peak occurs at time

$$t_{\text{peak}} = \frac{\ln \frac{t_o}{\tau}}{1/\tau - 1/t_o}.$$

Trapezoidal Ramp Example

Using a linear ramp for the source excitation, a sample SPICE simulation is shown in Figure 2.14, where the schematics are shown in inset. For this problem $\tau = 0.1$ns, so the 1ns rise time appears somewhat slow, and the effect of the discontinuity should be small. In Figure 2.14a, the source is located sufficiently far from the inductor so that it has finished switching by the time the pulse returns. The pulse at point 1 is roughly square since the incident edge is triangular. The peak amplitude of the reflected pulse can be estimated using (2.47) to be

$$v_1^- \approx \frac{1}{2}\frac{10 \times 10^{-9}}{50}\frac{1}{1 \times 10^{-9}} = 0.1\text{V},$$

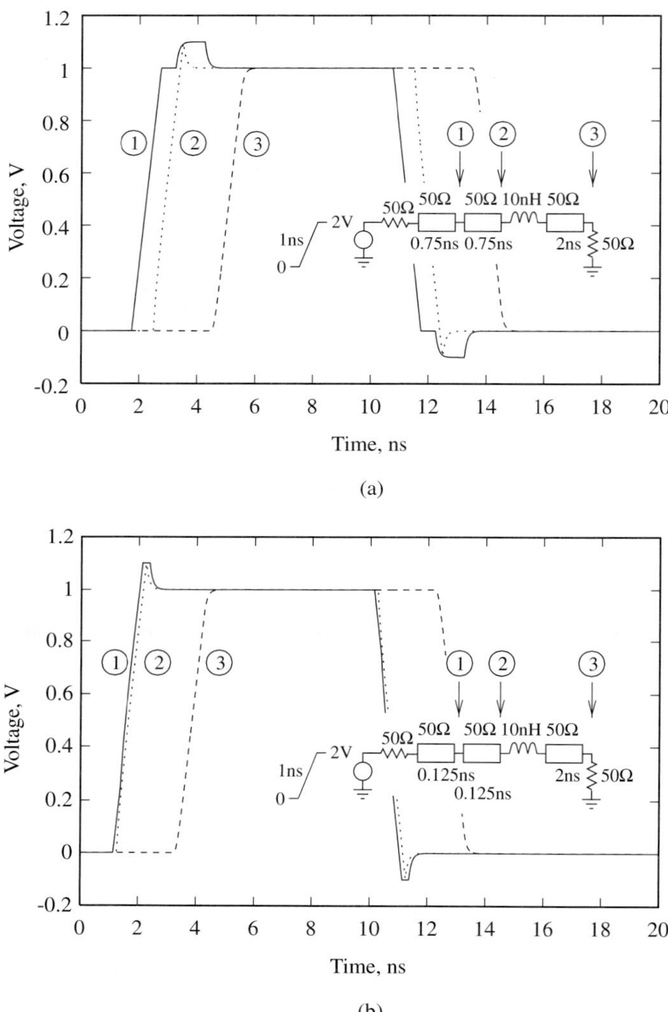

Figure 2.14. Reflected and transmitted waveforms at a series inductance discontinuity: (a) source far from the discontinuity, (b) source near to the discontinuity.

which closely matches the simulation. The voltage swing is 1V because the 2V source evenly divides between the source and transmission line impedances. The pulse width of 1ns is approximately equal to the edge rate. For the transmitted waveform observed at point 3, there is no significant change in the edge rate, but rounding of the corners is observable.

In Figure 2.14b, the source is close to the inductor so that the initial edge is still rising when the reflected pulse returns. The reflected pulse width and amplitude are similar to the far case in Figure 2.14a, but the reflected pulse voltage superimposes onto the rising edge and is not as easily observed. In this case, some of the pulse is "hidden" within the edge itself, where it helps the signal to rise more quickly.

These two cases show how signal integrity is improved when the edge rate lengthens compared to the delays in the circuit. Slower edges produce less noise and allow the noise to be hidden in the rise/fall times if the interconnects are short enough.

2.4.4 Shunt Capacitance

For a wave incident from the left of the shunt capacitance discontinuity in Figure 2.10b, the voltage and current are

$$v_1 = v_1^+ + v_1^-$$

and

$$i_1 = \frac{1}{Z_o} v_1^+ - \frac{1}{Z_o} v_1^- .$$

On the right of the capacitor, assume that the line is long so there is no reflected wave, then the voltage and current there are

$$v_2 = v_2^+$$

and

$$i_2 = \frac{1}{Z_o} v_2^+ .$$

Applying Kirchhoff's current law yields

$$\frac{1}{Z_o}v_1^+ - \frac{1}{Z_o}v_1^- = C\frac{dv_2^+}{dt} + \frac{1}{Z_o}v_2^+, \tag{2.52}$$

while continuity of the voltage requires

$$v_1^+ + v_1^- = v_2^+. \tag{2.53}$$

Eliminating v_1^+ yields

$$v_1^- = -\tau\frac{dv_2^+}{dt}, \tag{2.54}$$

where $\tau = \frac{CZ_o}{2}$ is the time constant of the discontinuity.

Approximate Solution for Weak Reflections

For small capacitances causing weak reflections, the incident wave passes through almost unchanged, then $\frac{dv_2^+}{dt} \approx \frac{dv_1^+}{dt}$ and

$$v_1^- \approx -\tau\frac{dv_1^+}{dt}. \tag{2.55}$$

The reflected wave is a pulse with width approximately equal to the edge rate of the incident wave. The pulse amplitude is negative for low-to-high transitions and positive for high-to-low.

Exact Solutions and Duality

Eliminating v_1^- between (2.52) and (2.53) yields

$$v_1^+ = v_2^+ + \tau\frac{dv_2^+}{dt}.$$

This equation is exactly the same as that for the series inductance in (2.48), so the shunt capacitor behaves exactly the same as the series inductance but with a different time constant. Similarly, the reflected wave in (2.54) is the same as that in (2.46) except for a sign change. The shunt capacitor reflects the same pulse as the series inductor but with opposite sign and different time constant. The series inductance and shunt capacitance are duals, a reasonable result since inductance encapsulates the effects of the magnetic field while those for the electric field are captured by capacitance.

Cascades of discontinuities are not covered by the mathematical framework above, but it is interesting to note that since the pulses reflected from shunt capacitive and series inductive discontinuities are of opposite sign, they cancel when $\tau_C = \tau_L$, which occurs when $Z_o = \sqrt{L/C}$.

Trapezoidal Ramp Example

A sample SPICE simulation is shown in Figure 2.15, where the edge rate of 1ns is slow compared to the 0.1ns time constant of the shunt capacitor. In Figure 2.15a, the source is located sufficiently far from the capacitor so that it has finished switching by the time the pulse returns. The pulse at point 1 is roughly square since the incident edge is triangular. The peak amplitude of the reflected pulse can be estimated using (2.55) to be

$$v_1^- \approx -\frac{1}{2}\, 4 \times 10^{-12}\, 50\, \frac{1}{1 \times 10^{-9}} = -0.1\text{V},$$

which closely matches the simulation. The pulse width is approximately equal to the edge rate. For the transmitted waveform observed at point 3, there is no significant change in the edge rate, but rounding of the corners is observable.

In Figure 2.15b, the source is close to the capacitor so that the initial edge is still rising when the reflected pulse returns. The reflected pulse width and amplitude are similar to the fast case in Figure 2.15a, but the reflected pulse voltage superimposes onto the rising edge and is not as easily observed. In this case, some of the pulse is "hidden" within the edge itself, where it inhibits the rise of the signal.

2.4.5 Impedance Step

The general rule of thumb is to keep all transmission lines at the same characteristic impedance; otherwise, reflections are generated with amplitudes given by the reflection coefficient in (2.40). For series inductance and shunt capacitance, the frequency-dependent impedance of these discontinuities causes frequency-dependent

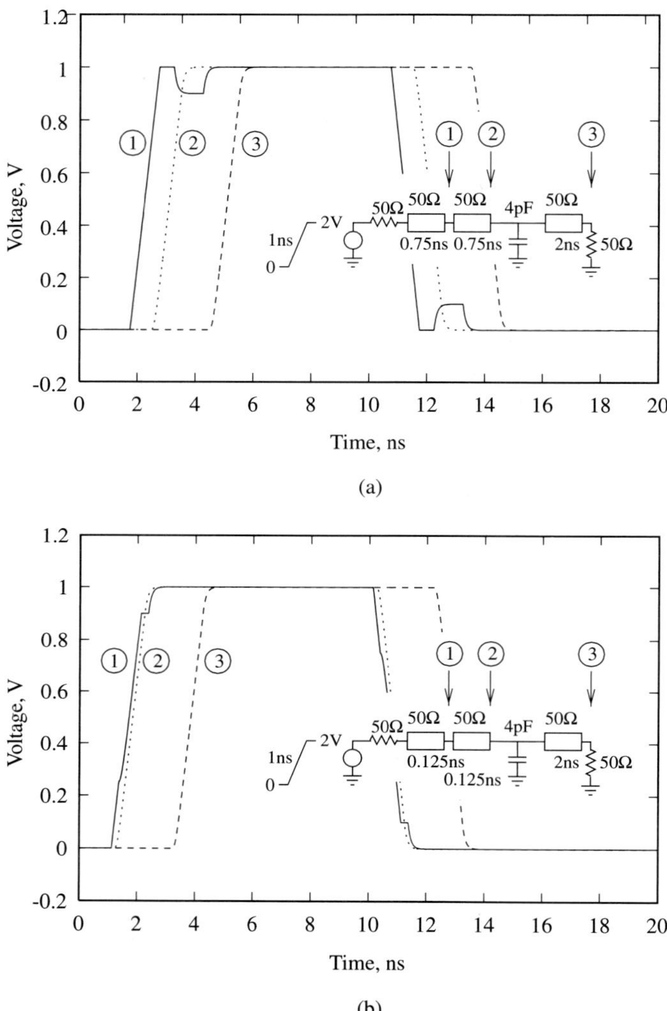

Figure 2.15. Reflected and transmitted waveforms at a shunt capacitance discontinuity. (a) Fast edge produces a reflected pulse with width approximately equal to the edge rate. (b) Slow edge at the source produces a truncated pulse with modified rise time.

reflection coefficients and the generation of narrow spikes with widths similar to the edge rate. In contrast, the characteristic impedance of transmission lines can be essentially constant over very broad bandwidths, so the reflection coefficient can be strongly frequency-independent. In these cases, a fraction of the entire waveform can be peeled off and reflected back down the line.

Assume a signal propagates down a transmission line with characteristic impedance Z_1 and meets a discontinuity where the line impedance changes to Z_2 thereafter. When the lines are long or terminated, then (2.40) holds and two cases must be considered. When $Z_1 > Z_2$, the reflection coefficient is negative and the reflected wave is a negative copy of the incident wave with reduced amplitude. When $Z_1 < Z_2$, the reflected wave is a positive copy. The incident and reflected waves superimpose.

Voltage is continuous at the discontinuity, so the signal continues onto the second transmission line with peak amplitude based on the total voltage on the first line. When the incident and reflected waves have the same sign, they add, and the voltage signal on the second transmission line is large. Conversely, a negative reflection coefficient produces a smaller transmitted voltage signal.

Example

Sample SPICE simulations for both cases are shown in Figures 2.16 and 2.17, where the schematics are shown in inset. See section 2.2.1 for a discussion on fast and slow edges. In Figure 2.16, the reflection coefficient is +0.1667, so the reflected wave has the peak amplitude 1V*0.1667=0.1667V. The incident and reflected waves add up to 1.1667V when they happen to be coincident; otherwise, just the 1V incident wave or 0.1667V reflected wave is observable. The transmitted wave is launched with the peak voltage of 1.1667V, so its slew rate increases because it transitions over a larger voltage in the same time as the incident wave.

In Figure 2.17, the reflection coefficient is -0.1667, so the reflected wave has the peak amplitude -0.1667V, and the transmitted wave has the peak amplitude -0.8333V. The slew rate of the transmitted wave decreases.

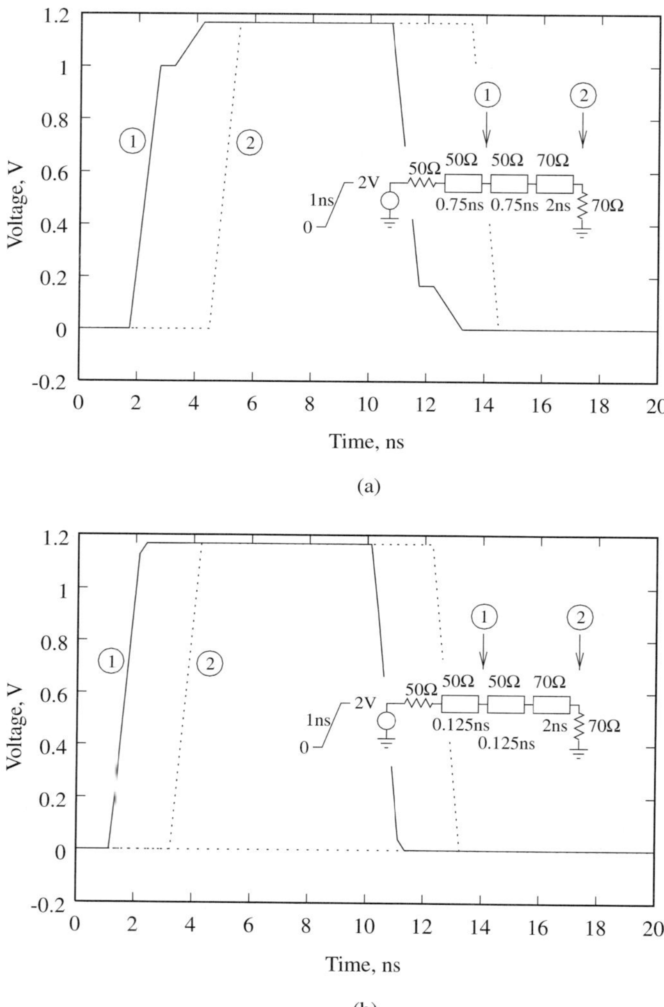

Figure 2.16. Reflected and transmitted waveforms for an impedance discontinuity to a larger impedance: (a) fast edge, (b) slow edge.

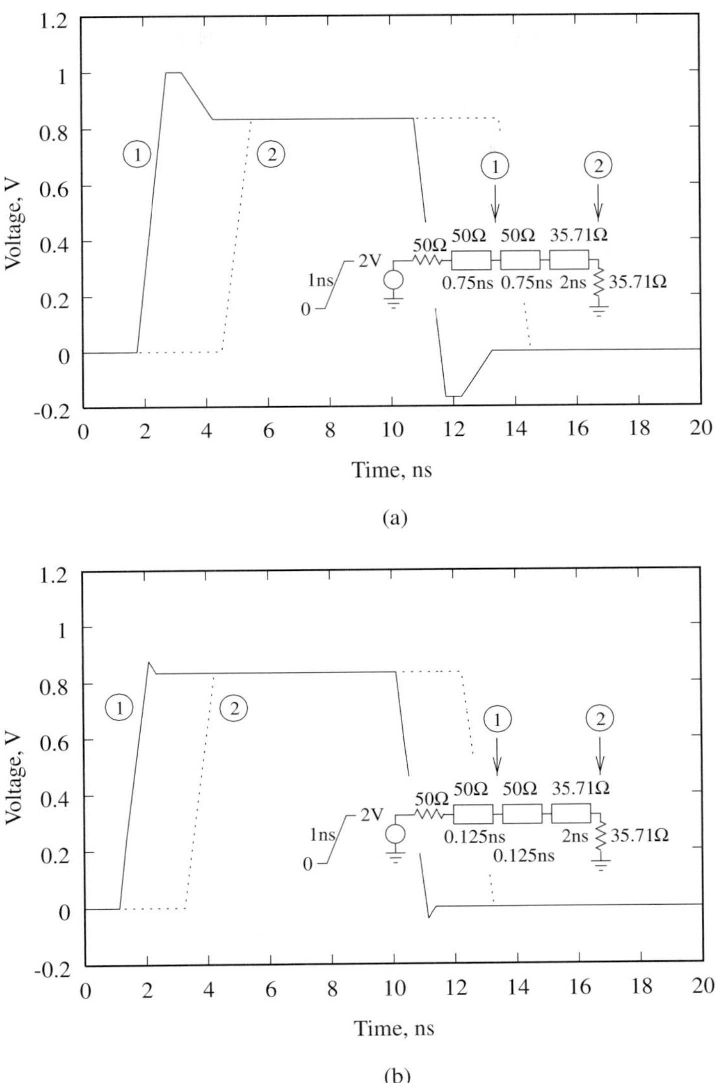

Figure 2.17. Reflected and transmitted waveforms for an impedance discontinuity to a smaller impedance: (a) fast edge, (b) slow edge.

2.5 Crosstalk

Leakage of a signal from one conductor to another is called *crosstalk*, and it can be induced through three coupling mechanisms: capacitive, inductive, and radiative. Radiative coupling is essentially a self-induced EMI disturbance and would be treated within an EMI design framework. This section is concerned with capacitive and inductive coupling.

2.5.1 Capacitive Crosstalk

All conductors have some capacitance between them, and when sufficiently close, the capacitance can become large enough to couple significant energy from one line (the aggressor, or active line) to another (the victim, or passive line). The capacitance allows displacement current to cross the gap and inject into the victim line. Since the impedance is equal looking both ways up and down the line, the current splits equally and sends waves propagating in each direction. The coupling is sketched in Figure 2.18, where capacitance is distributed along the length of two transmission lines. The mutual capacitance per-unit-length is c_m. The problem in general requires

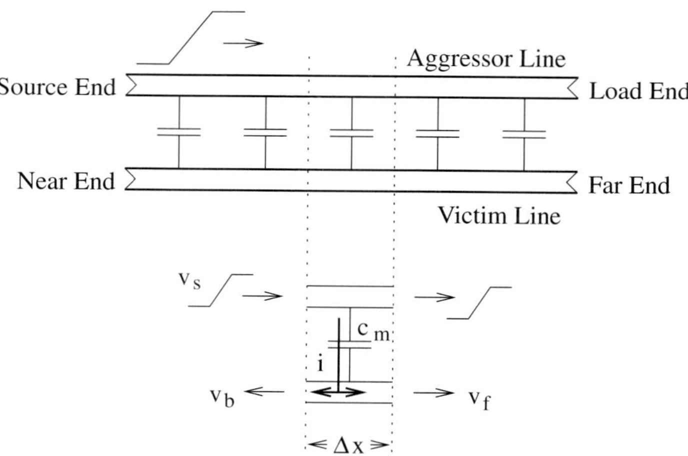

Figure 2.18. Sketch of capacitive coupling producing crosstalk.

the simultaneous solution of the coupled differential equations representing the two lines. Approximations can be applied to show the basic behavior and to derive useful formulas suitable for weak coupling.

A very short length of the coupled transmission lines is shown in cutout in Figure 2.18. As the voltage wave passes on the aggressor line, it injects current onto the victim line through displacement current, then this current splits evenly in each direction, due to symmetry. Kirchhoff's current law applied to the victim line yields

$$\frac{v_b}{Z_o} + \frac{v_f}{Z_o} = c_m \Delta x \frac{dv_s}{dt},$$

where it is assumed that, due to weak coupling, the voltage across the mutual capacitor is v_s, the amplitude of the voltage wave on the aggressor line. The voltage is continuous, so $v_b = v_f$, and

$$v_f = v_b = \frac{1}{2} Z_o c_m \Delta x \frac{dv_s}{dt}. \tag{2.56}$$

The aggressor wave creates pulses that have widths about equal to the edge rate and that propagate in opposite directions on the victim line. A low-to-high transition on the aggressor line produces positive pulses on the victim line, while a high-to-low transition produces negative pulses.

Once a crosstalk signal has been launched on the victim line, it too can create crosstalk back onto the aggressor, where it can upset the waveform there and complicate further computations of crosstalk. When this secondary crosstalk is negligible, then the coupling is said to be weak; otherwise, the coupling is strong. Simple formulas for crosstalk rely on weak coupling.

The aggressor and forward victim waveforms travel together towards the load and the far end, respectively. Therefore, at each increment on the lines, the aggressor edge can add to the victim pulse, and the victim pulse grows in magnitude the farther down the lines the waves travel. At the far end, for line length d, the total noise for weak coupling from (2.56) is

$$v_{FE} = \frac{1}{2} Z_o c_m d \frac{dv_s}{dt}, \tag{2.57}$$

which matches the capacitive portion of the same result in (10.36), derived using coupled mode theory. The noise at the far end—far-end noise (FEN)—is a single pulse with width approximately equal to the edge rate of the signal on the aggressor.

The aggressor and backward victim waveforms travel in opposite directions, so the overlap where the aggressor can inject current is only one-half the rise time, Δt. After this time period, the pulse travels unchanged to the near end. However, these pulses are generated continuously, so the near end receives them for at least 2×TOF, when the last pulse generated at the far end propagates back to the near end.

The interaction distance in (2.56) is $\Delta x = v_p \frac{1}{2} \Delta t$, where v_p is the phase velocity. To simplify the result, assume that the aggressor edge is triangular, then $dv_s/dt = v_o/\Delta t$, where v_o is the peak voltage. Substituting these into (2.56) yields the near-end noise (NEN) for weak coupling as

$$v_{NE} = \frac{1}{4} Z_o c_m v_p v_o.$$

However, $Z_o v_p = \sqrt{\ell/c}/\sqrt{\ell c} = 1/c$, where c is the capacitance per-unit-length of either line, so

$$v_{NE} = \frac{1}{4} \frac{c_m}{c} v_o.$$

Expressing as a coupling coefficient yields

$$K_{NE} = \frac{1}{4} \frac{c_m}{c}, \tag{2.58}$$

where $K_{NE} = v_{NE}/v_o$. This result matches the capacitive portion of the same result in (10.31), derived using coupled mode theory.

In summary, capacitive crosstalk results in a short pulse at the far end and a long signal at the near end. The far-end signal grows in amplitude with longer lines, while near-end noise grows in width with longer lines. The crosstalk noise is positive for low-to-high transitions and negative for high-to-low. The results are summarized in Figure 2.19.

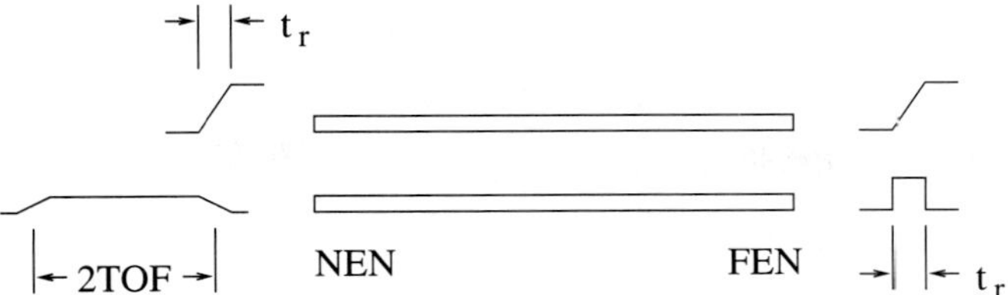

Figure 2.19. Summary of noise waveforms for weak capacitive coupling.

2.5.2 Inductive Crosstalk

The closed loops formed by two signal lines are coupled by mutual inductance, which causes a crosstalk voltage to be generated on the victim line due to changes in the current on the aggressor line according to $v = m\frac{di}{dt}$. In contrast to the capacitive case, where current is injected into the victim line, the net change in current is zero; the aggressor line can only drive current along the victim line. As a result, the forward and backward crosstalk have opposite polarities. The coupling is sketched in Figure 2.20, where mutual inductance, m, is distributed along the length of the line. Similar to the case for capacitive crosstalk, the problem in general requires

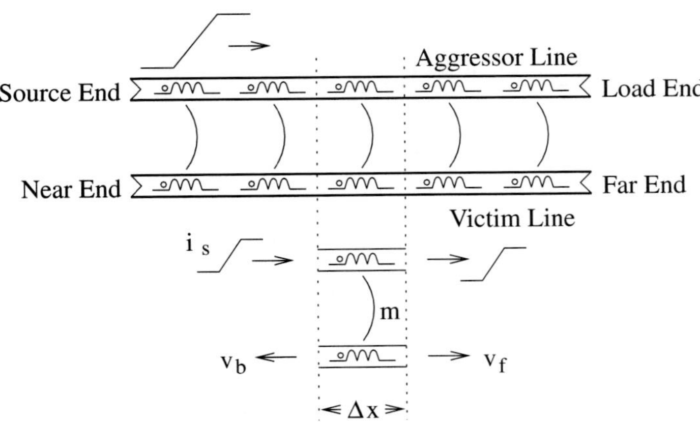

Figure 2.20. Sketch of inductive coupling producing crosstalk.

the simultaneous solution of the coupled differential equation representing the two lines. Approximations can be applied to show the basic behavior and to derive useful formulas suitable for weak coupling.

A very short length of the coupled transmission lines is shown in cutout in Figure 2.20. As the current wave passes on the aggressor line, it induces a series voltage on the victim line through the mutual inductance. Kirchhoff's voltage law applied to the victim line yields

$$v_b = m\Delta x \frac{di_s}{dt} + v_f.$$

The currents are continuous, so $v_b/Z_o = -v_f/Z_o$. Eliminating v_f and substituting $i_s = v_s/Z_o$ yields the backward wave amplitude as

$$v_b = \frac{1}{2}\frac{m}{Z_o}\Delta x \frac{dv_s}{dt},$$

and similarly for the forward wave

$$v_f = -\frac{1}{2}\frac{m}{Z_o}\Delta x \frac{dv_s}{dt}.$$

The aggressor wave creates pulses that have widths about equal to the edge rate and that propagate in opposite directions on the victim line. A low-to-high transition on the aggressor line produces a positive backward pulse and a negative forward pulse on the victim line, while a high-to-low transition reverses the signs.

At this point, the inductive crosstalk derivation follows that of capacitive crosstalk. The forward noise travels with the aggressor wave and picks up amplitude continuously. The backward noise only picks up noise for half the edge rate since it and the aggressor travel in opposite directions. The results for inductive crosstalk are

$$v_{FE} = -\frac{1}{2}\frac{m}{Z_o} d \frac{dv_s}{dt} \tag{2.59}$$

and, assuming a triangular edge on the aggressor waveform,

$$v_{NE} = \frac{1}{4}\frac{m}{\ell} v_o.$$

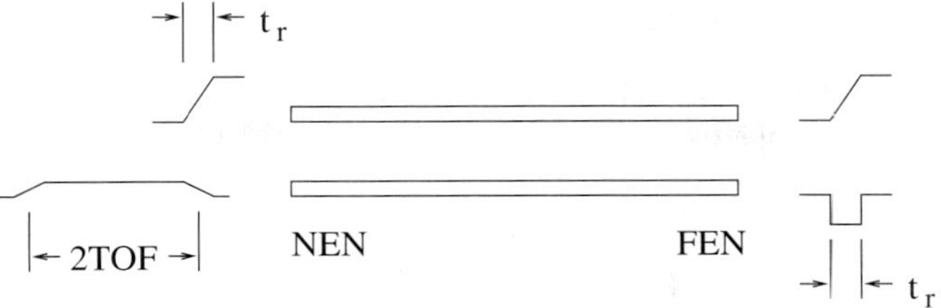

Figure 2.21. Summary of noise waveforms for weak inductive coupling.

Expressing the near-end noise as a coupling coefficient yields

$$K_{NE} = \frac{1}{4}\frac{m}{\ell}. \tag{2.60}$$

Note that these results match (10.36) and (10.31), derived using coupled mode theory.

In summary, inductive crosstalk is very similar to the capacitive crosstalk: a short pulse at the far end and a long signal at the near end. The difference is in the signs, where forward inductive and capacitive crosstalk have opposite signs. The results are summarized in Figure 2.21.

2.5.3 Total Crosstalk

In the general case, capacitive and inductive crosstalk are simultaneously present. Summing (2.57) and (2.59) yields

$$v_{FE} = \frac{1}{2}d\left(Z_o c_m - \frac{m}{Z_o}\right)\frac{dv_s}{dt}, \tag{2.61}$$

which applies for weak coupling. An implicit assumption is that the phase velocities on the two coupled lines are equal. Because of the opposite polarities for capacitive and inductive crosstalk, far-end noise enjoys some cancellation. Far-end noise is eliminated when $Z_o c_m - \frac{m}{Z_o} = 0$, or

$$\frac{c_m}{c} = \frac{m}{\ell},$$

a balanced condition where both coupling terms represent the same fraction of the aggressor line's per-unit-length parameters.

For near-end noise, summing the capacitive crosstalk from (2.58) and the inductive crosstalk from (2.60) yields

$$K_{NE} = \frac{1}{4}\left(\frac{c_m}{c} + \frac{m}{\ell}\right). \tag{2.62}$$

Cancellation is not possible, so near-end crosstalk always exists.

Note that the cancellation of far-end crosstalk enables the creation of directional couplers. With far-end cancellation, a sample of a signal moving left-to-right only appears at the left port. Similarly, a signal moving right-to-left produces a signal only at the right port. Such a directional coupler can be used to create network analyzers, which operate on the principal of directional traveling waves. From the theory, good performance requires equal phase velocities on two lines, plus careful balance of capacitive and inductive crosstalk.

In digital systems, many closely spaced lines exist with a mix of crosstalk components. For capacitive crosstalk, lines shield each other so c_m decreases rapidly for victim lines farther from the aggressor. Therefore, capacitive crosstalk is typically strong to a line's neighbors but weak to other lines. For inductive crosstalk, magnetic fields are not shielded by normal metals, so m decreases slowly and inductive crosstalk is strong for many of a line's neighbors. Summing these two results, the typical crosstalk behavior is that coupling to neighbors is both inductive and capacitive, with capacitive crosstalk often dominating, while coupling to farther lines is mostly inductive.

2.6 Topology

A point-to-point connection between two components yields the highest signal integrity, but most circuits require several components to be connected with a single net. A net connecting three or more pins is a multidrop net. For low-speed operation, just about any topology that completes the connection is acceptable. To achieve high-speed operation, the layout of the net becomes important. Several topologies for interconnecting components are sketched in Figure 2.22.

The daisy chain simply stitches from one circuit to the next and is suitable for low-speed nets. With no termination (other than any driver source impedance), each drop introduces a shunt capacitance to ground, and the stubs between drops support complex multiple reflections. Daisy chains work well when the delay of the entire length of the net is short compared to the rise time; then the net acts as a capacitive load and the driver can be scaled accordingly.

The near-end cluster or star cluster can be very effective but requires delicate balance to achieve good performance for fast edges. Consider the example in Figure 2.23, where a single stub is connected in the middle of a 10ns line for a short rise-time pulse. As the stub length increases, the ringing period increases and the ringing amplitude takes longer to settle. The voltage dividers, due to the source impedance and the impedance step at the stub, can be seen within the first 12ns on the near-end plot. When the stub reaches the same length as the main line length beyond the branch, the waveform quality improves dramatically, although multiple reflections are still required to reach the full line value. Star clusters require equal stub length and loading.

Just as for the daisy chain, if the edge rate is sufficiently long, then the line lengths in the star cluster are not important. In Figure 2.24, the star cluster of Figure 2.23 is resimulated with a slow 30ns edge, which is long enough for the source edge to propagate to the far-end load and back plus 10ns, which is one round trip TOF on the 5ns far-end section. The deviations from the perfect ramp are primarily isolated to the edge's ramp time with almost no ringing later.

In general, stubs create two principal opportunities for loss of signal quality. First, the branch in the line creates an impedance discontinuity by placing the two lines in parallel. The impedance discontinuity will appear as a simple voltage divider for the shortest round trip TOF to the closest discontinuity. Second, reflections from the unmatched terminations will return to the branching point and reflect again,

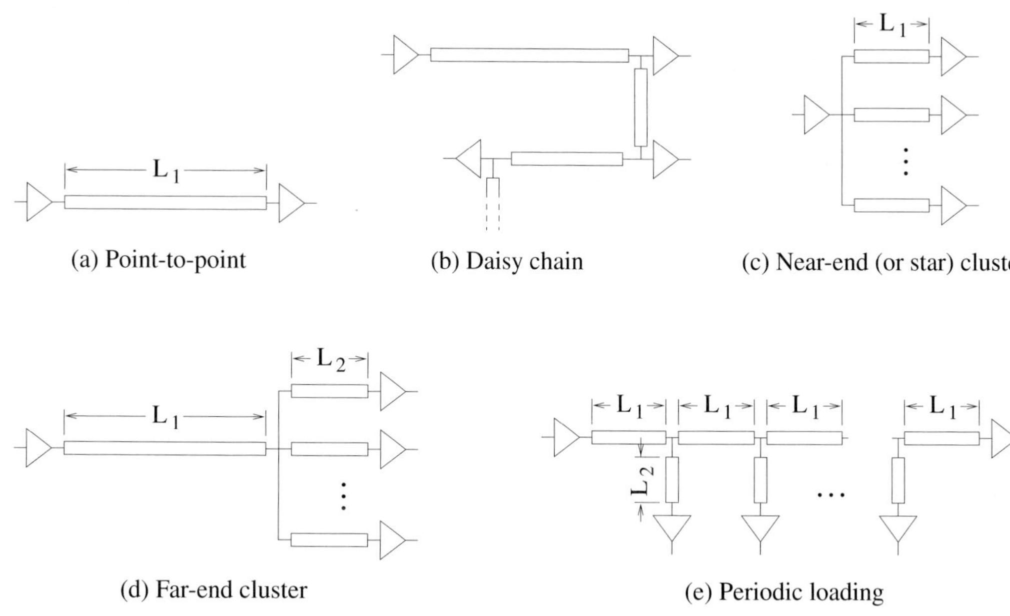

Figure 2.22. A few topologies for connecting components.

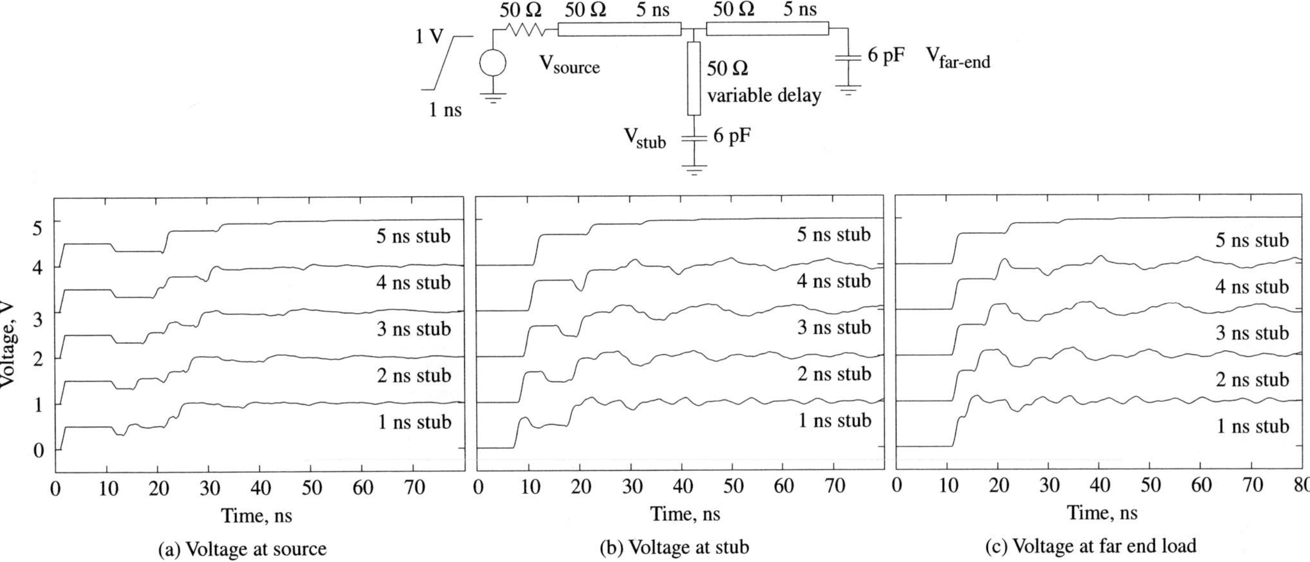

Figure 2.23. Example SPICE simulations of the effects of stubs with a fast 1ns edge. Note the 1V offsets added to separate the waveforms.

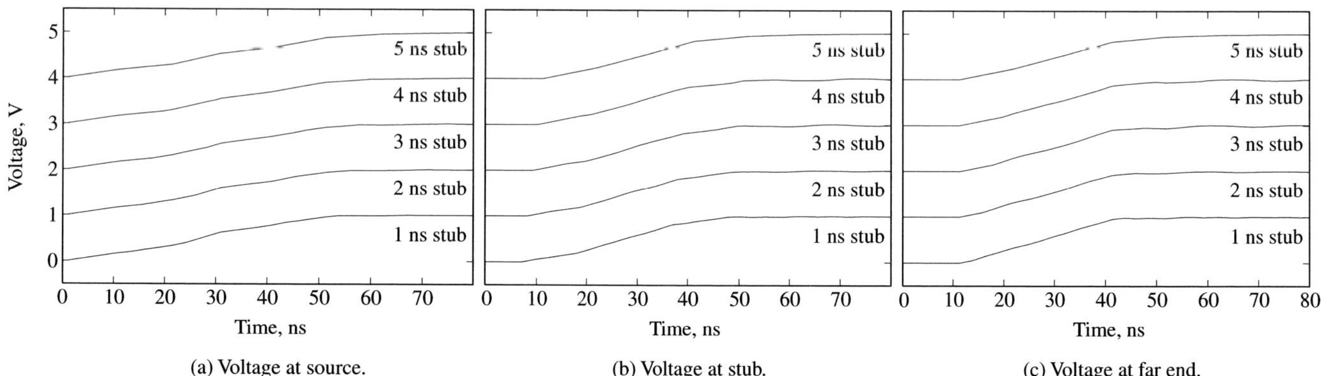

(a) Voltage at source.

(b) Voltage at stub.

(c) Voltage at far end.

Figure 2.24. Example SPICE simulations of the effects of stubs with a slow 30ns edge. Note the 1V offsets added to separate the waveforms.

causing multiple reflections. The sum of all of these interactions can be quite complex and will in general require detailed simulation.

The far-end cluster improves signal integrity by limiting the length of the far-end stubs. By keeping the lengths of stubs short compared to the edge rate, the far-end cluster behaves like a single capacitive load.

Finally, good signal integrity can be achieved by periodically loading a transmission line with additional capacitance. A straight transmission line can be tapped along its length, as shown in Figure 2.22e, to create a multidrop net well suited to creating an expansion bus. Assuming that the stubs and the distance between them are short compared to the signal rise time, then the composite structure acts as an artificial transmission line with characteristic impedance lower than the unloaded trunk transmission line. The inductance is unaffected, but the capacitive loading from the stubs and associated receivers adds to the capacitance of the trunk to pull down the characteristic impedance. If the total capacitive load is C_L, then the distributed loading is C_L/d, where d is the length of the trunk line. The loaded characteristic impedance and phase velocity are then

$$Z_L = \sqrt{\frac{\ell}{c + C_L/d}} = \frac{Z_o}{\sqrt{1 + C_L/(cd)}} = \frac{Z_o}{\sqrt{1 + C_L/C}},$$

and

$$v_L = \frac{1}{\sqrt{\ell(c + C_L/d)}} = \frac{v_p}{\sqrt{1 + C_L/(cd)}} = \frac{v_p}{\sqrt{1 + C_L/C}},$$

where $C = cd$ is the total capacitance of the trunk line, and Z_o and v_p are the unloaded values for the trunk line. Note how the loading capacitance only lowers impedance and phase velocity. To avoid reflections from an impedance step, the loaded impedance must be matched to other system impedances. High-performance expansion bus specifications include limits on stub lengths and capacitive loading per slot to ensure impedance match and signal integrity.

Wiring Rules

In a large system, detailed simulation of every net is too time consuming. Precharacterization of several topologies can provide a library of acceptable layouts, called *wiring rules*, that are guaranteed to produce acceptable signal quality and delay. A system designed using such a library is then correct by construction. Any generic topology, such as one from Figure 2.22, can be considered for characterization for wiring rules.

Detailed simulations over process corners with each variable combination are required, so the effort to construct a usable set of wiring rules can be considerable. Metrics from the simulation may include delay, rise and fall time, overshoot, undershoot, crosstalk, settling time, and SSN. After appropriate metrics and their bounds have been selected, a large number of simulations can be performed to find the range of driver strengths and loadings, characteristic impedances, and line lengths that fit within the bounds. The set of valid values are fitted to equations, and these equations become the wiring rules.

To demonstrate, the simple circuit shown in inset in Figure 2.25 has fixed driver impedance and rise time, load capacitance, and transmission line delay. What range of characteristic impedance is allowable so that the voltage at the load rises to 0.75V within 3ns? Running several cases, it can be seen that impedances above 40Ω are allowed. If a limit for overshoot is also specified, then an upper limit on characteristic impedance can be established.

2.7 Simultaneous Switching Noise

Ideally, every signal would be provided with separate and isolated power and ground connections. For packaged digital components, such dedicated pins drive up packaging costs by increasing pin counts. Common practice is to share several ground and power pins among all of the signal pins, and the performance of the system is strongly related to the ratio of the number of signal pins to the number of ground and/or power pins. For example, an 8:1 signal-to-ground ratio is common, where the number of power pins equals the number of ground pins. Very high-speed interfaces require ratios as small as 2:1, while very low-performance interfaces may need only one power pin and one ground pin.

Shared ground and power pins enable coupling from one signal line to another. Because the power distribution inductance is small, the coupled noise is small if few drivers are switching at any given time. However, the level of coupled noise increases with the number of switching outputs, and when the whole bus is switching, the noise level can become excessive. For this reason, this noise mechanism is called *simultaneous switching noise* (SSN). It is also known as *delta-I noise* since it results from the rate-of-change of current across the package inductance. SSN is treated in detail in chapter 3.

2.8 System Timing

Prior sections have discussed in detail many mechanisms that cause degradation in digital signaling. If signal integrity is sufficiently reduced, then the performance of

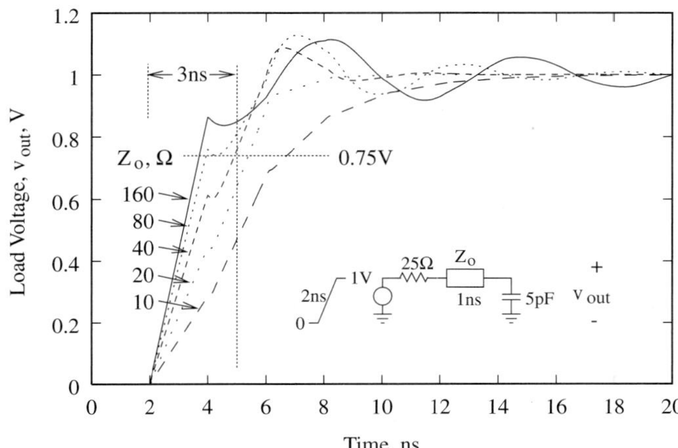

Figure 2.25. Example simulations to generate a wiring rule for allowed transmission line impedance given delay, driver impedance, and load.

the digital system will suffer. The maximum clock rate of a digital system is related to the signaling specifications, driver and receiver performance, and interconnect length. Eye diagrams offer a convenient method to verify that signaling specifications are met while providing estimates of skew.

2.8.1 Maximum Clock Rate

The maximum clock rate of a synchronous system is of primary interest, and the waveforms in Figure 2.26 show the major components that set the upper limit on the clock rate: driver propagation delay (t_p), TOF on the interconnect, and the setup and hold times at the receiver. Note that the driver's rise/fall time is subsumed into the driver's propagation delay; a longer rise time causes the driver's output to reach the receiver's trigger point later, so delay is increased.

Assuming that the data transitions and is sampled on the rising clock edge, then before the rising clock edge, it must be true that

$$P > t_p + \text{TOF} + t_S.$$

Since the clock rate is inversely proportional to frequency, the maximum clock rate is

$$f_{max} < \frac{1}{t_p + \text{TOF} + t_S}. \tag{2.63}$$

After the rising clock edge, it must be true that

$$t_H < t_p + \text{TOF}. \tag{2.64}$$

So the maximum clock rate is given by (2.63) under the condition that (2.64) is true.

Normally, short driver delay is desired to help maximize the system clock frequency. However, for very short and fast interconnects, the TOF is very small and longer t_p may be needed to satisfy the receiver hold time.

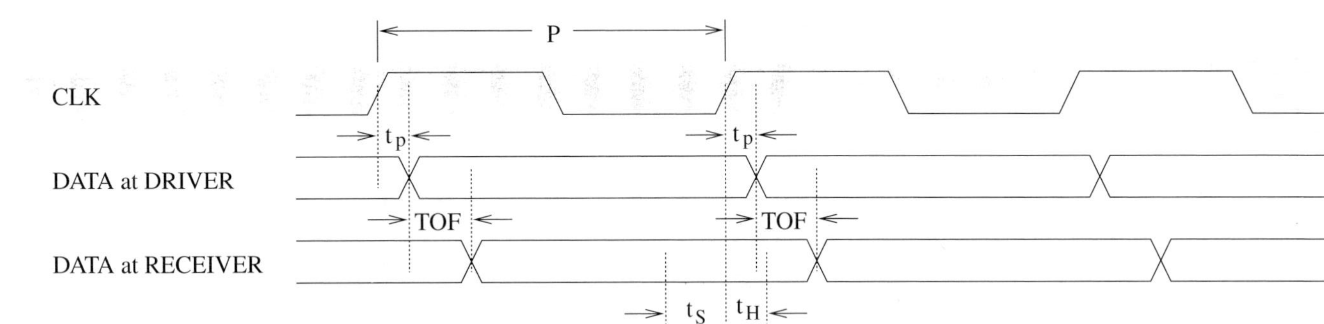

Figure 2.26. Timing diagram for determining maximum clock rate.

2.8.2 Eye Diagrams

Signaling specifications must be met for every signal at every clock cycle. For a long stream of bits, it can be difficult to tell if the signal meets the specifications. To facilitate analyses, the bits can be superimposed to create a single plot from which all of the signaling specifications can be checked.

Consider the clocked data presented in Figure 2.27, where it is assumed that the data is sampled on the rising edge of the clock. The data associated with each rising edge can be cut out and plotted together, as shown for a few of the edges. A good plot results when the waveform is plotted for one clock period before and after the edge, then the full data bit, plus half the one before and half one after, are captured. Superimposing all of the bits then builds an eye diagram. For multiconductor interconnects, the eye diagrams for each line can be superimposed to examine the signal specifications for the whole interconnect.

As an example, consider the schematic in Figure 2.28, where the driver and

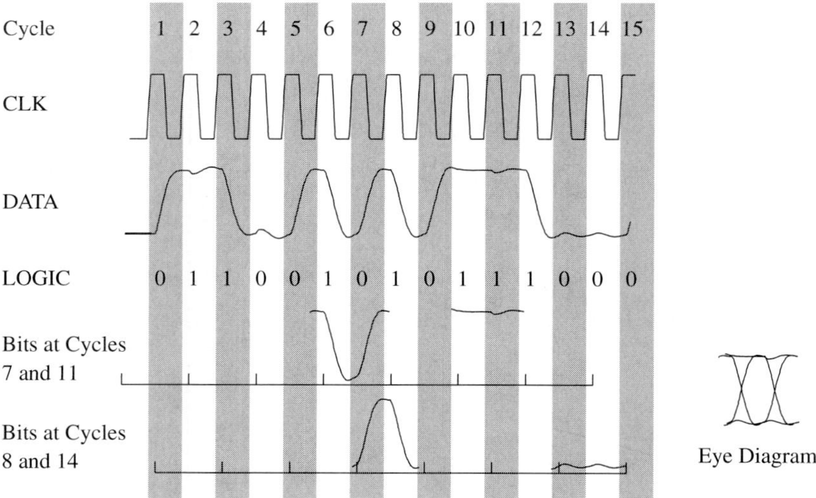

Figure 2.27. Eye diagrams are constructed by overlapping the individual bits in a bit stream.

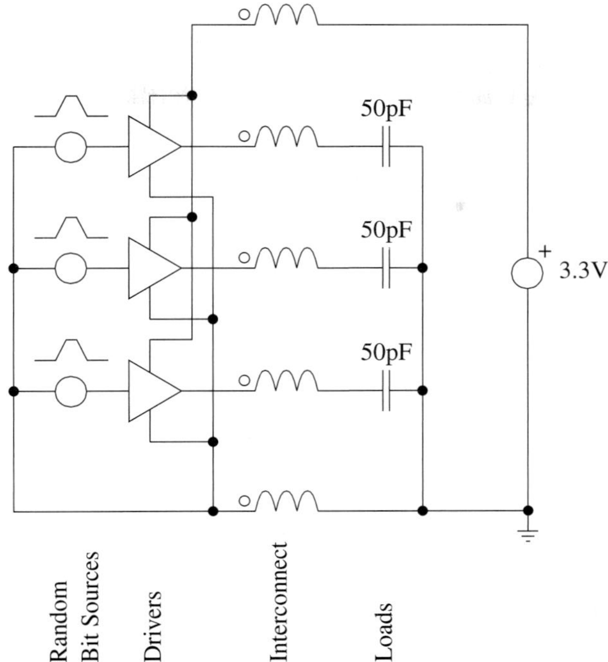

Figure 2.28. Schematic for demonstration of eye diagrams from SPICE simulations.

interconnect models are those appearing in the netlist in Appendix C. Each driver is driven by a random bit stream to simulate a digital signaling sequence. SSN and crosstalk introduced by the package degrade the signal quality at the receiver, modeled here by simple capacitance. In Figure 2.29, the data is driven at a 125MHz clock rate, and the resulting eye diagram for all three signal lines taken together is very clean. As the clock rate is increased to 250MHz in Figure 2.30, the eye diagram begins to show significant impact due to finite edge rate and skew. By 500MHz, shown in Figure 2.31, the eye diagram begins closing down. Note how as the clock frequency rises, the waveforms do not trace each other as well.

One bit of data is often called a *symbol*. In a single stream of data, if the waveform of one bit is not completely settled by the time of the clock transition

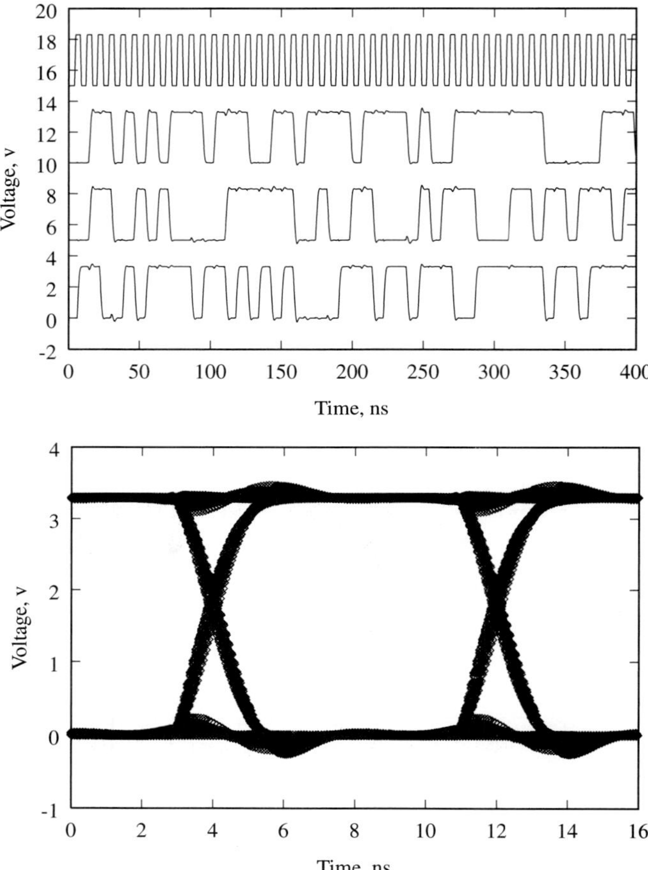

Figure 2.29. Clock and output waveforms (with 5V offsets) and eye diagram of the outputs for the circuit in Figure 2.28 at 125MHz.

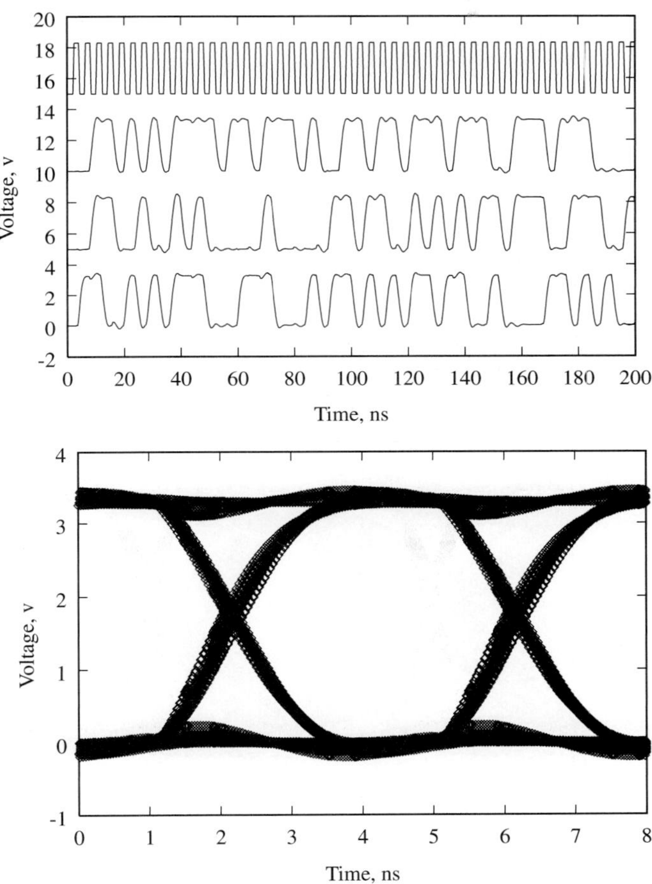

Figure 2.30. Clock and output waveforms (with 5V offsets) and eye diagram of the outputs for the circuit in Figure 2.28 at 250MHz.

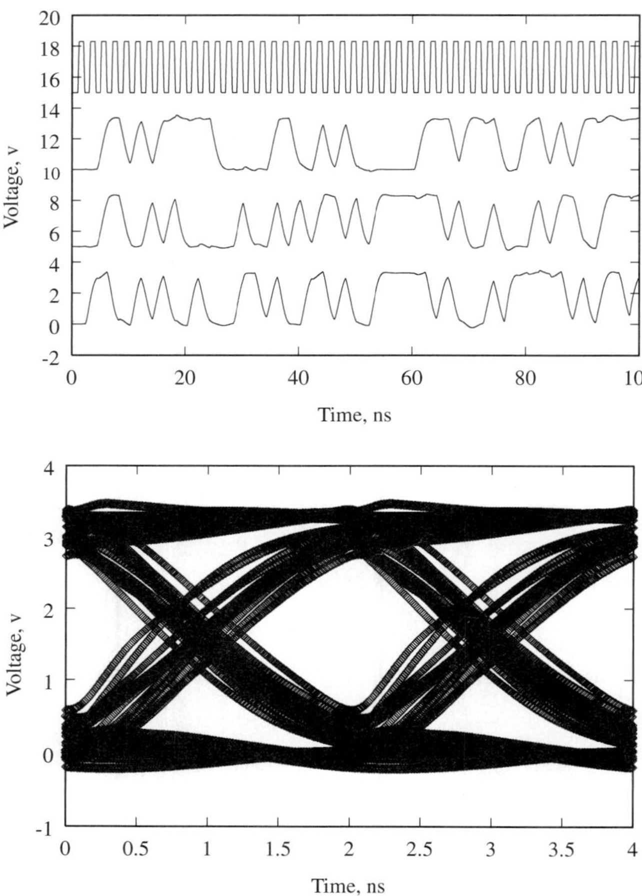

Figure 2.31. Clock and output waveforms (with 5V offsets) and eye diagram of the outputs for the circuit in Figure 2.28 at 500MHz.

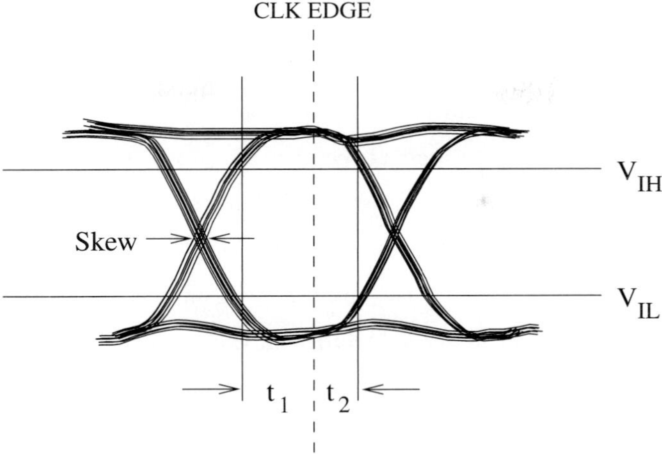

Figure 2.32. Eye diagrams help determine if signals satisfy signaling specifications.

for the next bit, it will affect the shape of the following bit, leading to lower eye quality. The effect is readily apparent in Figure 2.31. When one symbol affects the next, *intersymbol interference* (ISI) is said to occur.

Signal quality only really matters at receivers, where the signal is interpreted according to the logic specifications. Consider the representative eye diagram shown in Figure 2.32. The input specification is superimposed onto the eye diagram. Assuming that the signal is sampled at the clock edge, then the receiver specifications will be satisfied when $t_1 > t_S$ and $t_2 > t_H$, the setup and hold times, respectively.

2.8.3 Skew, Jitter, and Margin

The timing diagram in Figure 2.26 implies that the clock and data edges fall at precise times. As can be seen in the eye diagrams in Figures 2.29 through 2.31, data edges fall along a range of times due to the influences of SSN, crosstalk, TOF variations, and other effects. The uncertainty in the arrival time of a signal edge is the *signal skew*, and all sources of skew must be taken into account to ensure reliable operation of the system. *Clock skew* refers specifically to skew from all sources on the clock line in a synchronous system.

Clock generators inherently produce some variation in the timing of clock edges at their outputs, and this variation is called *clock jitter*. In a system driven by a single clock generator, all components see the same variation in the timing of the clock edges, so the timing budget is not directly affected by clock jitter. However, jitter does effectively reduce the available clock cycle by occasionally producing one clock edge late and the next one early, and the lower effective clock cycle does affect the timing budget. So clock jitter is included in the timing budget, not because of the uncertainty in the clock edge, but because of the uncertainty in the clock period.

Skews and jitter can be incorporated into the system clock estimates in (2.63) and (2.64), with the precise representation depending on how the skew is broken down. Assuming that the driver propagation delay and the receiver setup and hold times are all worst-case from the manufacturer's data sheet, then with TOF skew, clock skew, clock jitter, and margin, the maximum clock rate is given by

$$f_{max} < \frac{1}{t_p + \mathrm{TOF} + t_S + \Delta t_{\mathrm{TOF}} + \Delta t_{clock} + \Delta t_{jitter} + \Delta t_{margin}}$$

assuming that

$$t_H < t_p + \mathrm{TOF} + \Delta t_{\mathrm{TOF}} + \Delta t_{clock} + \Delta t_{jitter} + \Delta t_{margin}$$

holds true.

If eye diagrams are not available, skew can be estimated from peak noise and the signal slew rate. In-phase noise pushes a waveform higher, so the rising edge arrives at the load earlier. Similarly, out-of-phase noise pushes a waveform lower for later arrival of rising edges. Therefore, additive noise directly causes skew in the timing of the rising and falling edges. The effect of delay caused by noise is often referred to as *pushout*. Assuming that the edges are approximately linear, the slew rate is $\Delta v/\Delta t$. The skew is then approximately the noise voltage times the inverse of the slew rate:

$$\Delta t_{\mathrm{TOF}} = v_{\mathrm{noise}} \frac{\Delta t}{\Delta v}. \qquad (2.65)$$

For example, in Figure 2.31, the peak-to-peak noise at the center of the eye is approximately 1V, while the slew rate is 3.3V/2ns. Applying these numbers to (2.65) yields about 0.6ns of skew compared to about 0.7ns from the eye diagram.

Note from (2.65) that a steeper waveform edge induces less skew for a given amount of noise. On the other hand, sharp edges induce noise through additional crosstalk and SSN. Therefore, there is a tradeoff between noise and timing skew. Because digital signaling operates with noise margins, there is no benefit to reducing noise below a certain level. Since lowering the noise through reduced edge rate pays a penalty in timing, an optimal system will utilize the fastest edge rates consistent with an acceptable noise level.

Skew estimates from peak noise voltage can also be used to help set noise budgets. Assuming worst-case additive skew, then the allowed skew from the timing budget can be allocated to noise sources such as crosstalk and SSN to help resolve design options like routing density and package style. Because detailed eye diagram simulations are time-consuming to set up and run, they may be more appropriate in later design confirmation stages.

2.8.4 Dual Data Rate

In some systems, data is sampled on both the rising and falling edges of the clock, and these systems are said to be *dual data rate* (DDR). One of the main advantages of DDR is that the bandwidth of the clock is halved for a given data rate, so DDR is an excellent choice for high-speed interconnects. However, to latch the data with a rising edge, the clock must be delayed by 90° with a delay-locked loop (DLL) or doubled in frequency with a phase-locked loop (PLL), so extra circuitry is required at the receive end. Timing considerations are unchanged for DDR except for the halving of the period plus the addition of jitter for the DLL or PLL.

2.9 Exercises

1. Show that

$$v^+ = \frac{v + Z_o i}{2}$$

and

$$v^- = \frac{v - Z_o i}{2}$$

on a lossless transmission line.

2. Show that the inductance per-unit-length of the parallel-plate transmission line in Figure 2.33 is

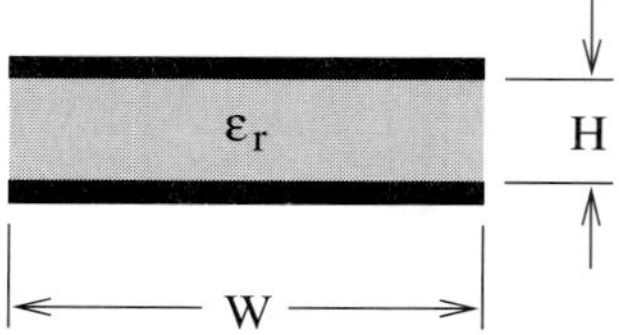

Figure 2.33. Cross section of a parallel plate transmission line.

$$\ell = \mu_o \frac{H}{W}$$

when $W \gg H$. Show that the characteristic impedance is

$$Z_o = \sqrt{\frac{\mu_o}{\epsilon_r \epsilon_o} \frac{H}{W}}.$$

3. (a) Using the notation in Figure 2.34, show that the input impedance of a loaded lossy transmission line is

$$Z_{\text{in}} = Z_o \frac{Z_L + Z_o \tanh(\gamma d)}{Z_o + Z_L \tanh(\gamma d)}. \tag{2.66}$$

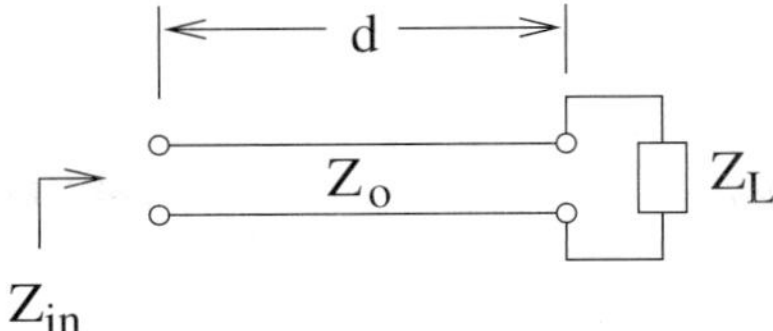

Figure 2.34. Notation for a loaded transmission line.

(b) Show that the input impedance of a loaded lossless transmission line is

$$Z_{\text{in}} = Z_o \frac{Z_L + \jmath Z_o \tan(\beta d)}{Z_o + \jmath Z_L \tan(\beta d)}. \tag{2.67}$$

4. (a) Describe the signal integrity effect of a series capacitor in terms of frequency-dependent impedance and reflection coefficients.

   (b) Set up and run a SPICE simulation to demonstrate the effects of a series 10pF capacitor on the reflected and transmitted signal. Use a 50Ω system, 1ns rise and fall times, 7ns wide pulse, and choose the transmission line lengths to be long compared to the edge rate.

5. Assume that a capacitor is inserted in series into a long transmission line. Set up and solve for the exact reflected, transmitted, and total voltages for an incident unit step, assuming the circuit is initially at rest. Find an approximate solution for small capacitance. Show that the time constant for the discontinuity is $\tau = 2CZ_o$.

6. Repeat problem 4 for a shunt inductance of 25nH. Is this an effective way to block the transmitted signal?

7. Assume that a long transmission line is shorted to ground in the middle with a shunt inductor. Set up and solve for the exact reflected, transmitted, and total voltages for an incident unit step, assuming the circuit is initially at

rest. Find an approximate solution for small inductance. Show that the time constant for the discontinuity is $\tau = 2L/Z_o$.

8. Describe the signal integrity at the load for the circuit in Figure 2.4b as the driver impedance Z_o varies from 1Ω to 2500Ω. Assume a receiver with 6pF of capacitive loading, a 50Ω transmission line with 10ns of delay, and a driver pulse of 1ns rise and fall, 7ns width, and 1V peak.

9. How are forward and backward crosstalk affected when

   (a) The characteristic impedances of the two transmission lines are different, but the phase velocities are the same?

   (b) The characteristic impedances of the two transmission lines are equal, but the phase velocities are different?

10. (a) Find the largest driver impedance that supports first incidence switching on a long 70Ω transmission line with a small capacitive load.

    (b) Repeat for second incidence switching.

    Use a 5V source swing and TTL logic high (2V) receiver threshold with no margin.

11. Discuss the impact of using a driver impedance lower than necessary for first incidence switching on timing, SSN, crosstalk, and power dissipation. (A single SPICE example may be used for the discussion.)

12. Sketch 35ns of the waveform at points 1 and 2 for the schematic in Figure 2.7, modified as follows: source voltage rises from 0 to 3V in 2ns, source impedance is 150Ω, transmission line impedance is 75Ω with 10ns of delay, and the load capacitance is small.

13. In source synchronous systems, the clock is edge-aligned and routed with the data to clock the data into the receivers. This system eliminates clock skew, driver delay, and TOF from the timing budget, enabling higher clock rates.

Derive the setup and hold time requirements for a DDR source synchronous system.

14. Show that far-end skew caused by crosstalk is approximately independent of edge rate. Assume weak coupling and trapezoidal edges.

15. (a) Show that the input impedance of a lossless transmission line shorted at the far end is given by

$$Z_{\text{in}} = \jmath Z_o \tan(\beta d).$$

Draw a sketch of Z_{in} for βd ranging from 0 to 3π and label the frequency bands where the transmission line looks inductive and where it looks capacitive. At what frequencies does this shorted transmission line look like an open circuit?

(b) Show that the input impedance of a lossless transmission line open-circuited at the far end is given by

$$Z_{\text{in}} = -\jmath Z_o \cot(\beta d).$$

Draw a sketch of Z_{in} for βd ranging from 0 to 3π and label the frequency bands where the transmission line looks inductive and where it looks capacitive. At what frequencies does this opened transmission line look like a short circuit?

(c) Repeat for a transmission line terminated in a resistance equal to the characteristic impedance of the line.

16. Find the transmission line equations by applying Kirchhoff's voltage and current laws to the differential transmission line section in Figure 2.35. How does the result compare to (2.3) and (2.4)?

17. Let $\tau = z + \nu t$ and find the direction of travel, velocity, and characteristic impedance of a lossless transmission line for an arbitrary waveform.

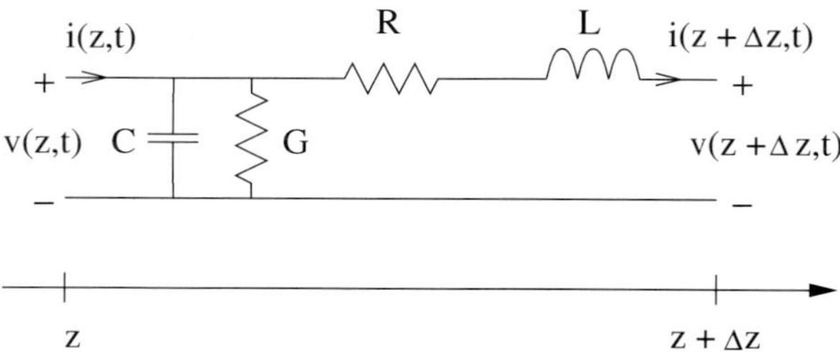

Figure 2.35. Lumped model of a short length of a transmission line. Note the reversal in component order from Figure 2.1.

18. Assume that a voltage source is connected in series at the junction between two lossy transmission lines. Find the reflection and transmission coefficients for the junction. What is the condition for matched impedance ($\Gamma = 0$)?

19. A 1V 50Ω driver with a 1ns edge rate drives a transmission line with $\ell =$ 289nH/m, $c =$ 115pF/m, and $r =$ 20Ω/m loaded at the far end by 5pF. For the lossy and lossless cases, plot the waveforms at the source and load for a low-to-high transition for line lengths of 10cm, 30cm, 90cm, and 270cm. When is the lossless assumption valid?

20. Determine a wiring rule for the variable length line in Figure 2.23 such that the ringing is limited to ±20% overshoot and undershoot.

21. For the star and daisy chain topologies in Figure 2.36,use SPICE simulations to determine which is better in terms of delay, skew, overshoot/undershoot, and power dissipation. Use multicycle simulation and eye diagrams, if available.

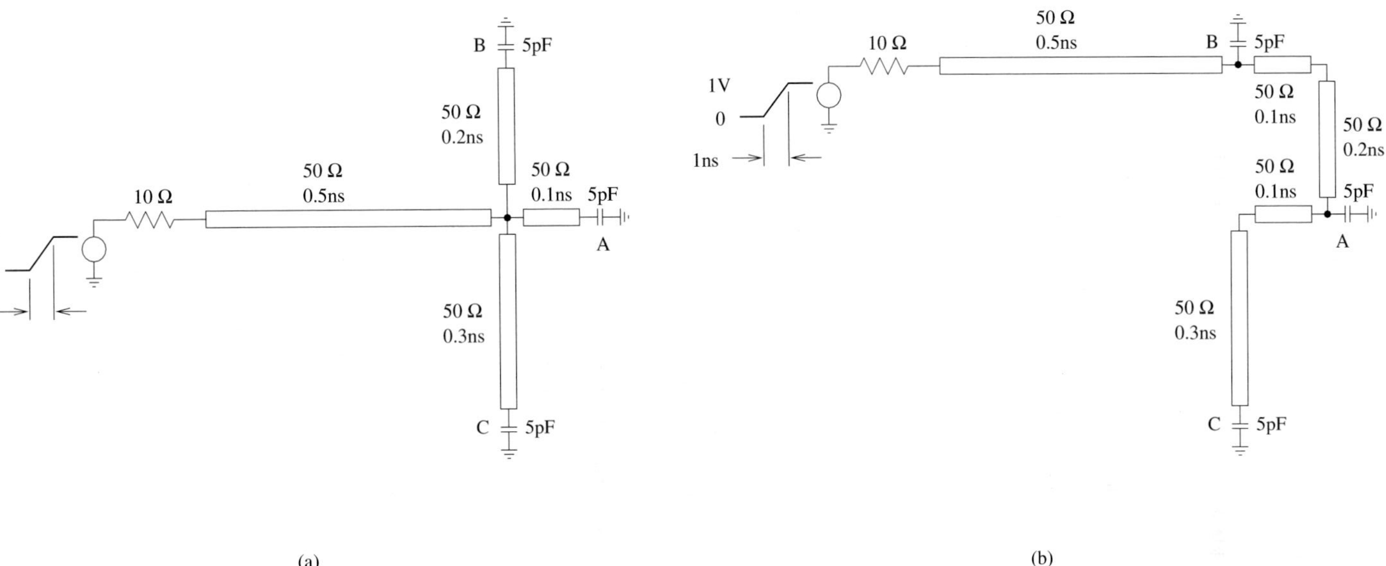

Figure 2.36. Two routing schemes to connect three loads to a single driver: (a) star, (b) daisy chain.

Chapter 3

SIMULTANEOUS SWITCHING NOISE

Simultaneous switching noise (SSN) refers to noise generated in a digital system due to rapid changes in voltage and current caused by many circuits switching at the same time. It can be a severe problem for fast and/or wide buses on packaged devices, where the fast-changing currents produce large voltages on the package power and ground pins. For this reason, SSN is also referred to as SSO, for *simultaneously switching outputs*.

SSN is not a random noise process, such as thermal noise or one-shot noise, so SSN is deterministic and can be modeled and simulated. A major goal of detailed system simulation is to characterize SSN to look for violations of noise margins and for effects on timing.

SSN is generated in circuits through many mechanisms, where each has a different mitigating strategy. On-chip and on-module SSN, generated by signaling between drivers and receivers located on the same package, can be minimized through interconnect parasitic reduction and/or incorporation of bypass capacitors. Off-chip and off-module SSN, generated by signaling between drivers and receivers in separate packages, generally requires parasitic reduction, although bypass capacitors have some effect. SSN induced from crosstalk can be minimized by increased line spacing or by multilayered substrates using planes to isolate routing layers.

SSN is principally due to package and socket inductance in the power distribution system. High-performance parts are almost always mounted on PCBs with solid power and ground planes, so power distribution up to the part is excellent. The weak link is then the connection between the die and the planes: the package and any sockets. The power and ground inductances in packages and sockets can usually be treated as lumped elements, and this chapter makes that assumption.

The huge variety of available logic styles and termination schemes makes impossible an exhaustive coverage of all combinations. For the following discussion, CMOS rail-to-rail unterminated signaling is assumed where all power supply parasitics are assumed to result from a package and/or socket. Logic signaling involves the charging and discharging of capacitances by the signal line. As shown in Figure 3.1, logic 0 or 1 is signaled with the appropriate charging and discharging of two capacitors by using the signal line to force the voltage at their junction. Driving the signal line to zero charges the upper capacitance while discharging the lower, as in Figure 3.1a. Conversely, driving the signal line to $+V$ discharges the upper capacitance while charging the lower, as in Figure 3.1b. The paths taken by these charging and discharging currents create unwanted voltages on power distribution inductances according to $L\frac{di}{dt}$.

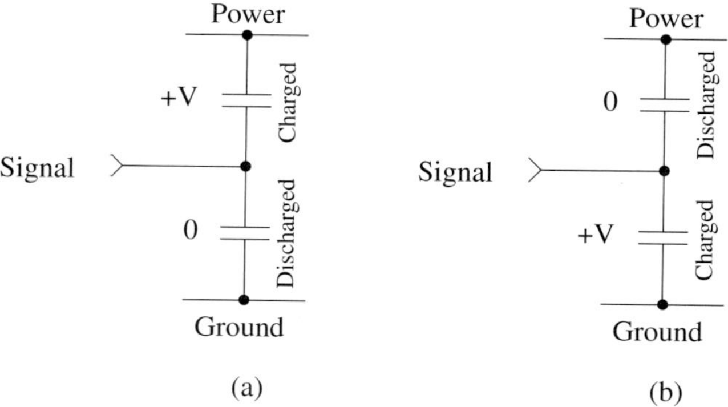

Figure 3.1. CMOS rail-to-rail unterminated signaling: (a) logic 0, (b) logic 1.

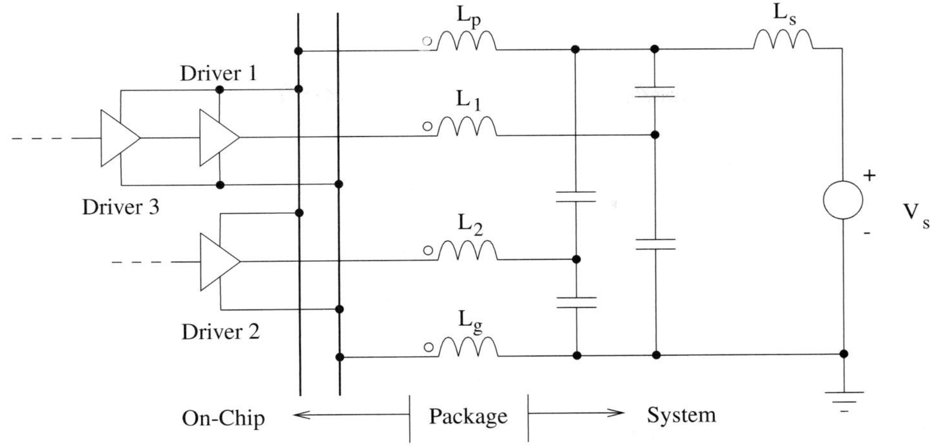

Figure 3.2. Canonical schematic for investigating SSN.

3.1 Origins of SSN

The origins of SSN can be investigated with the help of the simple schematic in Figure 3.2, which includes all of the components involved with SSN. The schematic shows a system consisting of a digital chip with two off-chip drivers (Drivers 1 and 2) and one on-chip driver (Driver 3) that connect to the system through a package. The package model consists of partial self-inductances, which are shown, and partial mutual inductances, which are not. The system loads for the off-chip drivers are represented by capacitances.

There are two major cases to consider: on-chip switching and off-chip switching. For on-chip switching, Driver 3 signals to Driver 1. For off-chip switching, Drivers 1 and 2 signal to the system receivers represented by capacitive loads. The effects of the package inductance are different for each case.

3.1.1 On-Chip Switching

For on-chip switching, Driver 3 must charge and discharge the input capacitances of Driver 1. For the high-to-low transition, the current paths are shown in Figure 3.3.

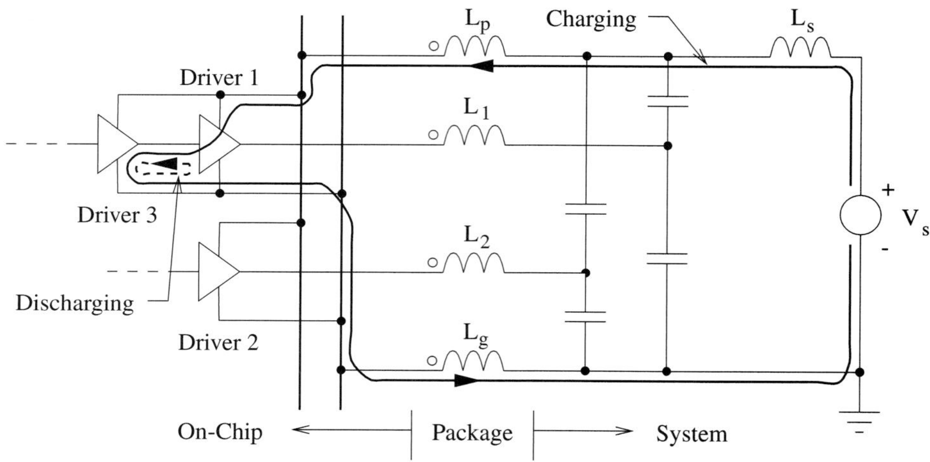

Figure 3.3. Current paths for signaling a change from logic 1 to logic 0 between an on-chip driver and an on-chip receiver.

For discharging the lower capacitance, the currents are wholly contained on-chip and generate no interconnect noise. For charging the upper capacitance, the currents flow through the package and supply inductances and induce voltages according to $L\frac{di}{dt}$. For the low-to-high transition, the upper capacitance is discharged while the lower capacitance is charged, and from the interconnect and supply point of view, the current paths are identical.

The current flows through both the package power and ground partial inductances, but not through the signal partial inductances. Since the current flow through L_p and L_g is in opposite directions, the partial mutual inductance reduces the inductance in the loop. The inductance in this case is $L = L_p + L_g - 2M_{pg}$, where M_{pg} is the partial mutual inductance between L_p and L_g.

Due to voltages induced on the package and supply inductances, the on-chip voltage is reduced from v_s. The on-chip voltage is given by $v_{\text{chip}} = v_s - L\frac{di}{dt} - L_s\frac{di}{dt}$. The drop in on-chip voltage is called *supply droop*, *rail collapse*, or *supply compression*. While the on-chip voltage initially drops in value, at later times it will typically ring around v_s with a damped profile. The reduction in supply levels slows on-chip circuits and reduces the maximum clock speed.

Several strategies are apparent for minimizing rail collapse. The basic strategy is to reduce the induced noise by decreasing either inductance or the rate of change in the current. The options are:

1. Slow down the on-chip drivers to reduce di/dt. This option is not practical when maximum clock rate is required.

2. Reduce the power supply inductance. High-speed design generally requires closely spaced power and ground planes to obtain the least inductance from the power supply to the circuits.

3. Reduce the package partial self-inductances for the power and ground paths. Lower inductance can be achieved by adding power and ground pins and by shortening the length of the power and ground path. Power and ground planes may also reduce the partial self-inductances.

4. Increase the package partial mutual inductance between the power and ground partial self-inductances. Power pins should be located as close to ground pins as possible. Power and ground pins are ideally manipulated in pairs. Closely spaced power and ground planes may also produce large partial mutual inductance.

5. Bypass the power supply inductance. Adding capacitors to the circuit enables rapidly changing current components to be supplied close to the circuits. The slowly changing current still comes from the power supply. This strategy works as long as the bypass capacitors can be recharged between switching events. The current paths for bypass capacitors are shown in Figure 3.4. Note that the load capacitances for the off-chip drivers provide some bypass capacitance, but the total value of capacitance is typically very small.

6. Bypass the package inductance. Bypass capacitance can also be placed on the package or on the chip to supply rapidly changing current even closer to the

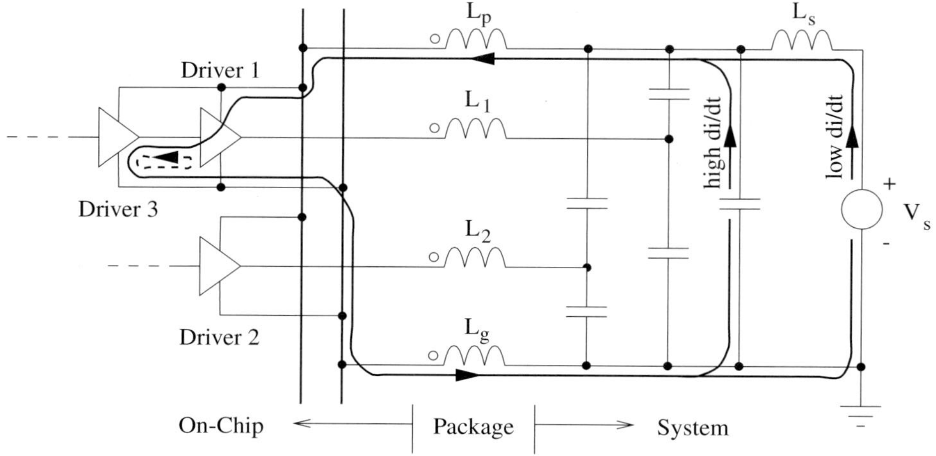

Figure 3.4. On-chip switching current paths for high-to-low transition with bypassed power supply inductance.

drivers. Rail collapse can be reduced to very small levels by bypassing the package.

7. Eliminate the package inductance entirely by directly attaching the chip to the system card, called *direct chip attach* (DCA). This is a very high-performance option that raises issues in other areas like testing, shipping, handling and assembly, and reliability.

In practice, a combination of these strategies are used with the cost of implementation regulating the relative weight of each. Low-cost approaches could use QFP packaging with surface-mount bypass capacitors located on the motherboard. High-cost, high-performance systems could use flip-chip BGA with integral power and ground planes for minimal inductance, along with bypass capacitors mounted both on the package itself and on the motherboard.

3.1.2 Off-Chip Switching

For signaling from on-chip to off-chip, the on-chip drivers must pass signals through the package (and socket, if present) to charge and discharge off-chip capacitance. Tracking the current paths demonstrates noise sources and suggests remedial strategies.

For switching high-to-low, the current paths are shown in Figure 3.5. With respect to the chip, the current flows in on the signal lines and out on the ground connection. Note that the power connection is not used. For the low-to-high case, the currents flow in on the power connection and out on the signal lines. Because the current loops for on-chip and off-chip signaling are different, these two aspects of the interconnect are treated separately.

As shown in Figure 3.6, a voltage is generated on the ground inductance according to

$$v_{\mathrm{gb}} = (L_1 + L_g - 2M_{1g})\frac{di}{dt}.$$

Because this voltage is positive with respect to the off-chip (system) ground, the on-chip ground moves up during a switching event. The offset in reference voltage between on-chip and off-chip grounds is called *ground bounce.* For low-to-high transitions, the inrush of current causes power bounce where the on-chip power reference moves down with respect to off-chip power.

Ground bounce affects on-chip power integrity. The on-chip supply voltage is $V_{\mathrm{s}} - v_{\mathrm{gb}}$, so ground bounce directly causes rail collapse. Rail collapse due to off-chip signaling can be reduced by:

1. Reducing the driver edge rates. The slowest edges that still meet timing produce the least noise.

2. Reducing the loop inductance of the package. The partial self-inductances can be reduced and/or the partial mutual inductance can be increased.

3. Adding on-chip and/or on-package bypass capacitance. The current paths sketched in Figure 3.7 show that on-chip bypass enables the power connec-

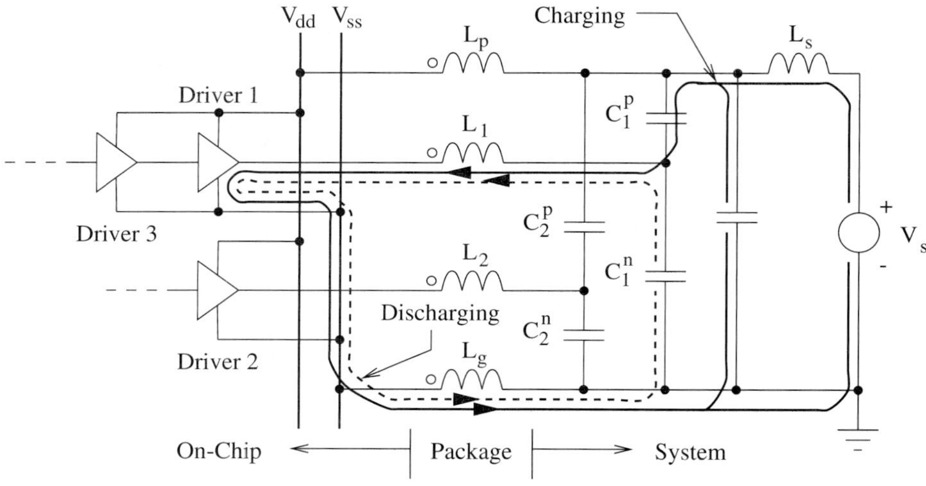

Figure 3.5. Driver 1 current paths for switching an off-chip receiver from high to low.

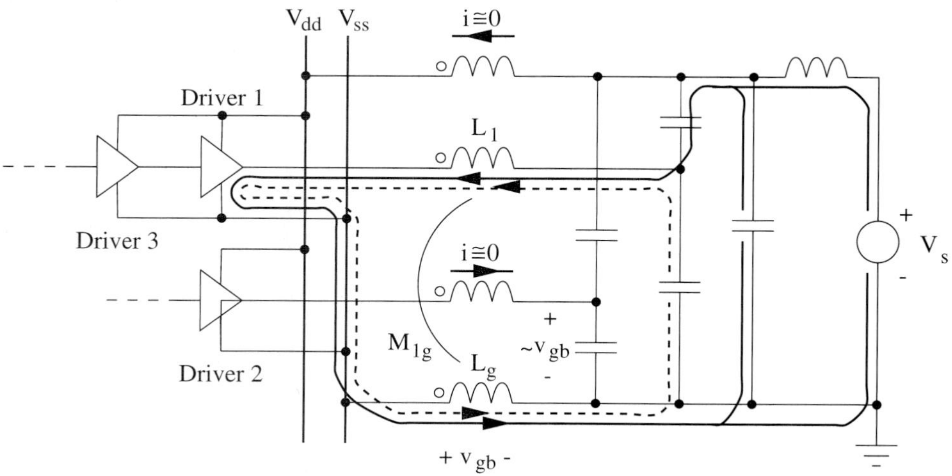

Figure 3.6. Ground bounce and quiet line noise holding low for high-to-low transitions.

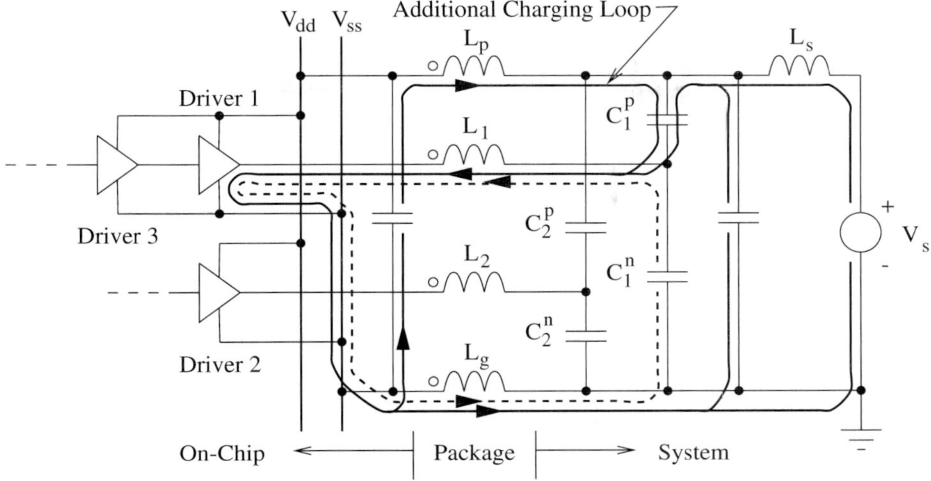

Figure 3.7. Off-chip switching current paths with on-package or on-chip capacitance.

tion to be used in addition to the ground connection. The extra path enables noise to be shared between the power and ground connections. Bypass capacitance in the system, located off-chip and off-package, does not help reduce ground bounce generated on the package due to off-chip signaling. By improving the power supply, system bypass enables faster edge rates and potentially more noise. Power bounce similarly creates rail collapse and can be addressed through the same techniques.

Ground bounce also affects off-chip signal integrity. Driver 2 holding low in Figure 3.6 is, in effect, signaling the on-chip ground potential to the off-chip receiver, which interprets the signal with respect to the off-chip ground potential. Ground bounce confuses this exchange by causing $v_{\rm gb}$, instead of zero, to be signaled to the receiver. This unwanted signal is called *quiet line noise*. If the quiet line noise is sufficiently large, the receiver can interpret the voltage as a logic 1 instead of a logic 0, and a false signal results. Quiet line noise decreases as ground bounce is reduced. In addition, quiet line noise is the ground bounce after filtering by the load capacitance, so large capacitive loads tend to see less quiet line noise. Faster systems require less capacitance, so these are more susceptible to quiet line noise. Quiet line noise is also referred to as *transmitted noise* since the noise is transmitted from on-chip to off-chip.

Just as ground bounce causes quiet line noise for signals holding low, power bounce causes quiet line noise for signals holding high. Simulations and measurements should check both cases against the signaling specification.

Since quiet line noise occurs during signal transitions after the clock edge, synchronous circuits are generally not sensitive to it as long as the waveforms are settled before the next clock edge. However, asynchronous circuits, such as interrupts and resets, are susceptible to quiet line noise at any time.

Uniqueness

It is important to emphasize that the real physical effects that affect system operation are rail collapse and quiet line noise. Strictly speaking, ground bounce is the voltage drop across the ground inductance; however, the ground inductance is actually a partial inductance (see chapter 5), and partial inductance is not unique. Therefore, *ground bounce is not unique.* Similarly, power bounce is not unique. In contrast, rail collapse and quiet line noise *are* unique because they are dependent on closed-loop inductance, which is unique. Measurements and simulations should always target rail collapse and quiet line noise. For additional information on this topic, see section 5.10.

3.1.3 Example SPICE Simulations

To demonstrate the effects of SSN, SPICE simulations based on the schematic of Figure 3.2 are presented. The circuit includes three signal lines rather than two so that an active signaling line can be simulated along with lines holding high and low. The receivers are modeled as 25pF capacitors to power and ground for bus-like loading levels. The total AC load of 50pF must be split if the effects of bypass capacitors are to be investigated. When the total capacitive loading is applied only to ground, perfect bypass is assumed.

The sample circuit is provided in full in Appendix C. While all of the magnitudes are selected to be realistic, this circuit cannot exhibit the full range of behavior that can be expected. Any observations of noise and noise mitigation cannot be considered to be general or directly applicable to any other circuit.

The voltage at the capacitive loading of the actively switching line is shown in Figure 3.8 for several bypass schemes. To facilitate comparisons between the waveforms, a 4V running offset is applied. With no bypass, the active waveform exhibits overshoot and some ringing. Comparison with the other waveforms shows that the edge rate is slowed. Adding off-chip bypass reduces these effects to almost imperceptible levels, so the bulk of these effects are due to the power supply inductance. With on-chip bypass, overshoot and edge rate degradation disappears, but higher

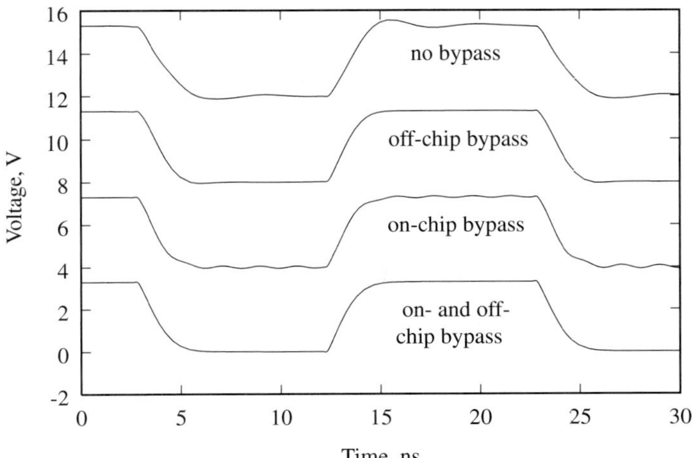

Figure 3.8. Example SPICE simulations of the voltage at an off-chip capacitive load generated by an actively switching packaged on-chip driver. The waveforms are separated by 4V offsets.

frequency ringing is induced due to the interaction between the on-chip capacitance and the package and source inductance. With both on-chip and off-chip bypass, the active waveform is very clean.

The rail voltage on-chip is shown in Figure 3.9, where a 1V running offset is used to separate the waveforms. With no bypass, the deviation from the 3.3V supply level is quite severe. Adding off-chip bypass capacitance reduces, but does not eliminate, the noise because of the noise induced on the package inductance. On the other hand, on-chip bypass by itself almost completely eliminates the rail noise. With sufficient on-chip bypass, off-chip bypass makes no discernable difference. It is interesting to note that the off-chip bypass is 1000 times larger than the on-chip bypass, but cannot do a fraction of the job.

The voltage at the capacitive loading of the driver holding low is shown in Figure 3.10, where a 0.3V running offset is used to separate the waveforms. With no bypass, the quiet line deviates from the expected voltage of zero by $\approx \pm 0.1$V. This low level will not cause false switching of receivers, but then, only one driver

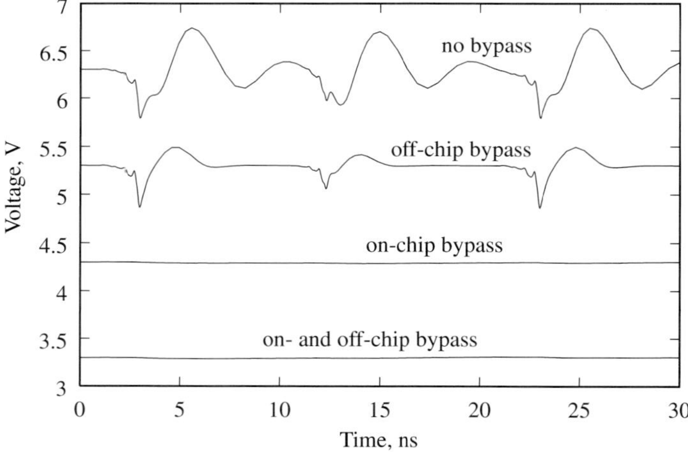

Figure 3.9. Example SPICE simulations of the on-chip power supply voltage (the rail voltage, or $V_{dd} - V_{ss}$) for a packaged device with active off-chip drivers. The waveforms are separated by 1V offsets.

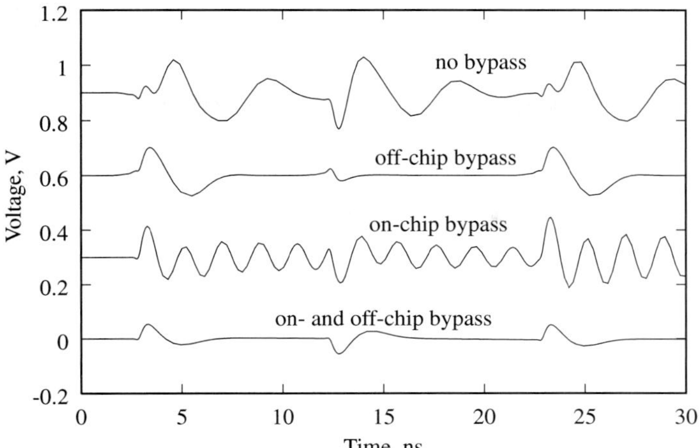

Figure 3.10. Example SPICE simulations of the voltage at an off-chip capacitive load generated by a packaged on-chip driver holding low. The waveforms are separated by 0.3V offsets.

is switching. When tens to hundreds of drivers switch simultaneously, the noise can become substantial. With off-chip bypass, the noise waveform is modified, but the peak excursion is only marginally reduced. The same is true for on-chip bypass. The quiet line holding-low noise is only substantially reduced when both on-chip and off-chip bypass capacitance are used.

The voltage at the capacitive loading of the driver holding high is shown in Figure 3.11, where a 1V running offset is used to separate the waveforms. With no bypass, the quiet line deviates from the supply voltage by $\approx \pm 0.3$V, which is much larger than that for the holding-low noise due to noise introduced on the supply inductance. With off-chip bypass, this noise is minimized so that the holding-high noise drops to $\approx \pm 0.1$V, which is similar to the holding-low noise. The on-chip bypass also effectively bypasses the supply noise inductance but does not achieve lower quiet line noise. Inclusion of both on- and off-chip bypass capacitance minimizes the holding-high noise to a level and waveform similar to the holding-low noise.

Note that the bypass capacitances essentially eliminate the rail noise but only

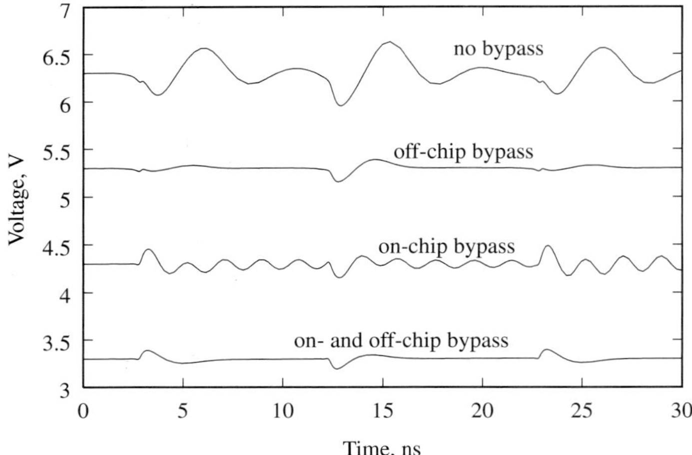

Figure 3.11. Example SPICE simulations of the voltage at an off-chip capacitive load generated by a packaged on-chip driver holding high. The waveforms are separated by 1V offsets.

cuts the quiet line noise by about a factor of 2. For on-chip signaling, rail noise can be reduced to arbitrarily low levels with sufficient bypass. For off-chip signaling, signals must traverse the package, but the bypass capacitance enables the power and ground paths to be used together for lower inductance. In this example, symmetry cuts the inductance almost in half, leading to a drop in SSN of about half.

3.2 Effective Inductance

Rapidly changing currents passing through packages, sockets, and connectors generate switching noise according to

$$V_{\text{SSN}} = L_{\text{eff}} \frac{di}{dt}, \tag{3.1}$$

where L_{eff} is the effective inductance of the path through which the current passes. The effective inductance varies according to the path. For on-chip SSN, the path is the loop formed by the power and ground leads. For off-chip SSN, there are two

L_{eff}s: one for low-to-high transitions, where the path is the loop formed by the power and signal leads, and the other for high-to-low transitions, where the path is the loop formed by the signal and ground leads. Usually, L_{eff} for on-chip switching is very small (as low as 80pH for a C4-CBGA), while the L_{eff}s for off-chip switching are asymmetrical and much larger. The asymmetry occurs because there are often more ground leads than power leads.

The off-chip L_{eff}s also vary depending on how many off-chip drivers are simultaneously switching and in what direction. If some drivers are switching low-to-high, there is some cancellation of noise with those switching high-to-low. The two worst cases occur when all of the drivers are switching low-to-high or high-to-low.

The on-chip L_{eff} plus the two worst-case off-chip L_{eff}s form a set of three metrics that encapsulate much of the performance of a package (or socket, or connector, etc.). Higher L_{eff} means higher SSN. Comparisons between packages can be meaningfully made by computing and comparing L_{eff}s. In contrast, comparisons of lead partial self-inductance are not meaningful and can easily lead to improper conclusions (such as that one package is "better" than another when it is really not).

3.2.1 Computing Worst-Case L_{eff}s

Given a partial inductance matrix for a package (or socket, or edge connector, etc.), effective inductances can be computed with the simple matrix operations derived in chapter 4, section 4.5.2. Since the power leads are connected in parallel both on the chip and on the system board, they can be paralleled together to obtain a single effective power lead. Similarly, the ground lead can be paralleled to obtain a single effective ground lead.

The worst-case off-chip switching scenario occurs when all signals swing either low-to-high or high-to-low. In these cases, it can be assumed (with some introduction of error) that if all of the simultaneously switching drivers maintain the same voltage on each signal throughout the transition, then the leads with simultaneously switching drivers can be paralleled together. While this assumption is strong, there is little analytical alternative, and it does provide for reasonable agreement with detailed simulations and measurements.

Once the paralleling operations are completed, the result is a reduced 3×3 partial inductance matrix:

$$L_{\text{paralleled}} = \begin{bmatrix} L_s & M_{sp} & M_{sg} \\ M_{sp} & L_p & M_{pg} \\ M_{sg} & M_{pg} & L_g \end{bmatrix},$$

where L_s represents the partial self-inductance of all of the signal leads paralleled together, L_p represents the paralleled power leads, and L_g represents the paralleled ground leads. The on-chip effective inductance is the inductance of the loop formed by the power and ground leads, and is given by $L_{\text{eff,P}} = L_p + L_g - 2M_{pg}$. The off-chip effective inductance for low-to-high SSN is given by the loop formed between the power and signals, and is given by $L_{\text{eff,LH}} = L_p + L_s - 2M_{sp}$. Finally, the off-chip effective inductance for high-to-low SSN is given by the loop formed between the signals and ground, and is given by $L_{\text{eff,HL}} = L_s + L_g - 2M_{sg}$.

The three L_{eff}s form a metric that enables comparison between components. For example, assume that two packages are fully characterized, leading to $L_{\text{eff,P1}} =$ 0.08nH, $L_{\text{eff,LH1}} =$ 0.3nH, and $L_{\text{eff,HL1}} =$ 0.25nH for the first package, and for the second package $L_{\text{eff,P2}} =$ 0.1nH, $L_{\text{eff,LH2}} =$ 0.25nH, and $L_{\text{eff,HL2}} =$ 0.25nH. The following conclusions would be reasonable. The first package will produce less rail noise than the second by virtue of its lower $L_{\text{eff,P}}$; the second package will produce less quiet line noise for low-to-high transitions since $L_{\text{eff,LH1}} > L_{\text{eff,LH2}}$; and the two packages will produce about equal quiet line noise high-to-low. Which package is better? It depends on whether rail collapse or quiet line noise is more important for a given application.

3.3 Off-Chip SSN Dependencies

For off-chip switching, SSN dependencies can be investigated[1] by considering the effect of noise on MOSFET saturation current in CMOS totem-pole drivers. The low-to-high and high-to-low cases are similar, so just the high-to-low case is treated. A driver can source only so much current from the signal line to ground, so the rate of discharge of a capacitive load is restricted to the saturation current of the pull-down MOSFET. Note that the discharge rate is not the RC time constant of the capacitive load in series with the driver's pull-down impedance.

Without short-channel effects, the saturation current for a MOSFET is given by

$$i_{\text{sat}} = \frac{1}{2}K(V_{GS} - V_T - V_{\text{SSN}})^2, \tag{3.2}$$

where V_{GS} is the gate-to-source voltage, V_T is the process-dependent threshold voltage of the pull-down MOSFET, and $K = \frac{\mu C_{ox} W}{L}$. This equation is normally presented under zero noise conditions with $V_{\text{SSN}} = 0$. For a worst-case scenario with N drivers simultaneously switching, the total current would be N times higher. To compute the noise with the effective inductance using (3.1), the time rate of current is required. Approximating $di/dt \approx \Delta i/\Delta t$, where $\Delta i = N i_{\text{sat}}$ and $\Delta t = t_f$ is the fall time of the signal transition, then combining (3.2) and (3.1) yields

$$V_{\text{SSN}} = \frac{N L_{\text{eff}} K(V_{GS} - V_T - V_{\text{SSN}})^2}{2t_f}. \tag{3.3}$$

Solving (3.3) for V_{SSN} produces

$$V_{\text{SSN}} = (V_{GS} - V_T) + \frac{t_f}{N L_{\text{eff}} K}\left(1 - \sqrt{1 + 2(V_{GS} - V_T)\frac{N L_{\text{eff}} K}{t_f}}\right), \tag{3.4}$$

where L_{eff} is from the high-to-low switching case.

High-speed drivers are often designed to achieve a given impedance, such as 50Ω, to source match a transmission line. If it is assumed that the pull-down impedance is given as $Z = V_{DD}/i_{\text{sat}}$, then K can be estimated given the impedance and the

1. R. Senthinathan and J. L. Prince, "Simultaneous switching ground noise calculation for packaged CMOS devices," *IEEE Trans. Solid-State Circuits*, vol. 26, no. 11, Nov. 1991, pp. 1724–28.

supply voltage, V_{DD}. Applying this information to (3.2) under zero noise conditions, then

$$K = \frac{2V_{DD}}{Z(V_{GS} - V_T)^2}.$$

V_{GS} can usually be taken as equal to V_{DD}, and V_T is process-dependent (but very approximately equal to $V_{DD}/5$). Finally, the fall time, t_f, can be estimated, but the best approach is to simulate the edge rate of the driver using a SPICE or IBIS model with a typical load.

With good estimates of effective inductance, fall time, driver impedance, and threshold voltage, (3.4) can provide reasonably accurate estimates of SSN. In many cases, the formula can be used to rough out designs or noise mitigation strategies.

3.3.1 Feedback and Saturation

The SSN formula in (3.4) shows that SSN is not linearly dependent on any of the user-controlled quantities like power supply, number of switching drivers, effective inductance, or fall time. Without calculation, it can be difficult to estimate the effect of changes in the system. A feedback loop exists where SSN causes rail collapse, leading to less drive strength in the driver, and then lower noise: the lower noise then causes less rail collapse, more drive, and higher noise. The system reaches an equilibrium between rail collapse and SSN due to the feedback between rail collapse and driver strength.

Feedback causes saturation in SSN with increasing numbers of simultaneously switching drivers. As the number of simultaneously switching drivers increases, SSN increases at a sublinear rate. The general dependence of SSN on N is shown in Figure 3.12, where the linear dependence on N is maintained only when few drivers are switching. Saturation is typically a beneficial phenomenon as most systems rely on it to keep noise levels from exceeding margins.

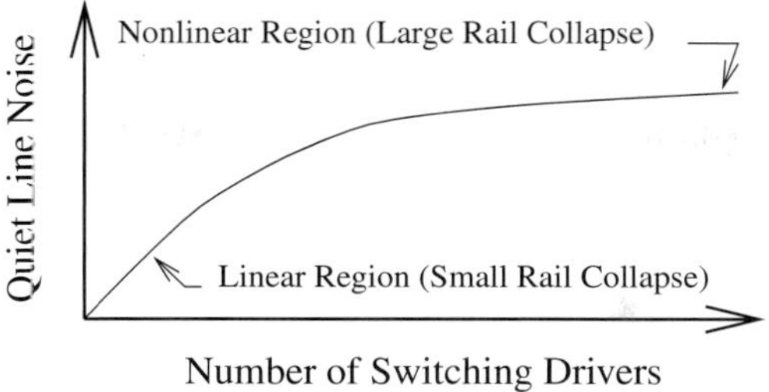

Figure 3.12. SSN saturates for high numbers of simultaneously switching drivers.

3.4 SSN-Induced Skew

SSN is noise, and like all other sources of noise, it can cause the movement of waveform edges, leading to skew. Skew from SSN is treated like that from any other source and is discussed in chapter 2, section 2.8.3.

3.5 Fast Simulation of Banks

A basic SSN simulation of an N-driver system would include an N-line interconnect model, N driver model instances, N driver loadings, plus a model of the power distribution system. For large N and complex driver models, this approach can become intractable due to the large number of elements in the circuit simulation. In addition, simulations with large numbers of driver instances may have difficulty converging due to nonlinearities and discontinuities in the driver transistor models.

The worst-case SSN event occurs when all drivers in a bank switch simultaneously in the same direction. A common occurrence is when all the lines on a wide microprocessor data bus switch from low to high (or vice versa). For these cases, the driver can be approximated to dramatically speed up computation time and perhaps eliminate convergence problems. The basic idea is to model one driver in detail and then to copy its behavior using current-controlled current sources (CCCS). A set of CCCSs that duplicate a driver is called a *current mirror*. The bank is then modeled by one detailed driver and many current mirrors, plus detailed drivers holding high and/or low to capture quiet line noise.

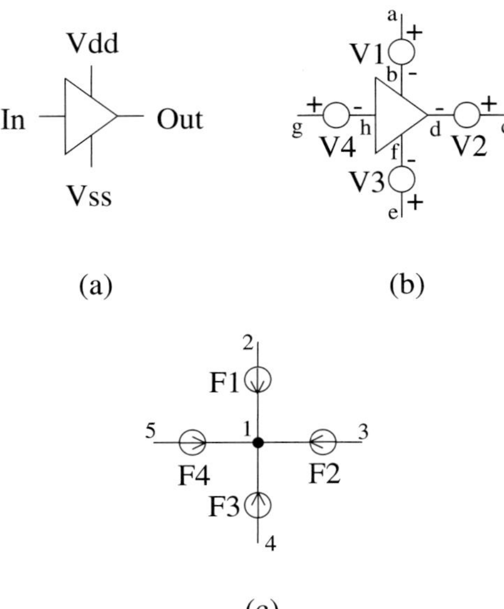

Figure 3.13. Driver and its current mirror: (a) original driver, (b) driver with 0V voltage sources used to detect current drawn by the driver node, (c) current mirror that duplicates the current draw of the real driver at some other place in the circuit.

The total effect of the current mirrors is to create a noise environment of an entire bank simultaneously switching, so the noise affects the behavior of the driver modeled in detail. Therefore, the all-important feedback mechanism is captured. While the use of current mirrors involves many approximations and limitations, as discussed below, the method provides very good results. Because of potential convergence difficulties or long runtimes, the use of current mirrors may be the only way to complete the simulation.

The implementation of a current mirror is demonstrated using SPICE elements. Assume that a driver has power, ground, input, and output leads as shown in Figure 3.13. The generalization to more complex drivers with additional leads is straightforward. The current drawn by the driver node is detected by 0V voltage sources placed on each branch of the node. Current-controlled current sources then duplicate these currents at another node, thereby duplicating the driver. A SPICE netlist to implement the current mirror looks like

```
        .
        .
        .
* The detailed SPICE driver model
Xdriver h d b f driver

* The current detectors
V1 a b 0
V2 c d 0
V3 e f 0
V4 g h 0

* The current mirror
F1 2 1 V1 1
F2 3 1 V2 1
F3 4 1 V3 1
F4 5 1 V4 1
        .
        .
        .
.subckt driver in out vdd vss
        .
        .
        .
.ends driver
```

Since nonlinear driver models are often slow to simulate, a bank of $N+1$ simultaneously switching drivers (in the same direction) can be rapidly simulated by

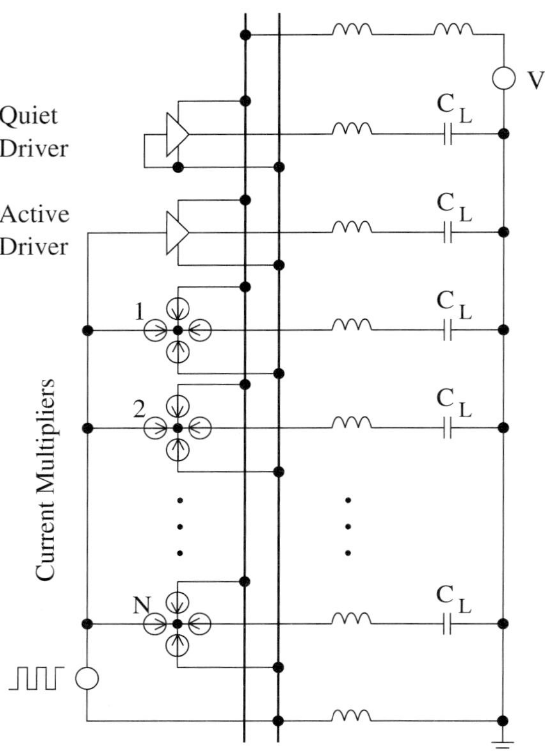

Figure 3.14. Switched bank with active driver mirrored by N current mirrors.

replacing N of the drivers with current mirrors, as shown in Figure 3.14, where two detailed driver instances simulate precise driver behavior while N current mirrors simulate the effects of many others.

Current mirrors can be used with any interconnect and loading as long as all loads on all drivers are nominally identical. With variance in the loadings, the active driver should see typical loading to average out the effects of large and small loading, which should still be reasonably similar. Current mirrors also force all of the drivers to switch at precisely the same time, so skew in driver switching time cannot be modeled.

As an example, the schematic of Figure 3.14, using socket inductance to intro-

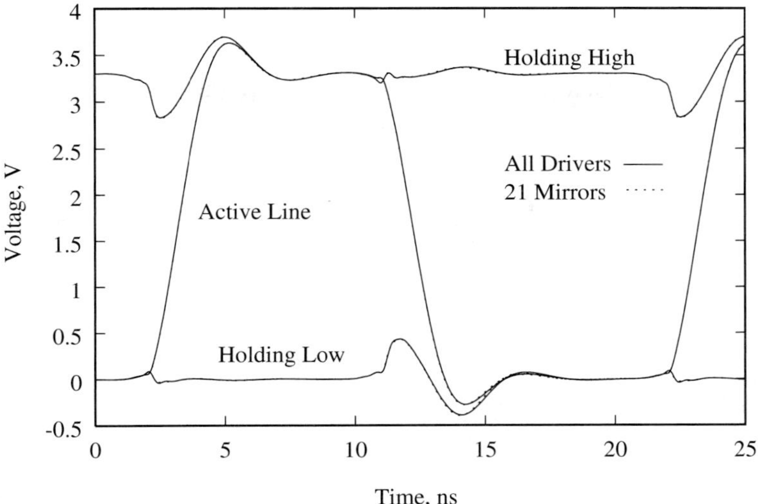

Figure 3.15. Active and quiet waveforms from a SPICE simulation using 24 detailed drivers compared to a simulation with 3 detailed drivers and 21 current mirrors.

duce SSN, is simulated with 24 drivers driving 50pF loads. One driver holds high and another low, leaving the other 22 actively switching low-to-high, then high-to-low. The system is simulated first with all 24 drivers modeled in detail with nonlinear MOSFET models, then a second simulation replaces 21 of the drivers with current mirrors that track the one remaining actively switching driver, so just 3 of the drivers are modeled in detail with MOSFETs. The results in Figure 3.15 show that the SPICE simulation with current mirrors compares very well to the simulation with all of the drivers modeled in detail.

3.6 Exercises

1. (a) Based on the schematic in Figure 3.3, sketch the current paths for on-chip signaling for the low-to-high transition.

   (b) Repeat with bypassed power supply.

   (c) Repeat with bypassed package and power supply.

2. (a) Based on the schematic in Figure 3.5, sketch the current paths for off-chip signaling for the low-to-high transition.

   (b) Repeat with bypassed power supply.

   (c) Repeat with bypassed package and power supply.

3. Sketch the current paths for SSN events for off-chip signaling with a terminated current-mode driver. How will the noise compare to that generated by a single-ended driver?

4. Set up and run SPICE simulations of SSN similar to those in section 3.1.3, based on the netlist in Appendix C, to investigate the effect of edge rates. Vary the edge rate of the driver and plot peak SSN vs. edge rate. What do the results imply about edge rates?

5. Estimate V_{SSN} using (3.4) for each edge rate in problem 4. How accurate are the estimates? Are the trends correct?

6. Derive a formula similar to (3.4) to estimate V_{SSN} for a packaged device with two buses with different driver strengths sharing a common ground.

7. What is the minimum number of current sources needed to implement the current mirror in Figure 3.13? Why?

Chapter 4

MULTIPORT CIRCUITS

Digital systems are interconnected with many parallel connections to increase the system throughput. Interconnect widths typically fall on byte boundaries, so interfaces are commonly 8-bit, 16-bit, 32-bit, 64-bit, and so on. Parity bits, at one per byte, can significantly increase width. For example, a typical address bus for a 32-bit microprocessor could include 32 bits for the address, 4 address parity bits, 64 data bits (double the 32-bit data word for double the throughput), 8 data parity bits, some control bits like CAS, RAS, and WE, plus CLK for a total of over 100 connections. For interconnect analysis, it is essential that all modeling and simulation techniques accommodate large numbers of coupled connections.

4.1 Z- and Y-Parameters

For circuit analysis, an interconnect model must relate voltage and current. For the general N-port circuit of Figure 4.1, the voltages and current are related by the $N \times N$ impedance matrix, $\overline{\overline{Z}}$, defined by

$$\begin{bmatrix} V_1 \\ V_2 \\ \vdots \\ V_N \end{bmatrix} = \begin{bmatrix} Z_{11} & Z_{12} & \cdots & Z_{1N} \\ Z_{21} & Z_{22} & \cdots & Z_{2N} \\ \vdots & \vdots & \ddots & \vdots \\ Z_{N1} & Z_{N2} & \cdots & Z_{NN} \end{bmatrix} \begin{bmatrix} I_1 \\ I_2 \\ \vdots \\ I_N \end{bmatrix},$$

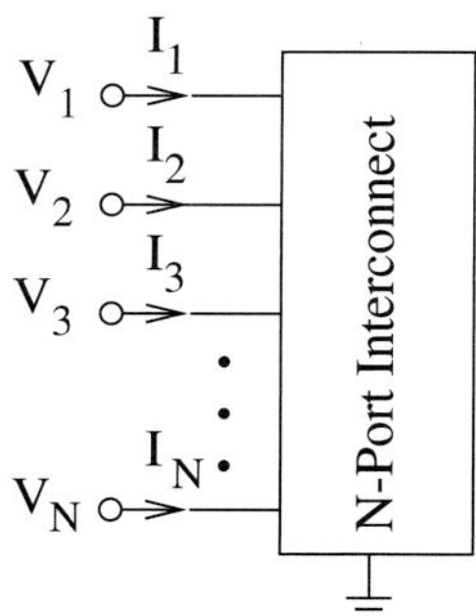

Figure 4.1. Voltage and current convention for a general N-port network.

where

$$\overline{V} = \begin{bmatrix} V_1 \\ V_2 \\ \vdots \\ V_N \end{bmatrix}$$

is a vector representing the voltage on each of the N lines, and $\overline{I}$ similarly represents the current on each of the N lines. Using a compact vector/tensor notation, then

$$\overline{V} = \overline{\overline{Z}}\,\overline{I}. \tag{4.1}$$

The important feature of (4.1) is that $\overline{\overline{Z}}$ is an $N \times N$ square matrix. Since reciprocity applies for the passive materials used in interconnects, $Z_{ij} = Z_{ji}$ and there are N self terms on the diagonal and $N(N-1)/2$ unique coupling terms off of the diagonal. When the coupling is negligible between two ports i and j, then Z_{ij} can be set to zero. Since interconnects are designed to pack many conductors into a small volume, many coupling terms are usually required. In some package styles, such as QFPs, often all of the off-diagonal terms are required. It is a rare event that only the self terms are needed.

Without loss of generality, the currents can be related to the voltages by the

$N \times N$ admittance matrix, $\overline{\overline{Y}}$, defined by

$$\begin{bmatrix} I_1 \\ I_2 \\ \vdots \\ I_N \end{bmatrix} = \begin{bmatrix} Y_{11} & Y_{12} & \cdots & Y_{1N} \\ Y_{21} & Y_{22} & \cdots & Y_{2N} \\ \vdots & \vdots & \ddots & \vdots \\ Y_{N1} & Y_{N2} & \cdots & Y_{NN} \end{bmatrix} \begin{bmatrix} V_1 \\ V_2 \\ \vdots \\ V_N \end{bmatrix},$$

or

$$\overline{I} = \overline{\overline{Y}}\,\overline{V}. \tag{4.2}$$

Like the impedance matrix, the admittance matrix is reciprocal. The admittance and impedance matrices are the inverse of each other: $\overline{\overline{Z}} = \overline{\overline{Y}}^{-1}$ and $\overline{\overline{Y}} = \overline{\overline{Z}}^{-1}$. If a datum node has not been defined, then the matrix inverse does not exist.

For the most general case, each element Z_{ij} of the impedance matrix is frequency-dependent. Special techniques may be required to accommodate frequency-dependent impedance matrices in time-domain simulators. If the bandwidth requirements of the simulation are modest, then the lumped approximation may hold, leading to frequency-independent matrix elements (to a good approximation). A similar discussion applies to the admittance matrix.

4.2 S-Parameters

At high frequencies, voltage becomes nonunique because of its path dependence. An alternative formulation for circuits can be based on the amplitudes and phases of voltage traveling waves (see chapter 2, section 2.1.3) scattered from the device under test (DUT). By separating signals into directional wave components, uniqueness problems with voltage are avoided. Scattering parameters (S-parameters) are convenient for two other reasons: laboratory equipment is commercially available covering as low as 300kHz to greater than 100GHz, and general-purpose electromagnetic simulators usually produce their results in the form of S-parameters. Familiarity with S-parameters is required to successfully apply high-frequency equipment and simulators. S-parameters can easily be converted into impedance and/or admittance parameters if needed, but modeling can proceed in S-parameters if desired.

4.2.1 Definition

Consider a black box with N ports providing access to the circuits inside. As sketched in Figure 4.2, at each port i a steady-state wave propagates towards the box on a transmission line with characteristic impedance Z_{oi}, and a wave also propagates away, having either been reflected by the black box or having been launched by the black box itself. By convention, the waves traveling towards port i are labeled as a_i, while the waves propagating away from the black box on port i are labeled as b_i. S-parameters are defined by relating the "reflected" b_i waves to the "incident" a_i waves according to

$$\begin{bmatrix} b_1 \\ b_2 \\ \vdots \\ b_N \end{bmatrix} = \begin{bmatrix} S_{11} & S_{12} & \cdots & S_{1N} \\ S_{21} & S_{22} & \cdots & S_{2N} \\ \vdots & \vdots & \ddots & \vdots \\ S_{N1} & S_{N2} & \cdots & S_{NN} \end{bmatrix} \begin{bmatrix} a_1 \\ a_2 \\ \vdots \\ a_N \end{bmatrix}. \tag{4.3}$$

Using a compact vector/tensor notation, then $\overline{b} = \overline{\overline{S}}\,\overline{a}$, where $\overline{\overline{S}}$ is the S-parameter matrix. If $b_i = 0$, then there is no reflected wave on the transmission line feeding port i and the port is said to be *matched*.

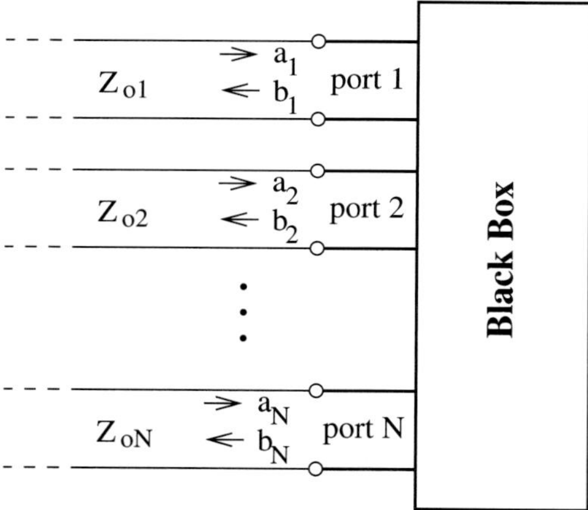

Figure 4.2. A black box circuit can be characterized by the waves incident to and reflected from the box's ports.

The voltage waves are related to the incident and reflected waves by the characteristic impedance of the transmission line connecting the port. If V_i^+ designates the incident voltage wave at port i, then

$$a_i = \frac{V_i^+}{\sqrt{Z_{oi}}},$$

where Z_{oi} is the (real) characteristic impedance of the transmission line connected to port i. Similarly, the reflected voltage wave is related to b_i by

$$b_i = \frac{V_i^-}{\sqrt{Z_{oi}}}.$$

The power propagating towards the black box on port i is

$$\begin{aligned} P_i^+ &= \frac{1}{2} V_i^+ \left(I_i^+\right)^* \\ &= \frac{1}{2} V_i^+ \left(\frac{V_i^+}{Z_{oi}}\right)^* = \frac{1}{2}\sqrt{Z_{oi}}\, a_i \left(\frac{\sqrt{Z_{oi}}\, a_i}{Z_{oi}}\right)^* \\ &= \frac{1}{2}|a_i|^2, \end{aligned} \tag{4.4}$$

for $Z_{oi} = Z_{oi}^*$, i.e., real. Similarly, $P_i^- = \frac{1}{2}|b_i|^2$. Therefore, by basing the definition of S-parameters on a and b rather than V^+ and V^-, calculations involving power avoid the need to know the characteristic impedances of the feed lines. For example, the gain of a two-port circuit for a signal entering port 1 and leaving port 2 is

$$\frac{P_2^-}{P_1^+} = \frac{\frac{1}{2}|b_2|^2}{\frac{1}{2}|a_1|^2} = \frac{|S_{21}a_1|^2}{|a_1|^2} = |S_{21}|^2,$$

assuming $S_{22} = 0$ (matched).

Measurements and simulations of S-parameters automatically incorporate the effects of the characteristic impedance of the lines driving the ports. For example, the reflection coefficient (S_{11}) of a resistor depends on the characteristic impedance of the transmission line driving it. For a 50Ω resistor, the reflection coefficient is $(50-75)/(50+75) = -0.2$ when measured from a 75Ω line, but it is $(50-50)/(50+50) = 0$ when measured from a 50Ω line. Since the effects of the transmission line characteristic impedances are "built into" the S-parameters, they are termed *unnormalized*. Section 4.4 shows that S-parameters can be normalized to any convenient characteristic impedance for easier use.

4.2.2 Circuit Calculations with S-Parameters

S-parameters can be incorporated into circuit analyses by relating the incident and reflected waves to the voltages and currents. First, Kirchhoff's voltage and current laws are used to find all voltage and current relationships. Second, the voltages and currents are eliminated with their traveling wave representations in (2.27) and (2.28), often with the convenient selection of $z = 0$. The voltage on port i is

$$V_i = V_i^+ + V_i^- = \sqrt{Z_{oi}}\, a_i + \sqrt{Z_{oi}}\, b_i,$$

while the current flowing into the port is

$$I_i = I_i^+ + I_i^- = \frac{V_i^+}{Z_{oi}} - \frac{V_i^-}{Z_{oi}} = \frac{a_i}{\sqrt{Z_{oi}}} - \frac{b_i}{\sqrt{Z_{oi}}}.$$

Finally, the equations are rearranged to allow the identification of the S-parameters.

As an example, consider the derivation of the two-port S-parameters for the series inductor in Figure 4.3. The voltage and current at port 1 are

$$\begin{aligned} V_1 &= a_1\sqrt{Z_{o1}} + b_1\sqrt{Z_{o1}} \\ I_1 &= a_1/\sqrt{Z_{o1}} - b_1/\sqrt{Z_{o1}}, \end{aligned}$$

while at port 2,

$$\begin{aligned} V_2 &= a_2\sqrt{Z_{o2}} + b_2\sqrt{Z_{o2}} \\ I_2 &= a_2/\sqrt{Z_{o2}} - b_2/\sqrt{Z_{o2}}. \end{aligned}$$

Applying Kirchhoff's voltage law around the loop formed by the inductor and the ends of the transmission lines yields

$$a_1\sqrt{Z_{o1}} + b_1\sqrt{Z_{o1}} = \jmath\omega L\left(a_1/\sqrt{Z_{o1}} - b_1/\sqrt{Z_{o1}}\right) + a_2\sqrt{Z_{o2}} + b_2\sqrt{Z_{o2}}.$$

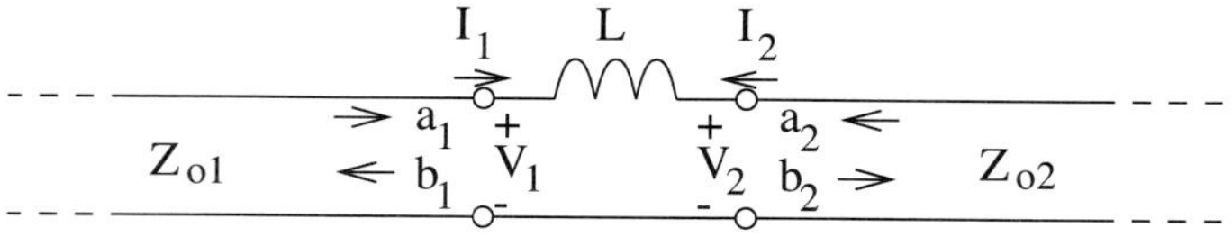

Figure 4.3. Setup for computing the S-parameters of a series inductor.

Conservation of charge requires that $I_1 = -I_2$, so

$$a_1/\sqrt{Z_{o1}} - b_1/\sqrt{Z_{o1}} = -a_2/\sqrt{Z_{o2}} + b_2/\sqrt{Z_{o2}}.$$

Eliminating b_2 yields

$$b_1 = \frac{Z_{o2} - Z_{o1} + \jmath\omega L}{Z_{o1} + Z_{o2} + \jmath\omega L} a_1 + \frac{2\sqrt{Z_{o1}Z_{o2}}}{Z_{o1} + Z_{o2} + \jmath\omega L} a_2,$$

while eliminating b_1 yields

$$b_2 = \frac{2\sqrt{Z_{o1}Z_{o2}}}{Z_{o1} + Z_{o2} + \jmath\omega L} a_1 + \frac{Z_{o1} - Z_{o2} + \jmath\omega L}{Z_{o1} + Z_{o2} + \jmath\omega L} a_2.$$

The S-parameters can be picked off by inspection according to the definition in (4.3), and the result is

$$\overline{\overline{S}} = \begin{bmatrix} \frac{Z_{o2}-Z_{o1}+\jmath\omega L}{Z_{o1}+Z_{o2}+\jmath\omega L} & \frac{2\sqrt{Z_{o1}Z_{o2}}}{Z_{o1}+Z_{o2}+\jmath\omega L} \\ \frac{2\sqrt{Z_{o1}Z_{o2}}}{Z_{o1}+Z_{o2}+\jmath\omega L} & \frac{Z_{o1}-Z_{o2}+\jmath\omega L}{Z_{o1}+Z_{o2}+\jmath\omega L} \end{bmatrix}.$$

Due to the symmetry in the circuit, the ports can be swapped by interchanging Z_{o1} and Z_{o2}. Note also that $S_{12} = S_{21}$, reflecting the reciprocity of a passive network, and that interchanging Z_{o1} and Z_{o2} has no effect on these.

4.3 Multiport Conversion Between S-, Y-, and Z-Parameters

Since the impedance and admittance matrices are related by inversion, only conversions involving S-parameters are needed. Two conversions are derived. First, the characteristic impedances of the transmission lines driving the ports are assumed to be equal. Second, this restriction is relaxed so that each transmission line can have a unique characteristic impedance.

It is assumed that the N-port impedance matrix is known and that the N-port S-parameter matrix is required. Transmission lines with a (real) characteristic impedance of Z_o are used to drive each port, so the voltages at the N-ports can be written as

$$\begin{aligned} V_1 &= \sqrt{Z_o}(a_1 + b_1) \\ V_2 &= \sqrt{Z_o}(a_2 + b_2) \\ &\vdots \\ V_N &= \sqrt{Z_o}(a_N + b_N). \end{aligned} \tag{4.5}$$

This can be compactly written as

$$\overline{V} = \sqrt{Z_o}(\overline{a} + \overline{b}). \tag{4.6}$$

Similarly, the currents can be written as

$$\overline{I} = \frac{1}{\sqrt{Z_o}}(\overline{a} - \overline{b}). \tag{4.7}$$

Substituting (4.6) and (4.7) into (4.1) yields

$$\sqrt{Z_o}(\overline{a} + \overline{b}) = \overline{\overline{Z}}\frac{1}{\sqrt{Z_o}}(\overline{a} - \overline{b}).$$

The definition for S-parameters, $\overline{b} = \overline{\overline{S}}\,\overline{a}$, can be used to eliminate $\overline{b}$ to get

$$Z_o(\overline{\overline{I}} + \overline{\overline{S}})\overline{a} = \overline{\overline{Z}}(\overline{\overline{I}} - \overline{\overline{S}})\overline{a},$$

where $\overline{\overline{I}}$ is the identity matrix. Then in general, the conversion from $\overline{\overline{S}}$ to $\overline{\overline{Z}}$ is given by

$$\overline{\overline{Z}} = Z_o(\overline{\overline{I}} + \overline{\overline{S}})(\overline{\overline{I}} - \overline{\overline{S}})^{-1}. \tag{4.8}$$

Note the similarity of (4.8) to the result from transmission line theory in equation (2.40), where the impedance is related to the reflection coefficient by $Z = Z_o(1 + \Gamma)/(1 - \Gamma)$. The conversion from $\overline{\overline{Z}}$ to $\overline{\overline{S}}$ is found from (4.8) as

$$\overline{\overline{S}} = (\overline{\overline{Z}} + Z_o\overline{\overline{I}})^{-1}(\overline{\overline{Z}} - Z_o\overline{\overline{I}}). \tag{4.9}$$

Once $\overline{\overline{Z}}$ is known, then $\overline{\overline{Y}} = \overline{\overline{Z}}^{-1}$. Dedicated formulas for $\overline{\overline{Y}}$ can be easily derived.

These conversions show that S-parameters can be converted directly into impedance or admittance parameters at any time, assuming that the necessary matrix inversions exist. In this sense, S-, Y-, and Z-parameters are equivalent, and conversions between them can be made as needed.

The restriction that all of the transmission lines driving the ports must have the same characteristic impedance can be relaxed to derive slightly more complex conversions. For a single port, $V = \sqrt{Z}(a+b)$, so the extension to multiple ports is

$$\overline{V} = \overline{\overline{k}}(\overline{a} + \overline{b}), \tag{4.10}$$

where

$$\overline{\overline{k}} = \begin{bmatrix} \sqrt{Z_{o1}} & 0 & \cdots & 0 \\ 0 & \sqrt{Z_{o2}} & \cdots & 0 \\ \vdots & \vdots & \ddots & \vdots \\ 0 & 0 & \cdots & \sqrt{Z_{oN}} \end{bmatrix}.$$

Similarly, the currents at the ports are given by

$$\overline{I} = \overline{\overline{k}}^{-1}(\overline{a} - \overline{b}). \tag{4.11}$$

Substituting (4.10) and (4.11) into (4.1) yields

$$\overline{\overline{k}}(\overline{a} + \overline{b}) = \overline{\overline{Z}}\,\overline{\overline{k}}^{-1}(\overline{a} - \overline{b}). \tag{4.12}$$

By the definition of S-parameters, $\overline{b} = \overline{\overline{S}}\,\overline{a}$, then $\overline{b}$ can be eliminated in (4.12) to obtain the conversion from Z to S as

$$\overline{\overline{S}} = (\overline{\overline{Z}}\,\overline{\overline{k}}^{-1} + \overline{\overline{k}})^{-1}(\overline{\overline{Z}}\,\overline{\overline{k}}^{-1} - \overline{\overline{k}}). \tag{4.13}$$

The result can be rearranged to find the conversion from $\overline{\overline{S}}$ to $\overline{\overline{Z}}$ as

$$\overline{\overline{Z}} = \overline{\overline{k}}(\overline{\overline{I}} + \overline{\overline{S}})(\overline{\overline{I}} - \overline{\overline{S}})^{-1}\overline{\overline{k}}. \tag{4.14}$$

4.4 Normalization of S-Parameters

S-parameters are dependent on the characteristic impedances of the transmission lines used to drive the ports; if the impedances change, the S-parameters change. It is often convenient to recompute S-parameters for different characteristic impedances, and it is standard practice to convert S-parameters to a form where all ports are driven from a transmission line of the same characteristic impedance, with 50Ω being the most common. S-parameters that have been recomputed to such a standard are said to be *normalized.* In practice, it is rare to encounter anything but normalized S-parameters. S-parameters measured with a calibrated 50Ω network analyzer are automatically normalized to 50Ω.

Assume that the original S-parameters are given by $\overline{\overline{S}}_1$ normalized by $\overline{\overline{k}}_1$ and that the desired S-parameters are $\overline{\overline{S}}_2$ normalized to $\overline{\overline{k}}_2$, where $\overline{\overline{S}}_1$, $\overline{\overline{k}}_1$, and $\overline{\overline{k}}_2$ are known. Since $\overline{\overline{Z}}$ is invariant to S-parameter normalization, the impedance matrices for $\overline{\overline{S}}_1$ and $\overline{\overline{S}}_2$, computed from (4.14), can be equated to obtain

$$\overline{\overline{k}}_1(\overline{\overline{I}}+\overline{\overline{S}}_1)(\overline{\overline{I}}-\overline{\overline{S}}_1)^{-1}\overline{\overline{k}}_1 = \overline{\overline{k}}_2(\overline{\overline{I}}+\overline{\overline{S}}_2)(\overline{\overline{I}}-\overline{\overline{S}}_2)^{-1}\overline{\overline{k}}_2.$$

Solving for $\overline{\overline{S}}_2$ yields

$$\begin{aligned}\overline{\overline{S}}_2 &= [\overline{\overline{k}}_1(\overline{\overline{I}}+\overline{\overline{S}}_1)(\overline{\overline{I}}-\overline{\overline{S}}_1)^{-1}\overline{\overline{k}}_1\overline{\overline{k}}_2^{\,-1}+\overline{\overline{k}}_2]^{-1}\\ &\quad [\overline{\overline{k}}_1(\overline{\overline{I}}+\overline{\overline{S}}_1)(\overline{\overline{I}}-\overline{\overline{S}}_1)^{-1}\overline{\overline{k}}_1\overline{\overline{k}}_2^{\,-1}-\overline{\overline{k}}_2].\end{aligned} \tag{4.15}$$

An interesting application for (4.15) is adjusting measured S-parameters for slight offsets in calibration standards. The short-open-load-through (SOLT) calibration technique relies on a precision load, typically 50Ω, to which the reference impedance is set. If the load value is off, then all of the S-parameters will also be off because any user will assume the "perfect" reference value. For example, a calibration standard may be trimmed to ±1%, so while the measured S-parameters could be normalized to 50.5Ω, any user would assume normalization to exactly 50Ω. A precision DC measurement would provide an accurate value for the load, and (4.15) could be applied to adjust the measured S-parameters from the slightly offset values to precisely 50Ω.

As a simple example of renormalization, consider a short section of 75Ω transmission line terminated in 50Ω. The reflection coefficient is

$$S_{11} = \frac{50-75}{50+75} = -0.2.$$

This S-parameter is normalized to 75Ω. Applying (4.15) to renormalize to 50Ω yields

$$\begin{aligned} S_{11_{\text{renorm}}} &= (\sqrt{75}(1-0.2)/(1+0.2)\sqrt{75/50}+\sqrt{50})^{-1} \\ &\quad (\sqrt{75}(1-0.2)/(1+0.2)\sqrt{75/50}-\sqrt{50}) \\ &= 0. \end{aligned}$$

The renormalized S-parameter is now zero and shows that the reflection from a matched termination is zero.

4.5 Matrix Reductions

When terminal characteristics at certain ports are known, the information can be built into the matrix to reduce its size. The two basic cases are when the excitation on a port is zero and when ports share excitation. Null excitation can occur when a port is short-circuited (zero voltage), open-circuited (zero current), or matched (zero incident wave). Common excitation can occur when ports are tied together (equal voltage, or paralleled leads).

4.5.1 Null Excitation

Consider the S-parameter matrix in (4.3). If port j is connected to a matched load, then $a_j = 0$, and column j in $\overline{\overline{S}}$ has no impact on $\overline{b}$. Therefore, there is no change in $\overline{b}$ if a_j is deleted from $\overline{a}$ and column j is deleted from $\overline{\overline{S}}$. Typically, there is no interest in b_j, so row j in $\overline{b}$ and $\overline{\overline{S}}$ can be deleted. Therefore, if port j is matched in the characteristic impedance of the system, then the S-parameter matrix can be simplified to take this into account by deleting row j and column j from the matrix.

For the impedance matrix in (4.1), the analogous excitation is that port j is open-circuited so that $I_j = 0$. Therefore, the impedance matrix can be simplified by deleting the rows and columns of ports that are open-circuited. Similarly, deleting the rows and columns of the admittance matrix from (4.2) simplifies the matrix for ports that are shorted to ground.

Note that the conversions in section 4.3 can be used to move a matrix in to and out of convenient forms for reduction. For example, if a port is to be shorted to ground for an impedance matrix, it can be inverted to admittance, processed, and then inverted back.

Row and column deletion can be efficiently programmed with a straightforward indexing scheme, but it is convenient for discussion to have a mathematical description for the process based on matrix multiplication. The jth row and column of an $N \times N$ matrix $\overline{\overline{B}}$ can be deleted to produce a new $(N-1) \times (N-1)$ matrix $\overline{\overline{B}}'$ through the operation

$$\overline{\overline{B}}' = \overline{\overline{A}}^T \overline{\overline{B}}\,\overline{\overline{A}}, \tag{4.16}$$

where $\overline{\overline{A}}$ is an $N \times (N-1)$ matrix formed by deleting column j of the $N \times N$ identity matrix. If several rows and columns are to be deleted from $\overline{\overline{B}}$, (4.16) can be applied just once with multiple columns deleted from $\overline{\overline{A}}$.

4.5.2 Common Voltage Excitation

When the voltages at two or more ports are forced to be equal, such as by shorting them together, the admittance matrix for the network can be processed to incorporate the information. The operation is useful to convert several parallel leads, such as multiple ground connections, into a single lead for faster simulation with the model.

The goal is to convert an $N \times N$ admittance matrix into a new $M \times M$ admittance matrix, where $M < N$. The voltages at each of the N-ports of the original admittance matrix, $\overline{\overline{Y}}_o$, are mapped to one of the M ports on the new admittance matrix, $\overline{\overline{Y}}_n$, using an $N \times M$ incidence matrix, $\overline{\overline{A}}$:

$$\overline{V}_o = \overline{\overline{A}}\,\overline{V}_n. \tag{4.17}$$

$\overline{\overline{A}}$ is constructed by inspection. The currents are related by

$$\overline{I}_n = \overline{\overline{A}}^T \overline{I}_o \tag{4.18}$$

to satisfy Kirchhoff's current law.

The construction of the incidence matrix is more easily demonstrated than described. Consider the setup in Figure 4.4. The incidence matrix for this example is

$$\overline{\overline{A}} = \begin{bmatrix} 1 & 0 & 0 \\ 1 & 0 & 0 \\ 0 & 1 & 0 \\ 0 & 1 & 0 \\ 0 & 0 & 1 \end{bmatrix}.$$

As can be seen, the voltages at the ports of the old matrix are simply assigned to the voltages at the ports of the new matrix.

The voltages and currents are related by the admittance matrix as

$$\overline{I}_o = \overline{\overline{Y}}_o \overline{V}_o \tag{4.19}$$

Substituting (4.17) into (4.19), multiplying by $\overline{\overline{A}}^T$, and then substituting (4.18) yields

$$\overline{I}_n = \overline{\overline{A}}^T \overline{\overline{Y}}_o \overline{\overline{A}} \overline{V}_n.$$

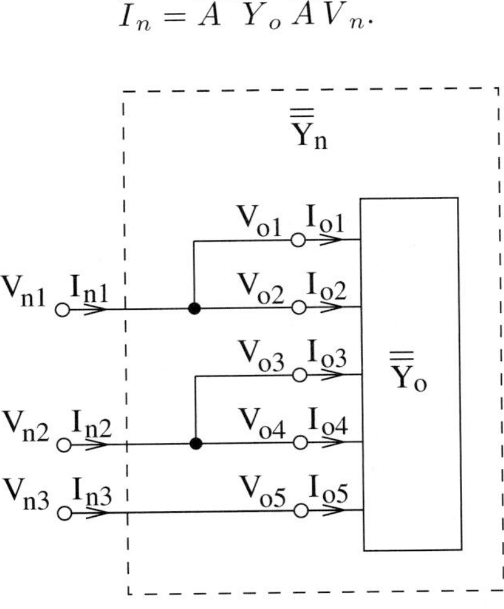

Figure 4.4. Example setup for conversion from one admittance matrix to another, with paralleled leads built-in.

The new reduced admittance matrix is then

$$\overline{\overline{Y}}_n = \overline{\overline{A}}^T \overline{\overline{Y}}_o \overline{\overline{A}}. \tag{4.20}$$

The conversions in section 4.3 can be used to move a matrix in to and out of convenient forms for reduction. The matrix operations in (4.20) simply add together the rows and columns of ports that are to be tied together, and an implementation based on this fact can be far quicker than actually performing the matrix multiplications.

For inductance matrices, (4.20) represents a generalization of the formula

$$L_{\text{total}} = \frac{1}{\frac{1}{L_1} + \frac{1}{L_2} + \cdots + \frac{1}{L_N}}$$

for paralleling inductors together. Note that the operation order is invert/add/invert. In (4.20), the same is performed to parallel leads of an inductance matrix: invert from impedance to admittance, add rows and columns, and invert back to impedance.

4.6 Exercises

1. Show that when $\overline{\overline{Z}}$ is symmetric, then $\overline{\overline{S}}$ is also symmetric. Recall that $\overline{\overline{A}}\,\overline{\overline{B}} \neq \overline{\overline{B}}\,\overline{\overline{A}}$ for general matrices $\overline{\overline{A}}$ and $\overline{\overline{B}}$, nor for symmetric matrices.

2. Show that the power dissipated in a multiport network is $P = \frac{1}{2}\overline{a}^T(\overline{\overline{I}} - \overline{\overline{S}}^T\overline{\overline{S}}^*)\overline{a}^*$, where $*$ denotes complex conjugate. Then for a lossless network, $\overline{\overline{S}}^T\overline{\overline{S}}^* = \overline{\overline{I}}$ or $\overline{\overline{S}}^* = (\overline{\overline{S}}^T)^{-1}$.

3. Show that the power dissipated in a multiport network is $P = \frac{1}{4}\overline{I}^T\left(\overline{\overline{Z}}^T + \overline{\overline{Z}}^*\right)\overline{I}^*$. Then for a lossless network, $\overline{\overline{Z}}^T + \overline{\overline{Z}}^* = 0$.

4. Show that the reflection coefficient looking into port 1 of a two-port network terminated with an arbitrary load impedance at port 2 is $\Gamma = S_{11} + \frac{S_{12}S_{21}\Gamma_L}{1-S_{22}\Gamma_L}$, where Γ_L is the reflection coefficient at the load.

5. Show that a conversion from $\overline{\overline{Y}}$ to $\overline{\overline{S}}$ is

$$\overline{\overline{S}} = (\overline{\overline{k}}^{-1} + \overline{\overline{Y}}\,\overline{\overline{k}})^{-1}(\overline{\overline{k}}^{-1} - \overline{\overline{Y}}\,\overline{\overline{k}}).$$

6. Show that a conversion from $\overline{\overline{S}}$ to $\overline{\overline{Y}}$ is

$$\overline{\overline{Y}} = \overline{\overline{k}}^{-1}(\overline{\overline{I}} - \overline{\overline{S}})(\overline{\overline{I}} + \overline{\overline{S}})^{-1}\overline{\overline{k}}^{-1}.$$

7. Multiport networks with equal numbers of input and output ports can be cascaded as shown in Figure 4.5. The $2N \times 2N$ S-parameter matrices can be divided into four $N \times N$ submatrices each as

$$\overline{\overline{S}}_1 = \begin{bmatrix} \overline{\overline{S}}_{1A} & \overline{\overline{S}}_{1B} \\ \overline{\overline{S}}_{1C} & \overline{\overline{S}}_{1D} \end{bmatrix}$$

and

$$\overline{\overline{S}}_2 = \begin{bmatrix} \overline{\overline{S}}_{2A} & \overline{\overline{S}}_{2B} \\ \overline{\overline{S}}_{2C} & \overline{\overline{S}}_{2D} \end{bmatrix}.$$

Show that the submatrices for the S-parameter matrix formed by the cascade of $\overline{\overline{S}}_1$ and $\overline{\overline{S}}_2$ are

$$\begin{aligned}
\overline{\overline{S}}_A &= \overline{\overline{S}}_{1A} + \overline{\overline{S}}_{1B}(\overline{\overline{I}} - \overline{\overline{S}}_{2A}\overline{\overline{S}}_{1D})^{-1}\overline{\overline{S}}_{2A}\overline{\overline{S}}_{1C} \\
\overline{\overline{S}}_B &= \overline{\overline{S}}_{1B}(\overline{\overline{I}} - \overline{\overline{S}}_{2A}\overline{\overline{S}}_{1D})^{-1}\overline{\overline{S}}_{2B} \\
\overline{\overline{S}}_C &= \overline{\overline{S}}_{2C}(\overline{\overline{I}} - \overline{\overline{S}}_{1D}\overline{\overline{S}}_{2A})^{-1}\overline{\overline{S}}_{1C} \\
\overline{\overline{S}}_D &= \overline{\overline{S}}_{2D} + \overline{\overline{S}}_{2C}\overline{\overline{S}}_{1D}(\overline{\overline{I}} - \overline{\overline{S}}_{2A}\overline{\overline{S}}_{1D})^{-1}\overline{\overline{S}}_{2B}.
\end{aligned}$$

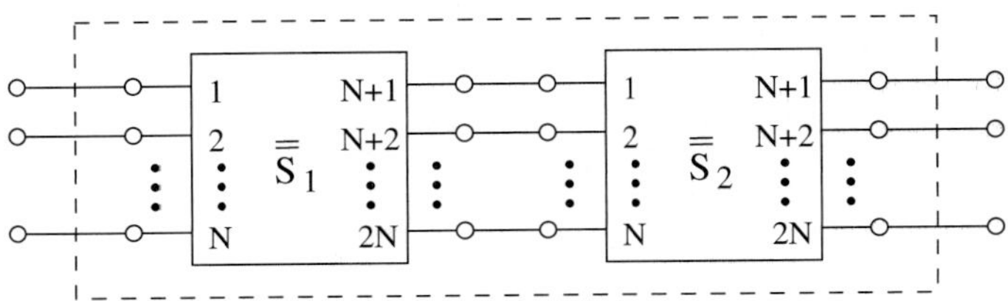

Figure 4.5. Cascaded multiport S-parameters.

8. Multiport networks with equal numbers of input and output ports can be connected in parallel as shown in Figure 4.6. The $2N \times 2N$ admittance matrices can be divided into four $N \times N$ submatrices each as

$$\overline{\overline{Y}}_1 = \begin{bmatrix} \overline{\overline{Y}}_{1A} & \overline{\overline{Y}}_{1B} \\ \overline{\overline{Y}}_{1C} & \overline{\overline{Y}}_{1D} \end{bmatrix}$$

and

$$\overline{\overline{Y}}_2 = \begin{bmatrix} \overline{\overline{Y}}_{2A} & \overline{\overline{Y}}_{2B} \\ \overline{\overline{Y}}_{2C} & \overline{\overline{Y}}_{2D} \end{bmatrix}.$$

Show that the admittance matrix formed by the parallel connection of $\overline{\overline{Y}}_1$ and $\overline{\overline{Y}}_2$ is

$$\overline{\overline{Y}} = \begin{bmatrix} \overline{\overline{Y}}_{1A} + \overline{\overline{Y}}_{2A} & \overline{\overline{Y}}_{1B} + \overline{\overline{Y}}_{2B} \\ \overline{\overline{Y}}_{1C} + \overline{\overline{Y}}_{2C} & \overline{\overline{Y}}_{1D} + \overline{\overline{Y}}_{2D} \end{bmatrix}.$$

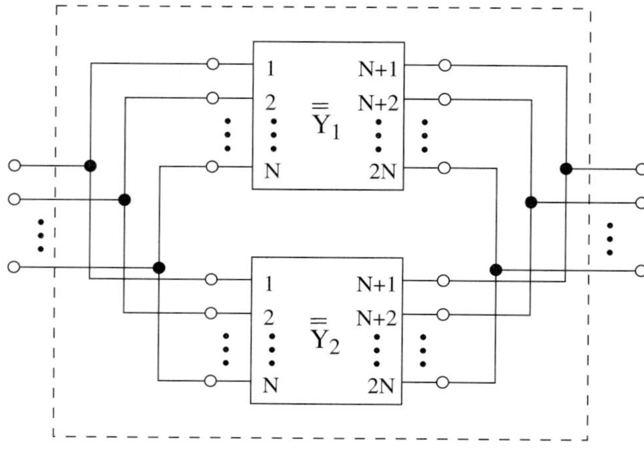

Figure 4.6. Parallel multiport admittance matrices.

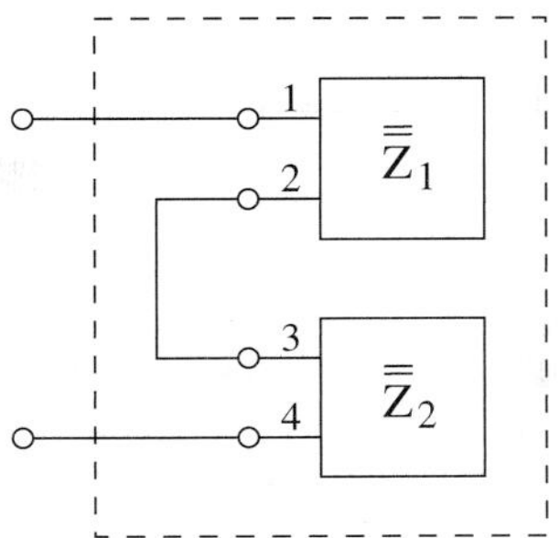

Figure 4.7. Series connection of two impedance matrices.

9. Show that the impedance matrix of the series connection of two 2×2 impedance matrices, as shown in Figure 4.7, is given by

$$\overline{\overline{Z}}_{\text{series}} = \begin{bmatrix} Z_{11} - Z_{12}Z_{21}/(Z_{22}+Z_{33}) & Z_{12}Z_{34}/(Z_{22}+Z_{33}) \\ Z_{21}Z_{43}/(Z_{22}+Z_{33}) & Z_{44} - Z_{34}Z_{43}/(Z_{22}+Z_{33}) \end{bmatrix}$$

if

$$\overline{\overline{Z}}_1 = \begin{bmatrix} Z_{11} & Z_{12} \\ Z_{21} & Z_{22} \end{bmatrix}$$

and

$$\overline{\overline{Z}}_2 = \begin{bmatrix} Z_{33} & Z_{34} \\ Z_{43} & Z_{44} \end{bmatrix}.$$

10. Given an $N \times N$ S-parameter matrix normalized to Z_o on each port, show that the new S-parameter matrix formed from adding to each port j a transmission line with characteristic impedance Z_o and phase shift $\theta_j = \beta \ell_j$ is

$$\overline{\overline{S}}_{new} = \begin{bmatrix} e^{-j\theta_1} & 0 & \cdots & 0 \\ 0 & e^{-j\theta_2} & \cdots & 0 \\ \vdots & \vdots & \ddots & \vdots \\ 0 & 0 & \cdots & e^{-j\theta_N} \end{bmatrix} \overline{\overline{S}} \begin{bmatrix} e^{-j\theta_1} & 0 & \cdots & 0 \\ 0 & e^{-j\theta_2} & \cdots & 0 \\ \vdots & \vdots & \ddots & \vdots \\ 0 & 0 & \cdots & e^{-j\theta_N} \end{bmatrix}.$$

Note that this conversion is frequency-dependent since β is frequency-dependent.

11. (a) Show that the admittance matrix of a lossless transmission line is

$$\overline{\overline{Y}} = \frac{\jmath y_o}{\sin(\beta\ell)} \begin{bmatrix} -\cos(\beta\ell) & 1 \\ 1 & -\cos(\beta\ell) \end{bmatrix}.$$

(b) Show that the admittance matrix of a lossy transmission line is

$$\overline{\overline{Y}} = \frac{y_o}{\sinh(\gamma\ell)} \begin{bmatrix} \cosh(\gamma\ell) & -1 \\ -1 & \cosh(\gamma\ell) \end{bmatrix}. \quad (4.21)$$

12. A 2cm section of transmission line is characterized at 1.79GHz with a 50Ω vector network analyzer to obtain

$$\overline{\overline{S}} = \begin{bmatrix} 0.322871 + \jmath 0.0216587 & 0.06333 - \jmath 0.0944073 \\ 0.06333 - \jmath 0.0944073 & 0.322871 + \jmath 0.0216587 \end{bmatrix}.$$

Find the line's characteristic impedance and effective dielectric constant.

13. Without using the result from chapter 9, equation (9.22), show that the admittance matrix for the π-network in Figure 4.8a is

$$\overline{\overline{Y}} = \begin{bmatrix} y_1 + y_2 & -y_2 \\ -y_2 & y_2 + y_3 \end{bmatrix}.$$

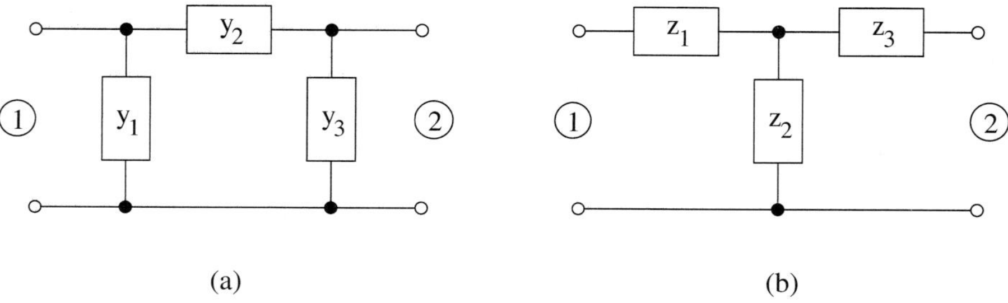

Figure 4.8. Networks for implementing 2 × 2 matrices: (a) π-network for admittance, (b) T-network for impedance.

14. Show that the impedance matrix for the T-network in Figure 4.8b is

$$\overline{\overline{Z}} = \begin{bmatrix} z_1 + z_2 & z_2 \\ z_2 & z_2 + z_3 \end{bmatrix}.$$

15. Show that the S-parameter matrix for a shunt capacitor inserted into a lossless transmission line is

$$\overline{\overline{S}} = \frac{1}{2 + \jmath\omega C Z_o} \begin{bmatrix} -\jmath\omega C Z_o & 2 \\ 2 & -\jmath\omega C Z_o \end{bmatrix}.$$

Chapter 5

INDUCTANCE

Because inductance leads directly to simultaneous switching noise (SSN) and crosstalk, it plays a central role in interconnect modeling and performance estimation in simulations of high-speed systems. Unfortunately, inductance is a remarkably deep subject, and it can often be difficult to apply. Because magnetic fields do not terminate, forming only closed loops, inductance is a long-range effect that couples together many individual lines of a multiconductor interconnect. Coupling requires the use of mutual inductance in interconnect models, and these can dramatically increase both model size and simulation runtime. As tempting as it may be, mutual inductance cannot be ignored.

As an example of the effect of neglecting mutual inductance, the SSN example from chapter 3, section 3.1.3 is recomputed without mutual inductance. The original and recomputed results are shown in Figure 5.1, where it can be seen that neglecting the mutual inductances causes as much as 100% error in the estimated SSN.

Maxwell's equations are the foundation of all classical work in electromagnetics and circuits. The concepts of inductance and capacitance help encapsulate the complex general electromagnetic field phenomenon of the full set of equations to the simpler circuit case involving Kirchhoff's voltage and current laws. Inductance relates magnetic flux to current, while capacitance relates the electric field to charge using voltage.

This chapter presents a thorough review of inductance and mutual inductance along with the application of the rigorous definitions to practical aspects of mod-

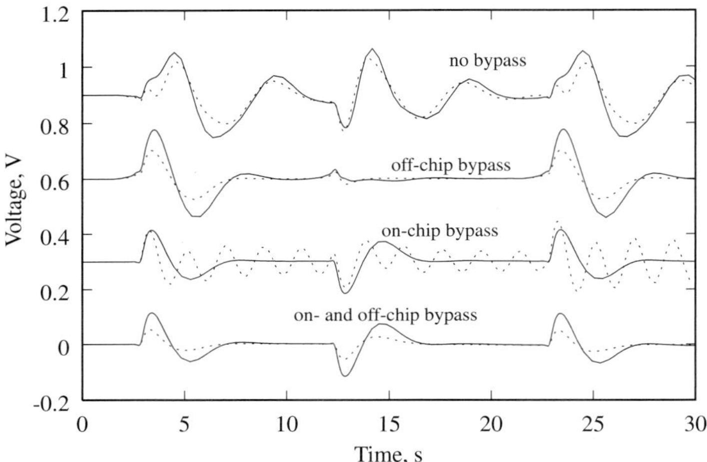

Figure 5.1. Recalculation of the plot in Figure 3.10, which is an example SPICE simulation of the voltage at an off-chip capacitive load generated by a packaged on-chip driver holding low. Dotted line indicates package model with mutual inductances. Solid line indicates package model without mutual inductance. The waveforms are separated by 0.3V offsets.

eling. The concept of partial inductance is explored in full. Common formulas and their uses are presented. Additional theoretical treatment investigates passivity, uniqueness, interconnect modeling, and reference.

5.1 Summary of an Electromagnetic Result

Maxwell's equations are quite flexible in their presentation, with several equivalent forms available. For the following discussion on inductance, the form utilizing the scalar potential and the magnetic vector potential is particularly useful. Since the derivation is available from innumerable sources,[1] just the results are presented here. Under the assumptions that the differential form of Maxwell's equations hold and that the material properties are uniform, then

1. See for example, C. Johnk, *Engineering Electromagnetic Fields and Waves*, 2nd ed. New York: Wiley, 1988, p. 546.

$$\overline{B} = \nabla \times \overline{A} \tag{5.1}$$

and

$$\overline{E} = -\nabla\Phi - \frac{\partial \overline{A}}{\partial t}, \tag{5.2}$$

where $\overline{A}$ is the magnetic vector potential, $\overline{B}$ is the magnetic flux density, Φ is the electric scalar potential, and $\overline{E}$ is the electric field strength. The complexity is embedded in the equations for $\overline{A}$ and Φ, which for quasi-static conditions appropriate for circuit analyses are given by

$$\overline{A}(\overline{r}) = \frac{\mu}{4\pi} \iiint_V \frac{\overline{J}(\overline{r}')}{|\overline{r} - \overline{r}'|} dV' \tag{5.3}$$

and

$$\Phi(\overline{r}) = \frac{1}{4\pi\epsilon} \iiint_V \frac{\rho_v(\overline{r}')}{|\overline{r} - \overline{r}'|} dV',$$

where $\overline{J}$ is the current density and ρ_v is the volume charge density.

Quasi-static fields do not appreciably change with frequency, a situation that occurs whenever the circuit is small compared to the wavelength of the electromagnetic field. For a given size circuit, there is an upper frequency limit under which quasi-static conditions prevail. Common rules of thumb are $d_{\max} < \lambda_{\min}/10$ and $d_{\max} < \lambda_{\min}/30$, where $d_{\max}$ is the largest circuit dimension and $\lambda_{\min}$ is the shortest wavelength of interest.

Circuits that are small for all frequency components of a signal appear to be lumped at a particular location. Circuits satisfying quasi-static conditions for all significant frequency components are said to be *lumped components* or *lumped elements*. While a component appears as lumped, the lumped approximation is said to apply. Therefore, *quasi-static* and *lumped approximation* are synonymous.

Circuit modeling and simulation deals with voltages and currents. Voltage is defined by

$$v = -\int \overline{E} \cdot \overline{d\ell}, \tag{5.4}$$

and current by

$$i = \int\int_S \overline{J}(\overline{r}) \cdot \overline{ds}.$$

While the voltage can be computed at any frequency, it only provides a unique answer for quasi-static fields. At high frequencies, the computed voltage depends on the path chosen for the path integral. In contrast, current is uniquely defined by the area, and its calculation is unambiguous at all frequencies.

While the lumped approximation holds, circuit analyses can use voltage and current. With the loss of uniqueness for voltage at high frequencies, circuit analyses must rely on more general concepts such as S-parameters.

5.2 Definitions of Inductance

The definition of inductance can be rigorously based on Maxwell's equations in integral form, and successful application of the inductance concept relies on careful and strict interpretation of the meanings of closed line integrals and surface integrals. Several definitions for inductance are possible: thin-wire for formula development, field-based for theoretical treatment, and energy-based for computer computation. Each of these definitions is derived in the following sections.

5.2.1 Thin Wire Definition

Many practical applications use long thin wires, and a definition for inductance specialized for this application is useful for deriving and cataloging formulas. In addition, the derivation clearly demonstrates the geometrical role of inductance.

Faraday's law in integral form is

$$\oint_C \overline{E} \cdot \overline{d\ell} = -\frac{d}{dt} \int\int_S \overline{B} \cdot \overline{ds}, \tag{5.5}$$

where $\oint_C \overline{E} \cdot \overline{d\ell}$ sums the tangential component of the electric field around a closed path given by contour C, and $\int\int_S \overline{B} \cdot \overline{ds}$ sums the normal component of the magnetic flux density through the area enclosed by C. These relationships are sketched in Figure 5.2. Substituting (5.1) into (5.5) and applying Stokes' theorem,

$$\iint_S (\nabla \times \overline{A}) \cdot \overline{ds} = \oint_C \overline{A} \cdot \overline{d\ell}, \tag{5.6}$$

true for any vector field $\overline{A}$, yields

$$\oint_C \overline{E} \cdot \overline{d\ell} = -\frac{d}{dt} \oint_C \overline{A} \cdot \overline{d\ell}.$$

Under quasi-static conditions, substituting (5.3) to eliminate $\overline{A}$ gives

$$\oint_C \overline{E} \cdot \overline{d\ell} = -\frac{d}{dt} \oint_C \left(\frac{\mu}{4\pi} \iiint_V \frac{\overline{J}(\overline{r}')}{|\overline{r} - \overline{r}'|} dV' \right) \cdot \overline{d\ell}.$$

For a circuit consisting of thin wires and small components, the current density vanishes for points off of the contour, C, and it travels in a direction tangential to the contour. Executing the volume integral yields the simplified form

$$\oint_C \overline{E} \cdot \overline{d\ell} = -\frac{d}{dt} \oint_C \left(\frac{\mu}{4\pi} \oint_{C'} \frac{i(\overline{r}')}{|\overline{r} - \overline{r}'|} \overline{d\ell}' \right) \cdot \overline{d\ell}, \tag{5.7}$$

where i is the current on the thin wire. According to the quasi-static assumption, the current is constant on the contour, and (5.7) can be rearranged into time-constant and time-varying parts as

$$\oint_C \overline{E} \cdot \overline{d\ell} = -\oint_C \left(\frac{\mu}{4\pi} \oint_{C'} \frac{1}{|\overline{r} - \overline{r}'|} \overline{d\ell}' \right) \cdot \overline{d\ell} \, \frac{di}{dt}. \tag{5.8}$$

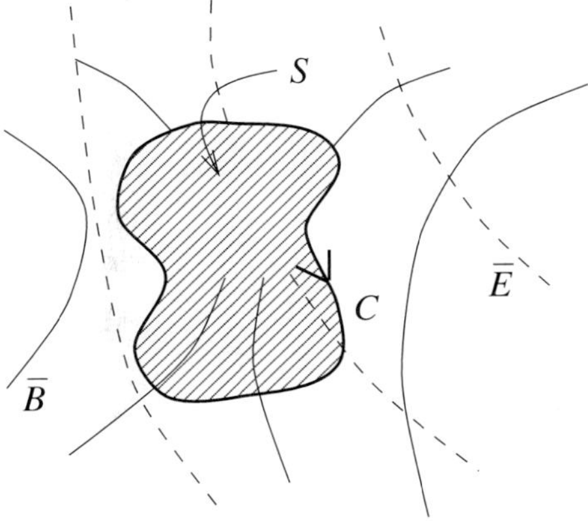

Figure 5.2. Contour and area sketch for interpretation of the integral relationships in Faraday's law.

The complexity of Maxwell's equations is then captured in a time-constant, geometrically-dependent factor called inductance, defined by

$$L = \frac{\mu}{4\pi} \oint_C \oint_{C'} \frac{\overline{d\ell}' \cdot \overline{d\ell}}{|\overline{r} - \overline{r}'|}, \tag{5.9}$$

and this formulation of L is called *Neumann's inductance formula.*

Circuits

The remaining electromagnetic quantity in (5.8) can be cast for circuits using the definition of voltage in (5.4). Breaking the closed contour C into N open segments C_i, as suggested in Figure 5.3 for $N = 8$, then

$$\begin{aligned} \oint_C \overline{E} \cdot d\ell &= \sum_{i=1}^{N} \int_{C_i} \overline{E} \cdot d\ell \\ &= -\sum_{i=1}^{N} v_i, \end{aligned} \tag{5.10}$$

where

$$v_i = -\int_{C_i} \overline{E} \cdot d\ell$$

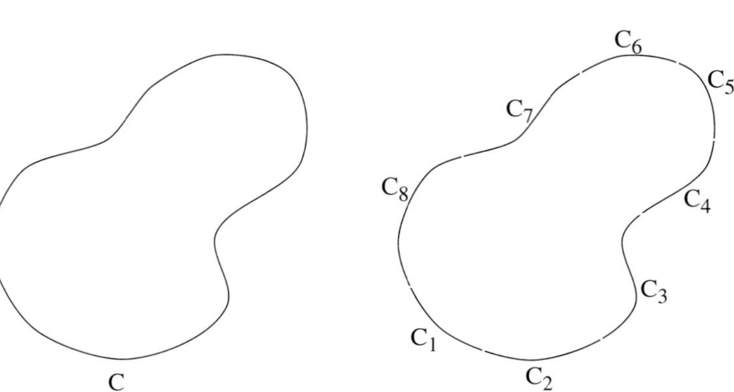

Figure 5.3. A closed contour can be broken into any number of open contours.

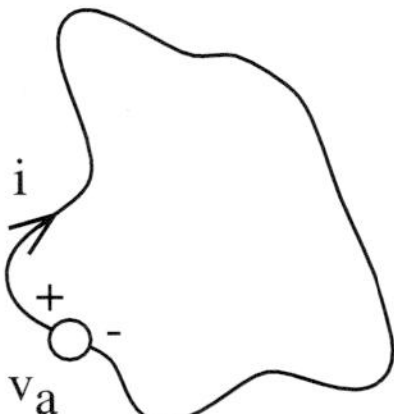

Figure 5.4. Prototypical circuit incorporating voltages from a source, resistive losses, and inductance.

are the voltage drops along each open contour segment, C_i. The sum of the voltage drops on all the segments is the voltage drop around the closed loop, C.

Incorporating (5.9) and (5.10) into (5.8) provides the classic voltage-current relationship for inductance as

$$v = L\frac{di}{dt}, \tag{5.11}$$

where v represents the voltage around a closed loop. So, a time-varying current produces a voltage on the closed loop, where the value is given by the time derivative of the current scaled by the inductance.

For the complete voltage-current relationship, consider the circuit shown in Figure 5.4, where a thin-wire loop is driven by a small voltage source, v_a. The resistive voltage drop is iR, where R is the total resistance of the thin wire, and the voltage generated on the loop due to inductance is $L\frac{di}{dt}$. From Kirchhoff's voltage law, the total of all voltages sums to zero, so $-v_a + iR + L\frac{di}{dt} = 0$, or

$$v_a = iR + L\frac{di}{dt}.$$

Loops vs. Wire

As shown by Faraday's law, the actual cause of the induced voltage is the time-varying magnetic flux density over a surface, but inductance allows a simplification and the use of the circuit quantity i. Since the current flows in the wire itself, this substitution can lead to confusion as to the source of the induced voltage and the

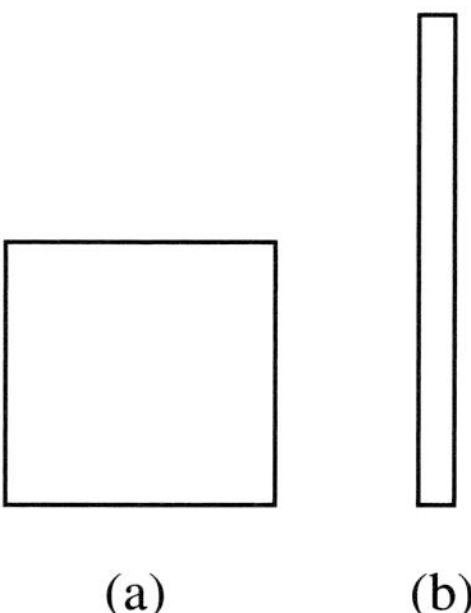

Figure 5.5. Two rectangular loops of wire with equal total wire lengths enclose different areas. The loop in (a) has much higher inductance than the loop in (b).

meaning of inductance. It is clear from the derivation that inductance is only defined for a closed contour: it is the closed loop formed by a wire that has inductance—not the wire itself.[2] A piece of a closed loop does not have inductance, so it is incorrect to assign inductance to a piece of wire. The geometry of the loop sets the inductance value; the wire itself has limited impact while the thin-wire assumption holds.

Consider the two wire loops in Figure 5.5. The two wires are the same length, but the inductances are radically different because the loop areas are very different. Inductance is minimized by reducing the loop area for a given wire length. One of the most effective methods for minimizing inductance is to route signals over a ground plane so that the loop area is just the length of the net times the dielectric thickness.

5.2.2 Field-Based Definitions

Equation (5.11) is valid in general, even though the formula for L in (5.9) is valid only for thin wires. To obtain a more general formula valid for all cases but requiring the use of field quantities (as all general solutions in electromagnetics must), the right-hand side of (5.11) can be equated to the right-hand side of Faraday's law in (5.5) to obtain

2. A common question in packaging is, What is the inductance of a wirebond? The correct answer is that wirebonds do not have inductance.

$$L\frac{di}{dt} = \frac{d}{dt}\iint_S \overline{B}\cdot\overline{ds}. \tag{5.12}$$

Integrating (5.12) with respect to time, and noting that the integration constant vanishes since fields do not exist without current in the circuit, then

$$L = \frac{\iint_S \overline{B}\cdot\overline{ds}}{i}. \tag{5.13}$$

In terms of field quantities, inductance is given by the total flux passing through a surface divided by the current magnitude in the circuit producing the magnetic flux.

A related field-based definition using the magnetic vector potential can be found by substituting (5.1) into (5.13) and then using (5.6) to obtain

$$L = \frac{\oint_C \overline{A}\cdot\overline{d\ell}}{i}. \tag{5.14}$$

5.2.3 Energy-Based Definition

While (5.13) is general, the reliance on the appropriate construction and application of a surface and the related surface integral can lead to difficulties in problem setup. Through the use of energy relationships, a field-based inductance formulation involving only volume integrals is possible. From circuit theory, the energy stored in an inductor is

$$\varepsilon = \frac{1}{2}Li^2,$$

while the result from electromagnetic field theory is

$$\varepsilon = \frac{1}{2}\iiint_V \overline{B}\cdot\overline{H}dV,$$

where $\overline{H}$ is the magnetic field strength. Equating these two definitions and solving for L yields

$$L = \frac{1}{i^2}\iiint_V \overline{B}\cdot\overline{H}dV. \tag{5.15}$$

Because of the relative simplicity of setup, (5.15) is often used to compute inductance in general-purpose electromagnetic simulators, where the computation of the fields over a volume is straightforward but manipulation of surface definitions is not.

5.3 Definition of Mutual Inductance

Systems will typically consist of several interacting circuits, so it is necessary to investigate the voltages induced on one circuit by the time-varying magnetic flux densities produced by another. The derivation and concepts developed in section 5.2 for inductance are extended in this section to investigate mutual inductance.

5.3.1 Thin Wire Definition

Consider a system consisting of two closed contours that are in close proximity so that the magnetic flux from one contour penetrates the area enclosed by the other. This system is sketched in Figure 5.6. The total magnetic flux density is the sum of that produced by each circuit, so $\overline{B} = \overline{B}_1 + \overline{B}_2$. Faraday's law applied to C_2 is

$$\oint_{C_2} \overline{E} \cdot \overline{d\ell} = -\frac{d}{dt} \iint_{S_2} \overline{B}_1 \cdot \overline{ds} - \frac{d}{dt} \iint_{S_2} \overline{B}_2 \cdot \overline{ds}. \tag{5.16}$$

Substituting (5.1) and applying Stokes' theorem from (5.6) yields

$$\oint_{C_2} \overline{E} \cdot \overline{d\ell} = -\frac{d}{dt} \oint_{C_2} \overline{A_1} \cdot \overline{d\ell} - \frac{d}{dt} \oint_{C_2} \overline{A_2} \cdot \overline{d\ell}.$$

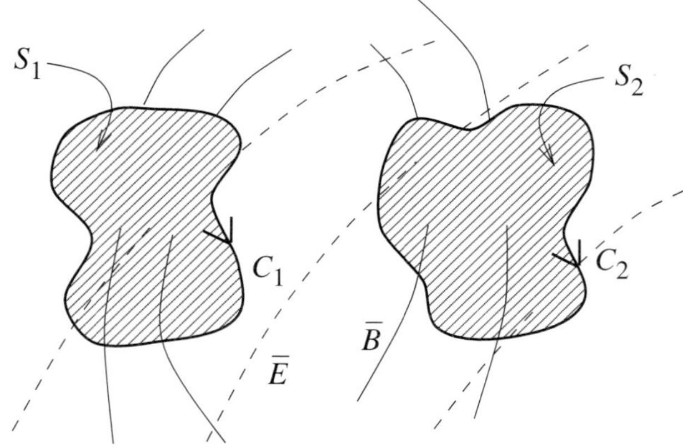

Figure 5.6. Contours and area sketches for interpretation of the integral relationships in Faraday's law for the case of mutual inductance.

Replacing the magnetic vector potentials from (5.3) gives

$$\oint_{C_2} \overline{E} \cdot \overline{d\ell} = -\frac{d}{dt} \oint_{C_2} \left(\frac{\mu}{4\pi} \iiint_V \frac{\overline{J_1}(\overline{r}')}{|\overline{r} - \overline{r}'|} dV' \right) \cdot \overline{d\ell}$$
$$-\frac{d}{dt} \oint_{C_2} \left(\frac{\mu}{4\pi} \iiint_V \frac{\overline{J_2}(\overline{r}')}{|\overline{r} - \overline{r}'|} dV' \right) \cdot \overline{d\ell}. \qquad (5.17)$$

For quasi-static operation with thin wires and small components, the current densities vanish off of the contours, and since the current travels in the same direction as the contours, then

$$\oint_{C_2} \overline{E} \cdot \overline{d\ell} = -\frac{d}{dt} \oint_{C_2} \left(\frac{\mu}{4\pi} \oint_{C_1} \frac{i_1(\overline{r}')}{|\overline{r} - \overline{r}'|} \overline{d\ell}' \right) \cdot \overline{d\ell}$$
$$-\frac{d}{dt} \oint_{C_2} \left(\frac{\mu}{4\pi} \oint_{C_2} \frac{i_2(\overline{r}')}{|\overline{r} - \overline{r}'|} \overline{d\ell}' \right) \cdot \overline{d\ell}. \qquad (5.18)$$

The currents i_1 and i_2 are constant on the contours according to the quasi-static assumption, so (5.18) can be rearranged into time-constant and time-varying parts as

$$\oint_{C_2} \overline{E} \cdot \overline{d\ell} = -M \frac{di_1}{dt} - L_2 \frac{di_2}{dt}, \qquad (5.19)$$

where

$$M = \frac{\mu}{4\pi} \oint_{C_2} \oint_{C_1} \frac{\overline{d\ell}' \cdot \overline{d\ell}}{|\overline{r} - \overline{r}'|} \qquad (5.20)$$

is the mutual inductance between contours C_1 and C_2 and is Neumann's formula for mutual inductance, and

$$L_2 = \frac{\mu}{4\pi} \oint_{C_2} \oint_{C_2} \frac{\overline{d\ell}' \cdot \overline{d\ell}}{|\overline{r} - \overline{r}'|}$$

is the self-inductance for contour C_2. Note that the only difference between self- and mutual inductance is in the contours of integration: for self-inductance, integration is over the same contour while for mutual inductance, integration is over both contours.

Circuits

Applying the definition of voltage from (5.4) to (5.19) provides a result completely in circuit quantities as

$$v_2 = M\frac{di_1}{dt} + L_2\frac{di_2}{dt}, \tag{5.21}$$

where v_2 is the voltage on the closed contour, C_2. Note that if the roles of the contours C_1 and C_2 are reversed, then the formula for mutual inductance in (5.20) remains unchanged. Therefore, inductance is reciprocal. The voltage-current relationship for the reversed case is

$$v_1 = M\frac{di_2}{dt} + L_1\frac{di_1}{dt}. \tag{5.22}$$

Equations (5.21) and (5.22) can be recast into matrix form as

$$\begin{bmatrix} v_1 \\ v_2 \end{bmatrix} = \begin{bmatrix} L_1 & M \\ M & L_2 \end{bmatrix} \begin{bmatrix} \frac{di_1}{dt} \\ \frac{di_2}{dt} \end{bmatrix}. \tag{5.23}$$

Generalizing this result for any number of contours yields

$$\overline{v} = \overline{\overline{L}}\frac{d\overline{i}}{dt}, \tag{5.24}$$

where $\overline{\overline{L}}$ is the inductance matrix. Since mutual inductance is reciprocal, then the inductance matrix is symmetric with $L_{ij} = L_{ji}$.

Loops vs. Wire

As in section 5.2, it is important to carefully note the origin of the self- and mutual inductance. It is the closed contours and not the wires that have inductance. The self-inductance relates the voltage produced on a closed loop to the magnetic flux generated by current on that same loop. The mutual inductance relates the voltage produced on a closed loop to the magnetic flux generated by current on another loop.

5.3.2 Field-Based Definition

Avoiding the thin-wire approximation, a more general formula for inductance can be found by equating the right-hand side of (5.21) to the negative of the right-hand side of (5.16) to obtain

$$M\frac{di_1}{dt} + L_2\frac{di_2}{dt} = \frac{d}{dt}\iint_{S_2} \overline{B}_1 \cdot \overline{ds} + \frac{d}{dt}\iint_{S_2} \overline{B}_2 \cdot \overline{ds}. \tag{5.25}$$

Integrating with a zero constant of integration yields

$$Mi_1 + L_2 i_2 = \iint_{S_2} \overline{B}_1 \cdot \overline{ds} + \iint_{S_2} \overline{B}_2 \cdot \overline{ds}, \tag{5.26}$$

from which it is apparent that

$$M = \frac{1}{i_1}\iint_{S_2} \overline{B}_1 \cdot \overline{ds}. \tag{5.27}$$

Mutual inductance is given by the flux generated by the current in loop 1 that is coupled into loop 2, then normalized by the current in loop 1. In other words, it is the flux linked into loop 2 that is the root cause of mutual inductance. For this reason, mutual inductance is often described as a *flux linkage*. Using this terminology, then self-inductance is caused by flux linkage from a loop to itself.

5.3.3 Energy-Based Definition

For many calculations, especially computer-based general-purpose simulators, an energy-based formulation for mutual inductance is simpler to implement and execute through the replacement of line and surface integrals with volume integrals. Equating the energy stored using circuit variables,

$$\begin{aligned} \varepsilon &= \frac{1}{2}M_{12}i_1 i_2 \\ &= \frac{1}{2}M_{21}i_2 i_1, \end{aligned} \tag{5.28}$$

to that using field variables,

$$\begin{aligned} \varepsilon &= \frac{1}{2}\iiint_V \overline{B}_1 \cdot \overline{H}_2 dV \\ &= \frac{1}{2}\iiint_V \overline{B}_2 \cdot \overline{H}_1 dV, \end{aligned} \tag{5.29}$$

yields a general field-based formula for mutual inductance:

$$\begin{aligned} M_{12} &= M_{21} \\ &= \frac{1}{i_1 i_2} \iiint_V \overline{B}_1 \cdot \overline{H}_2 dV \\ &= \frac{1}{i_1 i_2} \iiint_V \overline{B}_2 \cdot \overline{H}_1 dV. \end{aligned} \tag{5.30}$$

The subscripts on the field quantities $\overline{B}$ and $\overline{H}$ imply the use of superposition. If both loops are simultaneously driven with current, then just one set of $\overline{B}$ and $\overline{H}$ exist. Using superposition, loop 1 is driven to produce $\overline{B}_1$ and $\overline{H}_1$ while the source on loop 2 is open-circuited ($i_2 = 0$). These fields are stored and the process repeated with $i_1 = 0$ to obtain $\overline{B}_2$ and $\overline{H}_2$. These two solutions, picking either $\overline{B}_1$ and $\overline{H}_2$ or $\overline{B}_2$ and $\overline{H}_1$, are used in (5.30) to find the mutual inductance. Note that $\overline{B}_1$ and $\overline{H}_1$ alone determine the inductance of loop 1 according to (5.15), while $\overline{B}_2$ and $\overline{H}_2$ determine the inductance of loop 2. The extension to any number of loops is accomplished with a straightforward exercise in index manipulation.

5.3.4 Sign

However the mutual inductance is computed, its sign depends on the directions of the current flow. If the direction of one current is reversed, then the computed mutual inductance magnitude is unchanged but the sign flips. The user of a mutual inductance must be careful to set the sign properly.

5.4 Calculations with Neumann's Formula

While Neumann's formula is an outstanding vehicle for discussions on inductance, its direct use in computations is limited. The double integration requires that the positions in the denominator be colocated, causing the denominator to vanish and the integral to become undefined. The source of the difficulty is that, in contrast to real wires, the contour has zero cross-sectional area. Fortunately, it is possible to incorporate finite wire radius through a mutual inductance calculation.

To accommodate finite wire radius using Neumann's formula, the inductance of a loop is subdivided into two components: internal inductance, for the inductance

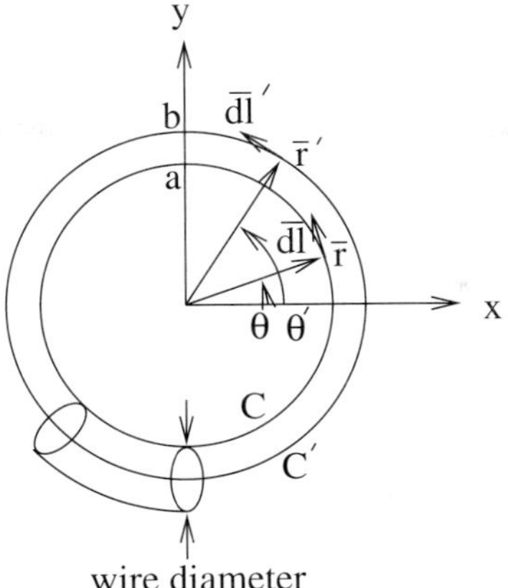

Figure 5.7. Concentric closed contours for computing external inductance using Neumann's formula for mutual inductance.

of the loop due to magnetic flux inside the wire; and external inductance, for the inductance of the loop due to magnetic flux outside the wire but still passing through the closed contour. The inductance is then just the sum of these two components.

Consider a thick, round wire formed into a circular loop. In Figure 5.7, a section of this loop is shown along with two thin wires located at the center of the thick wire (C') and along the inner edge (C). The important observation is that the external inductance of the thick wire is determined by the flux inside C, while the mutual inductance between loops C' and C is also determined by this flux (the flux linkage). Therefore, the external inductance of the thick wire can be computed as the mutual inductance between the thin wires. Neumann's formula for self-inductance is avoided. Note that the problem is not complete until the internal inductance is added.

5.4.1 Calculation of External Inductance for Loops of Thin Wires

The calculation of external inductance for thin wires is examined in this section by way of example, continuing the discussion on the inductance of a round loop of wire.

To find the external inductance of the loop through the calculation of the mutual inductance between C and $C^{'}$ in Figure 5.7, let

$$\begin{aligned}
\overline{r}^{'} &= b\cos\theta^{'}\hat{x} + b\sin\theta^{'}\hat{y} \\
\overline{r} &= a\cos\theta\hat{x} + a\sin\theta\hat{y} \\
\overline{d\ell}^{'} &= b\,d\theta^{'}(-\sin\theta^{'}\hat{x} + \cos\theta^{'}\hat{y}) \\
\overline{d\ell} &= a\,d\theta(-\sin\theta\hat{x} + \cos\theta\hat{y})
\end{aligned} \tag{5.31}$$

then

$$\begin{aligned}
|\overline{r} - \overline{r}^{'}| &= |(a\cos\theta - b\cos\theta^{'})\hat{x} + (a\sin\theta - b\sin\theta^{'})\hat{y}| \\
&= \sqrt{a^2 + b^2 - 2ab(\cos\theta\cos\theta' + \sin\theta\sin\theta')} \\
&= \sqrt{a^2 + b^2 - 2ab\cos(\theta - \theta')} \\
\overline{d\ell}^{'} \cdot \overline{d\ell} &= ab\cos(\theta - \theta^{'})\,d\theta^{'}\,d\theta.
\end{aligned} \tag{5.32}$$

Filling out Neumann's formula for mutual inductance in (5.20), then

$$M = \frac{\mu}{4\pi}\int_0^{2\pi}\int_0^{2\pi}\frac{ab\cos(\theta - \theta^{'})}{\sqrt{a^2 + b^2 - 2ab\cos(\theta - \theta^{'})}}d\theta d\theta^{'}. \tag{5.33}$$

From symmetry, set $\theta = 0$ and evaluate the integral over C by multiplying by 2π. Convert the integral on $C^{'}$ to the range 0 to π, then

$$\begin{aligned}
M &= \mu\int_0^{\pi}\frac{ab\cos\theta^{'}}{\sqrt{a^2 + b^2 - 2ab\cos\theta^{'}}}d\theta^{'} \\
&= \mu a\int_0^{\pi}\frac{\cos\theta^{'}}{\sqrt{1 - 2\left(\frac{a}{b}\right)\cos\theta' + \left(\frac{a}{b}\right)^2}}d\theta^{'} \\
&= 2\mu b\left[K\left(\frac{a}{b}\right) - E\left(\frac{a}{b}\right)\right],
\end{aligned} \tag{5.34}$$

where K and E are the complete elliptic functions of the first and second kind, respectively, and the integral formula used is

$$\int_0^{\pi}\frac{\cos x}{\sqrt{1 - 2p\cos x + p^2}}dx = \frac{2}{p}[K(p) - E(p)], \quad p^2 < 1, \tag{5.35}$$

where

$$K(k) = \int_0^{\frac{\pi}{2}} \frac{dx}{\sqrt{1 - k^2 \sin^2 x}} \tag{5.36}$$

and

$$E(k) = \int_0^{\frac{\pi}{2}} \sqrt{1 - k^2 \sin^2 x} dx. \tag{5.37}$$

For thin wires, $a/b \simeq 1$, so K and E can be approximated as

$$K(k) \simeq \ln\left(\frac{4}{\sqrt{1 - k^2}}\right) \tag{5.38}$$

and

$$E(k) \simeq 1. \tag{5.39}$$

Let the wire radius be defined as $r = b - a$, then

$$\sqrt{1 - \left(\frac{a}{b}\right)^2} = \sqrt{\frac{2r}{b} - \left(\frac{r}{b}\right)^2} \simeq \sqrt{\frac{2r}{b}} \tag{5.40}$$

since $\left(\frac{r}{b}\right)^2$ is negligible compared to $\frac{2r}{b}$ when $r \ll b$. Substituting (5.38), (5.39), and (5.40) into (5.34) yields a simplified formula for the external inductance of the round loop of thin wire:

$$L_{\text{external}} = M = \mu b \left[\ln\left(\frac{8b}{r}\right) - 2\right]. \tag{5.41}$$

5.4.2 Calculation of Internal Inductance for Round Wires

For internal inductance calculations, the thin-wire approximation does not apply, so Neumann's formula is inappropriate. Inductance must be calculated with a field-based formulation such as (5.15). For round wires under quasi-static conditions, Ampere's law in integral form,

$$\oint_C \overline{H} \cdot \overline{d\ell} = \iint_S \overline{J} \cdot \overline{ds} + \frac{d}{dt} \iint_S \overline{B} \cdot \overline{ds}, \tag{5.42}$$

can be used to find the magnetic fields inside the wire.

The cross section of a round wire is shown in Figure 5.8. Noting that $\overline{d\ell} = r d\phi \hat{\phi}$,

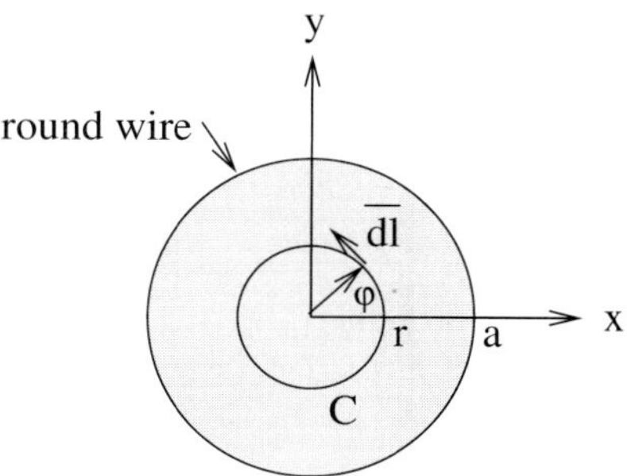

Figure 5.8. Cross section of a round wire with annotation for finding the magnetic fields inside the wire.

then $\overline{H} \cdot \overline{d\ell} = H_\phi(r) r d\phi$, where $\overline{H}$ has no dependence on ϕ due to symmetry. The current density is $\overline{J} = i/(\pi a^2)\hat{z}$. Neglecting the time-derivative for the quasi-static solution, then (5.42) can be evaluated as

$$\begin{aligned}
\oint_C \overline{H} \cdot \overline{d\ell} &= \iint_S \overline{J} \cdot \overline{ds} \\
\int_0^{2\pi} H_\phi(r) r d\phi &= \int_0^r \frac{i}{\pi a^2} 2\pi x dx \\
H_\phi(r) r \int_0^{2\pi} d\phi &= \frac{i}{\pi a^2} 2\pi \int_0^r x dx \\
H_\phi(r) &= \frac{i}{2\pi a^2} r.
\end{aligned}$$

With the constitutive relationship for magnetic fields, $\overline{B} = \mu \overline{H}$, then the magnetic flux density inside the wire is

$$B_\phi(r) = \frac{\mu i}{2\pi a^2} r. \tag{5.43}$$

Using this expression in (5.15) and integrating over the volume of the wire yields

$$L_{\text{internal}} = \frac{\ell}{i^2} \int_0^a \left(\frac{\mu i}{2\pi a^2} r\right)\left(\frac{i}{2\pi a^2} r\right) 2\pi r \, dr, \tag{5.44}$$

where the integrations over the length, ℓ, and the angular variations are already incorporated. Completing the computation yields

$$L_{\text{internal}} = \frac{\mu \ell}{8\pi}, \tag{5.45}$$

which provides the internal inductance of a straight round wire in isolation. Note that the value does not depend on the radius of the wire; all round wires of a given length have the same internal inductance.

Continuing the example of inductance calculation for the round loop of thin round wire, the inductance of the loop is the sum of the internal and external inductances. Adding (5.41) and (5.45) with $\ell = 2\pi b$ provides

$$L = \mu b \left[\ln \left(\frac{8b}{r} \right) - 7/4 \right]. \tag{5.46}$$

5.4.3 Frequency Dependence of Inductance

The fields external to a wire are weakly affected by the distribution of the current inside of the wire, so external inductance changes little with frequency while the quasi-static approximation holds. On the other hand, internal inductance is strongly affected by the current distribution. When the skin effect is well developed, the current resides on the surface of the wire and the internal inductance tends towards zero. Therefore, the low-frequency inductance of a loop of wire is the sum of the internal and external inductances, while at high frequencies with well-developed skin effects (but the quasi-static assumption must still hold), the inductance is just the external inductance because the internal inductance vanishes. The frequency dependence of inductance follows the trend sketched in Figure 5.9. For wires made from nonmagnetic materials, the low- and high-frequency limits typically differ by a few percent. Except for the breakdown of the quasi-static assumption, these trends are observable in chapter 7, Figure 7.7, where the inductance of a coaxial transmission line is computed with a PEEC method (see chapter 7, section 7.3).

Magnetic materials, such as Alloy 42, can generate very high internal inductance due to high relative permeability. The resulting high inductance does not necessarily lead to poor circuit performance. While magnetic materials can push up the internal inductance by orders of magnitude, skin effects shield the material at high frequencies. Since the effect of an inductor in a circuit is governed by its reactance, high inductance at low frequencies can still result in low reactance and minimal circuit effect.

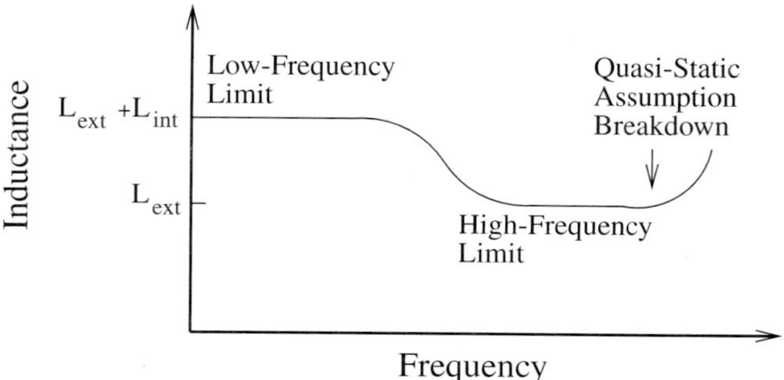

Figure 5.9. Frequency dependence of inductance.

5.5 Definition of Partial Inductance

Prior sections have established that self-inductance and mutual inductance are defined by loops of wire and not by the wires themselves (for thin wires). Although measurements must always be performed on closed loops—otherwise, no current flows and there is nothing to measure—*calculations* can be performed on pieces of closed loops. Calculation of the contributions to inductance from subsections of closed loops leads to the concept of partial inductance.

For thin wires, the inductance of a loop is given by (5.9), which is computed on a closed contour. This closed contour can be broken into any number of open contours, as suggested in Figure 5.3, then

$$\begin{aligned} L &= \frac{\mu}{4\pi} \sum_i \int_{C_i} \sum_j \int_{C_j'} \frac{\overline{d\ell}' \cdot \overline{d\ell}}{|\overline{r} - \overline{r}'|} \\ &= \sum_i \sum_j \left(\frac{\mu}{4\pi} \int_{C_i} \int_{C_j'} \frac{\overline{d\ell}' \cdot \overline{d\ell}}{|\overline{r} - \overline{r}'|} \right). \end{aligned}$$

Noting the similarity of the bracketed portion to the definitions of inductance in (5.9) and mutual inductance in (5.20), the partial self-inductance is defined as

$$L_i = \frac{\mu}{4\pi} \int_{C_i} \int_{C_i'} \frac{\overline{d\ell}' \cdot \overline{d\ell}}{|\overline{r} - \overline{r}'|}$$

and the partial mutual inductance as

$$M_{ij} = \frac{\mu}{4\pi} \int_{C_i} \int_{C_j'} \frac{\overline{d\ell}' \cdot \overline{d\ell}}{|\overline{r} - \overline{r}'|}. \tag{5.47}$$

A loop can have any number of partial self-inductances and partial mutual inductances, depending on how many sections into which the loop is broken. The number and configuration of pieces are selected for convenience in the solution of the problem at hand.

Calculations using partial self- and mutual inductances must carefully include all of the pieces needed to form a closed loop. Every piece of the loop has partial self-inductance plus partial mutual inductance to every other piece. Each contribution must be computed so that the sum for the entire closed loop produces the correct inductance according to (5.9).

The generalization of partial inductance to the mutual inductance between two loops is straightforward. Every piece of each loop has partial self-inductance plus partial mutual inductance to every other piece in all the loops, so several loops are treated in exactly the same manner as just one.

It is important to recognize that partial self- and mutual inductances are not physical quantities. As mathematical constructs, they cannot be measured. Certainly they should be used when convenient, but the focus should always remain with closed loops, the inductance of closed loops, and the mutual inductance between closed loops.

5.6 Formulas for Partial Self- and Mutual Inductance

Partial inductances are convenient when the geometry of the closed contour can be segmented into pieces for which formulas are already available. Using the thin-wire approximation and Neumann's formula, formulas for many useful structures can be derived. The following sections provide a few of these plus discussion on their use, while the Exercises section provides a few more results.

5.6.1 Partial Mutual Inductance between Two Parallel Wires

For two parallel straight wires, the setup for the partial mutual inductance calculation is shown in Figure 5.10. For use in (5.47), the various vectors are

$$\begin{aligned}
\overline{r} &= x\hat{x} \\
\overline{r}' &= x'\hat{x} + a\hat{y} \\
|\overline{r}' - \overline{r}| &= \sqrt{(x' - x)^2 + a^2} \\
\overline{d\ell} &= dx\hat{x} \\
\overline{d\ell}' &= dx'\hat{x},
\end{aligned} \tag{5.48}$$

then the partial mutual inductance is

$$\begin{aligned}
M_p &= \frac{\mu}{4\pi}\int_0^b\int_0^b \frac{dx\,dx'}{\sqrt{(x' - x)^2 + a^2}} \\
&= \frac{\mu b}{2\pi}\left[\ln\left(\frac{b}{a} + \sqrt{1 + \left(\frac{b}{a}\right)^2}\right) - \sqrt{1 + \left(\frac{a}{b}\right)^2} + \frac{a}{b}\right],
\end{aligned} \tag{5.49}$$

where the subscript p is used to emphasize that this is a partial mutual inductance. Two integral formulas useful in this solution are

$$\int \frac{du}{\sqrt{u^2 + a^2}} = \ln\left(u + \sqrt{u^2 + a^2}\right) \tag{5.50}$$

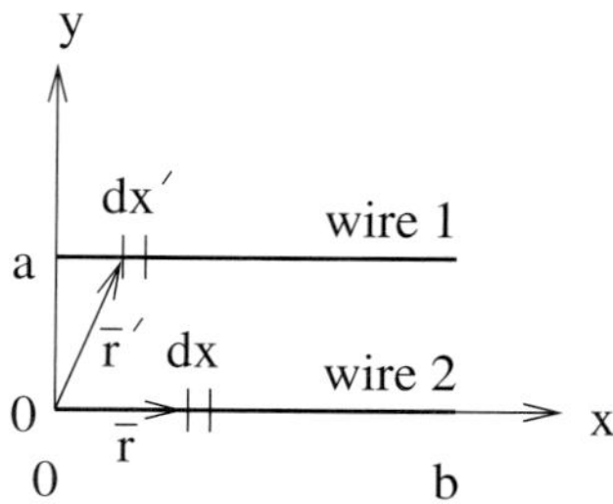

Figure 5.10. Setup for partial mutual inductance calculation for two parallel thin wires.

and

$$\int \ln\left(u + \sqrt{u^2 + a^2}\right) du = u \ln\left(u + \sqrt{u^2 + a^2}\right) - \sqrt{u^2 + a^2}. \tag{5.51}$$

For narrowly spaced wires with $b \gg a$, (5.49) can be approximated as

$$M_p = \frac{\mu b}{2\pi}\left[\ln\left(\frac{2b}{a}\right) - 1\right]. \tag{5.52}$$

The directions of the integrations imply that the current flow on both wires is in the same direction, so if the currents are actually moving in opposite directions, such as when one wire is acting as the return for another, then the sign of the partial mutual inductance must be reversed.

5.6.2 Partial Self-Inductance for a Round Wire

As discussed in section 5.4, Neumann's formula cannot be used to compute partial self-inductance. Instead, the partial self-inductance is found by computing the partial mutual inductance between a thin wire at the center of the (fat) round wire and one along its edge.

As shown in Figure 5.11, the flux linkage for the round wire is the same as for the two parallel thin wires, so (5.49) or (5.52) can be used to find the external partial self-inductance.

Since the self-flux linkage extends laterally to infinity, that is where the return currents are assumed to be. A realistic circuit will require a return path at a finite distance, and the return path's partial self-inductance along with all partial mutual inductances must be computed. The partial self-inductance formula derived in this way is said to *return at infinity.*

To the external partial self-inductance, the internal partial self-inductance from (5.45) must be added. For long thin wires with $b \gg a$, adding the internal to the external partial self-inductance yields

$$L_p = \frac{\mu b}{2\pi}\left[\ln\left(\frac{2b}{a}\right) - \frac{3}{4}\right]. \tag{5.53}$$

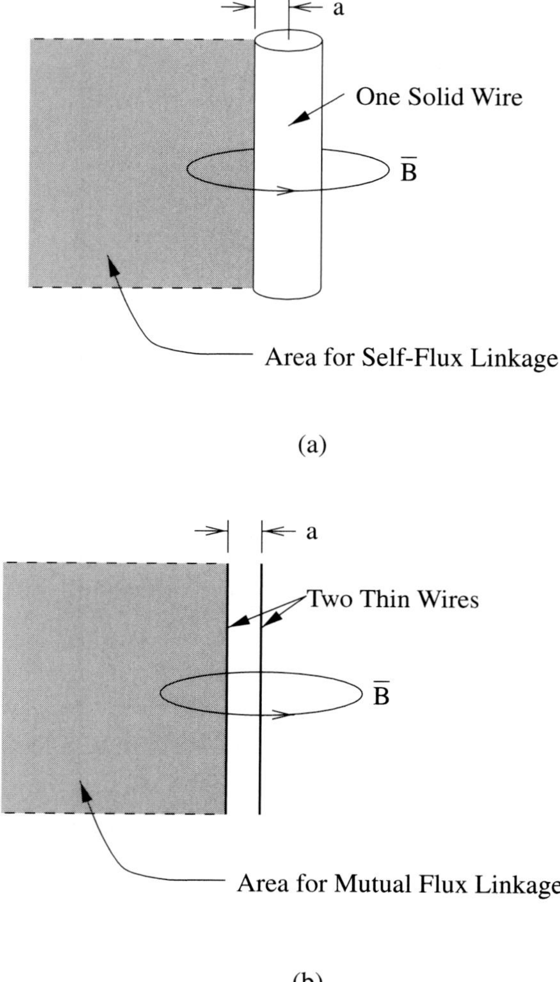

Figure 5.11. The flux linkage in (a) to find a partial self-inductance for a round wire is identical to that in (b) for the partial mutual inductance between two thin wires.

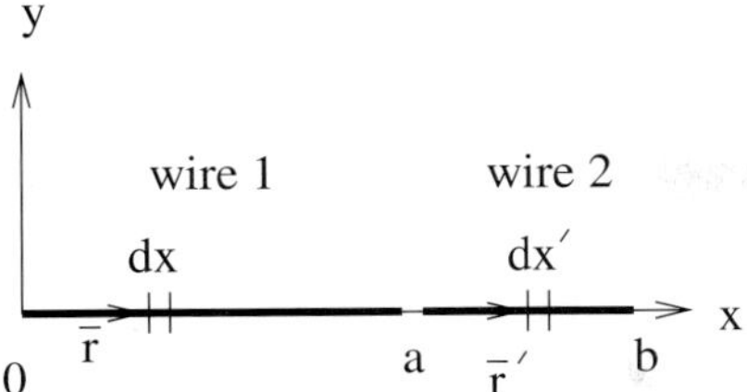

Figure 5.12. Setup for partial mutual inductance calculation for two colinear thin wires.

5.6.3 Partial Mutual Inductance of Two Colinear Wires

For two colinear thin wires, the setup in Figure 5.12 can be used to find the partial mutual inductance between the wires. The setup is simple, and Neumann's formula for this case is

$$\begin{aligned} M_p &= \frac{\mu}{4\pi}\int_0^a\int_a^b \frac{dx' dx}{x' - x} \\ &= \frac{\mu}{4\pi}\left[-a\ln a + b\ln b - (b-a)\ln(b-a)\right]. \end{aligned} \tag{5.54}$$

5.6.4 Assembling Solutions

The use of partial inductance to solve inductance problems requires four steps.

1. Break the closed loop into several sections.
2. Compute the partial self-inductance for each section.
3. Compute the partial mutual inductance between each pair of sections.
4. Assemble all results into a circuit simulation.

Since partial inductance is a mathematical construction, failure to completely apply the method can result in nonphysical results.

As an example, consider a transmission line constructed from two parallel thin wires. Assume that the line length, b, is long compared to the wire spacing, S, and that the wire radii are a_1 and a_2. The inductance of the complete circuit (neglecting end effects) with one wire as signal and the other as return is

$$\begin{aligned} L &= L_{p1} + L_{p2} - 2M_p \\ &= \frac{\mu b}{2\pi}\left[\ln\left(\frac{2b}{a_1}\right) - \frac{3}{4}\right] \\ &+ \frac{\mu b}{2\pi}\left[\ln\left(\frac{2b}{a_2}\right) - \frac{3}{4}\right] \\ &- 2\frac{\mu b}{2\pi}\left[\ln\left(\frac{2b}{S}\right) - 1\right] \\ &= \frac{\mu b}{2\pi}\left(\frac{1}{2} + \ln\left[\frac{S^2}{a_1 a_2}\right]\right). \end{aligned} \tag{5.55}$$

Examination of (5.53) reveals that the partial inductance is a nonlinear function of length with a $b \ln b$ dependence. The transmission line result in (5.55) is linearly dependent on the length of the line, so once the nonphysical partial inductances are correctly assembled into an inductance, the familiar per-unit-length dependence is recovered. Note that (5.55) is strongly dependent on length and weakly dependent on wire spacing and wire diameter. To reduce inductance, it is much more important to reduce interconnect length first, then reduce spacing or increase radius.

As another example, consider a simple computation regarding the partial self-inductance of a round wire. The partial self-inductance of a round wire 1cm long with a radius of 0.1mm from (5.53) is 9.097nH. The partial self-inductance of a wire half as long is 3.855nH. Note that $9.097 \neq 2 \times 3.855$. To correctly find the partial self-inductance of the long wire from two short wires, the partial mutual inductance between the two short wires must be included. From (5.54), the partial mutual inductance is 0.6931nH. The partial self-inductance of the long wire using the short wire results is $3.855 + 3.855 + 2 \times 0.6931 = 9.096$ nH, which is the same as that obtained for the long wire computed in one piece.

5.7 Circuit Symbols

The inductor circuit symbol represents both inductance and partial inductance in schematics, where the number of inductors present in each loop determines the type of inductance represented. Since a loop has just one inductance associated with it, if one inductor appears in the schematic of that loop, then the inductor represents the inductance. Otherwise, more than one inductor appears in the loop and the inductors represent partial inductances.

The distinction only matters in helping to ensure that mutual inductances are appropriately included. For inductance, mutual inductance is required only between loops, while for partial inductance, partial mutual inductance is required between each partial self-inductance, whether in the same loop or not.

The use of the inductor symbol for either inductance or partial inductance is shown in Figure 5.13. The inductors are shown polarized with the "o" to indicate where the mutual inductance formula assumes that the current enters. The polarity of one can be reversed if the sign of the mutual inductance is also reversed. Errors in polarity and/or mutual inductance sign can result in an active circuit due to violation of Lenz's law.

Depending on the circuit simulator, the mutual inductance may be entered directly in units of Henry or indirectly as a unitless value. SPICE uses the "k-factor" for mutual inductance calculated as

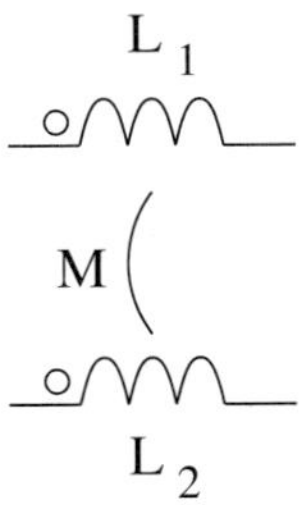

L_1

M

L_2

Figure 5.13. Demonstration of the use of the inductor symbol and mutual inductance.

$$k = \frac{M}{\sqrt{L_1 L_2}},$$

and k is bounded between ± 1. K-factors are convenient in that they indicate the strength of the mutual inductance, with magnitudes below about 0.1 being weak and above about 0.4 being strong. Note that unless a k-factor is very small, deleting a mutual inductance can lead to an active circuit when many loops are involved.

Much of circuit simulation is set up and run from schematic capture programs. However, many schematic capture programs, including top-of-the-line software from major vendors, do not accommodate mutual inductance on-screen in the same manner as other basic circuit components. It is important to recognize that this omission is due the difficulty of cleanly implementing mutual inductance in graphical schematic capture programs and not due to a lack of need for this basic circuit element. Work-arounds must be implemented in these cases so that mutual inductance can be included.

5.8 Modal Decomposition

Just as Fourier decomposition often leads to insight for time-varying signals by providing the frequency spectrum, general matrices can also be decomposed into constituent parts for further study. In general, an $N \times N$ matrix can be decomposed into N vectors, called *eigenvectors*, that are each weighted by a constant, called an *eigenvalue*. A primary property of interest is that eigenvectors, like the sinusoids of Fourier analysis, are orthogonal. Such modal decomposition enables extended analysis and study of matrices.

Inductance matrices are symmetric, due to reciprocity, real and square. For matrices with these properties, an extensive theoretical framework exists that can be directly used to develop a modal decomposition for inductance matrices. The modal decomposition enables the inductance matrix to be diagonalized, where all matrix values off of the diagonal are zero. Physically, the mutual inductances are eliminated. Once diagonalized, issues in simulation and passivity can be readily explored.

5.8.1 Diagonalization

The eigenvectors, $\bar{i}_k$, and eigenvalues, λ_k, of a square $N \times N$ real symmetric matrix, $\overline{\overline{L}}$, are determined from the solution of the equation

$$\overline{\overline{L}}\,\bar{i}_k = \lambda_k \bar{i}_k, \qquad k = 1, 2, \ldots, N. \tag{5.56}$$

All of the eigenvectors are linearly independent. If all of the eigenvalues are distinct, such as when the physical structure described by $\overline{\overline{L}}$ has no symmetry, then the eigenvectors are also orthogonal. In any case, the eigenvectors can be orthogonalized and normalized through the application of the Gram-Schmidt orthonormalization process.[3]

With orthonormal eigenvectors, an $N \times N$ matrix can be formed as

$$\begin{aligned}\overline{\overline{T}} &= \begin{bmatrix} \bar{i}_1 & \bar{i}_2 & \cdots & \bar{i}_N \end{bmatrix} \\ &= \begin{bmatrix} i_{1_1} & i_{2_1} & \cdots & i_{N_1} \\ i_{1_2} & i_{2_2} & \cdots & i_{N_2} \\ \vdots & \vdots & \ddots & \vdots \\ i_{1_N} & i_{2_N} & \cdots & i_{N_N} \end{bmatrix}.\end{aligned} \tag{5.57}$$

$\overline{\overline{T}}$ is an orthogonal matrix with the useful property that

$$\overline{\overline{T}}\,\overline{\overline{T}}^T = \overline{\overline{U}}, \tag{5.58}$$

where $\overline{\overline{U}}$ is the identity matrix and the superscript T denotes the matrix transpose. Using the orthogonality property of eigenvectors ($\bar{i}_m \cdot \bar{i}_n = 0$, $m \neq n$), $\overline{\overline{T}}$ diagonalizes $\overline{\overline{L}}$ as

$$\overline{\overline{T}}^T\,\overline{\overline{L}}\,\overline{\overline{T}} = \begin{bmatrix} \lambda_1 & 0 & \cdots & 0 \\ 0 & \lambda_2 & \cdots & 0 \\ \vdots & \vdots & \ddots & \vdots \\ 0 & 0 & \cdots & \lambda_N \end{bmatrix}. \tag{5.59}$$

3. See for example, C. R. Wylie, Jr. *Advanced Engineering Mathematics*, 3rd ed. New York: McGraw-Hill, 1966, p. 458.

With (5.58), (5.59) can be recast as

$$\overline{\overline{L}} = \overline{\overline{T}} \begin{bmatrix} \lambda_1 & 0 & \cdots & 0 \\ 0 & \lambda_2 & \cdots & 0 \\ \vdots & \vdots & \ddots & \vdots \\ 0 & 0 & \cdots & \lambda_N \end{bmatrix} \overline{\overline{T}}^T$$
$$= \overline{\overline{T}}\,\overline{\overline{\lambda}}\,\overline{\overline{T}}^T, \tag{5.60}$$

which shows that $\overline{\overline{L}}$ can be diagonalized into its eigenvalues by pre- and post-multiplication by $\overline{\overline{T}}$ and $\overline{\overline{T}}^T$. The benefit of diagonalization appears when (5.60) is folded into circuit analyses.

5.8.2 Circuit Theory

The voltage drop across an inductance matrix is related to the current by (5.24), repeated here as

$$\overline{v} = \overline{\overline{L}}\frac{d\overline{i}}{dt}.$$

Substituting the diagonalized form of the inductance matrix from (5.60) yields

$$\overline{v} = \overline{\overline{T}}\,\overline{\overline{\lambda}}\,\overline{\overline{T}}^T\frac{d\overline{i}}{dt}. \tag{5.61}$$

Since $\overline{\overline{L}}$ is not time-varying, then neither is $\overline{\overline{T}}$. Moving $\overline{\overline{T}}^T$ under the time derivative and using (5.58) yields

$$\overline{\overline{T}}^T\overline{v} = \overline{\overline{\lambda}}\frac{d}{dt}(\overline{\overline{T}}^T\overline{i}). \tag{5.62}$$

Modal voltages and currents can be defined as

$$\overline{v}_m = \overline{\overline{T}}^T\overline{v}$$
$$\overline{i}_m = \overline{\overline{T}}^T\overline{i}, \tag{5.63}$$

then the modal voltages are related to the modal currents by

$$\overline{v}_m = \overline{\overline{\lambda}}\frac{d\overline{i}_m}{dt}. \tag{5.64}$$

Since $\overline{\overline{\lambda}}$ is a diagonal matrix, (5.64) represents a *decoupled* set of N equations that can be solved independently. In contrast, (5.24) represents N *coupled* equations.

Direct solution of (5.24) involves the full and dense $\overline{\overline{L}}$ matrix with $N(N+1)/2$ terms. Solution of (5.64) involves just N terms. So a tradeoff exists: direct solution with many terms, or indirect solution with few terms but with pre and postprocessing required to transform into and out of the modal voltages and currents.

5.8.3 Manual Implementation

Inductance matrices can be incorporated into circuit simulation in terms of modal voltages and currents through auxiliary circuits set up within the simulation. The original circuit is shown in Figure 5.14, where the voltage-current relationship is

$$\overline{v}^L - \overline{v}^R = \overline{\overline{L}}\,\frac{d\overline{i}}{dt}. \tag{5.65}$$

The eigenvectors of $\overline{\overline{L}}$ are computed, orthonormalized, and then assembled into $\overline{\overline{T}}$ according to (5.57). The eigenvalues are computed and stored.

At each time step of the simulation, the modal voltages are computed as

$$\overline{v}_m = \overline{\overline{T}}^T (\overline{v}^L - \overline{v}^R)$$

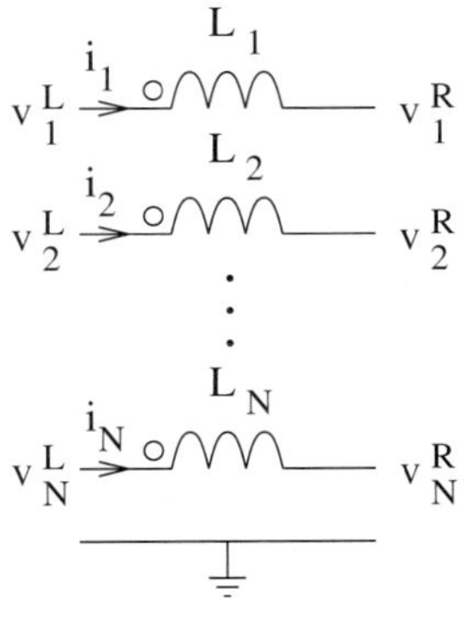

Figure 5.14. Voltage and current setup for an inductance matrix. For clarity, the mutual inductances are not shown.

and are used to drive N decoupled circuits that compute the modal currents from the modal voltages using (5.64). The circuits are shown in Figure 5.15, where the eigenvalues are the modal inductors. The modal currents are used to reconstruct the currents in the original circuit according to

$$\overline{i} = \overline{\overline{T}}\,\overline{i}_m.$$

The modal currents drive the circuit in Figure 5.16, which replaces the original circuit in Figure 5.14. The circuit simulator then steps from one time step to the next, solving the N auxiliary circuits along with any others present.

Direct implementation of the inductance matrix requires a circuit simulation with N^2 circuit elements. A manual implementation of a modal decomposition requires $N(2N+1)$ circuit elements, so simulation is slower with the modal decomposition. However, if the modal decomposition is hard-coded into the simulator, the number of circuit elements drops to $3N$ for a significant time savings. Note that the eigenvalue problem solution represents a small time penalty since it is solved just once when setting up a modal decomposition.

5.8.4 Passivity

The modal framework enables a direct investigation of the passivity of inductance matrices. It is important to check if an inductance matrix is passive because it is remarkably easy to generate one that is not. Active inductance matrices can cause nonconvergence of the circuit simulation, with computation of ever-growing voltage levels, and most dangerously, reasonable but incorrect results. A dense matrix completely filled out through a consistent electromagnetic simulation or measurement is not likely to be active; however, matrices constructed from partial results or by combinations of results generated from several techniques can easily be active. Numerical errors can cause an electromagnetic simulation to produce an active matrix.

The instantaneous power delivered to an inductance matrix is

$$p(t) = \overline{i}^T(t)\overline{v}(t).$$

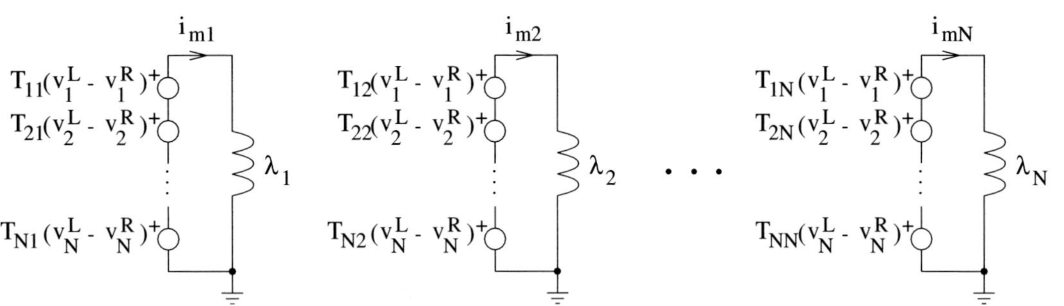

Figure 5.15. Auxiliary circuit setups for calculation of the modal currents from the modal voltages.

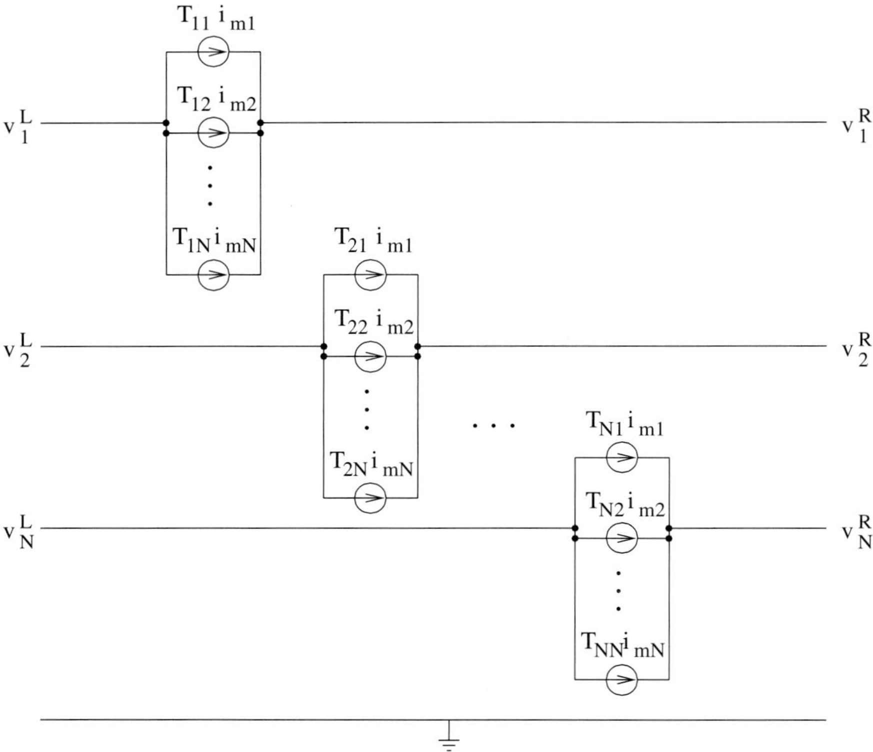

Figure 5.16. Replacement circuit for the original inductance matrix in Figure 5.14.

Substituting (5.24) to eliminate the voltage, then

$$p(t) = \bar{i}^T(t)\bar{\bar{L}}\frac{d\bar{i}(t)}{dt}.$$

The inductance matrix can be substituted with (5.60) to obtain

$$p(t) = \bar{i}^T(t)\bar{\bar{T}}\,\bar{\bar{\lambda}}\,\bar{\bar{T}}^T\frac{d\bar{i}(t)}{dt}. \tag{5.66}$$

Modal currents can be introduced using (5.63) to simplify the expression to

$$p(t) = \bar{i}_m^T(t)\bar{\bar{\lambda}}\frac{d\bar{i}_m(t)}{dt}.$$

Since $\overline{\overline{\lambda}}$ is diagonal, then the power simplifies to

$$p(t) = \sum_{k=1}^{N} \lambda_k i_{m_k}(t) \frac{di_{m_k}(t)}{dt}. \tag{5.67}$$

The total power is the sum of the powers in each mode.

For the inductance matrix to be passive, the total power delivered to the matrix from the start of time must be positive at any given time. Mathematically, this requirement is stated as

$$E(t) = \int_{-\infty}^{t} p(\tau) d\tau \geq 0 \ \forall \ t.$$

Integrating (5.67) yields

$$E(t) = \int_{-\infty}^{t} \sum_{k=1}^{N} \lambda_k i_{m_k}(\tau) \frac{di_{m_k}(\tau)}{d\tau} d\tau.$$

Interchanging the integral and summation, and assuming the currents are initially zero, then integration yields

$$E(t) = \frac{1}{2} \sum_{k=1}^{N} \lambda_k i_{m_k}^2(t).$$

Note that $i_{m_k}^2(t) \geq 0$ always, so for $E(t) \geq 0$ for any arbitrary time, in general

$$\lambda_k \geq 0 \ \forall \ k$$

is required. Therefore, the inductance matrix is passive only when all of its eigenvalues are greater than or equal to zero. However, a zero eigenvalue represents a shorted modal voltage, so in practice, positive eigenvalues are required.[4]

After an inductance matrix has been obtained from any source, it is good practice to compute its eigenvalues and confirm that all are positive. The appearance of any negative or zero eigenvalues disqualifies the matrix from use and indicates a problem in the methodology used to construct it.

4. The inductance matrix is Hermitian since it is real and symmetric. If all of its eigenvalues are greater than zero, then the matrix is positive definite.

5.9 Nonuniqueness of Partial Inductance

Section 5.5 shows that the inductance of a loop can be partitioned into N segments with N partial self-inductances and $N(N-1)/2$ partial mutual inductances between them. The cnly physical quantity is the inductance of the loop, so the partial inductances only have utility once they are combined to form a closed loop. Since the number of variables exceeds the number of constraints, partial inductance represents an under-determined system, and an infinite number of solutions exist.

As an example, consider the two partial self-inductors with the partial mutual inductor shown in Figure 5.17.If the ends are closed off to form a loop, then the inductance of the loop is

$$L = L_1 + L_2 - 2M.$$

Note that if any value is added to all three partial inductances, then the inductance of the loop is unchanged, so partial inductance is at least nonunique to within an additive constant.

The nonuniqueness of partial inductance can be shown mathematically to stem from the lack of a closed loop. Inductance is defined using the magnetic vector potential as

$$L = \frac{\oint_C \overline{A} \cdot \overline{d\ell}}{i}, \tag{5.68}$$

which is repeated here from section 5.2.2 for convenience. The loop can be subdivided as shown in Figure 5.3 to obtain

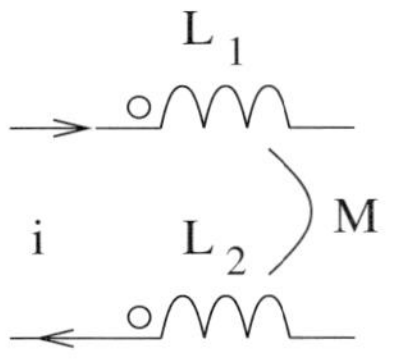

L1
M
i
L2

Figure 5.17. Two partial self-inductances and a partial mutual inductance.

$$L = \frac{\sum_{k=1}^{N} \oint_{C_k} \overline{A} \cdot \overline{d\ell}}{i}.$$

Then the partial inductance of section k is

$$L_k = \frac{\int_{C_k} \overline{A} \cdot \overline{d\ell}}{i}. \tag{5.69}$$

Since $\overline{A}$ is defined by $\overline{B} = \nabla \times \overline{A}$, $\overline{A}$ can have the gradient of a scalar field superimposed onto it without changing $\overline{B}$, then

$$\begin{aligned} \overline{B} &= \nabla \times (\overline{A} + \nabla\phi) \\ &= \nabla \times \overline{A} + \nabla \times \nabla\phi \\ &= \nabla \times \overline{A}, \end{aligned} \tag{5.70}$$

where $\nabla \times \nabla\phi = 0$ by vector identity. While $\overline{B}$ remains unchanged, L_k does not, and

$$\begin{aligned} L_k &= \frac{\int_{C_k} (\overline{A} + \nabla\phi) \cdot \overline{d\ell}}{i} \\ &= \frac{\int_{C_k} \overline{A} \cdot \overline{d\ell}}{i} + \frac{\int_{C_k} \nabla\phi \cdot \overline{d\ell}}{i}. \end{aligned} \tag{5.71}$$

Since ϕ is arbitrary, then L_k is not unique. Uniqueness is restored when the loop is closed since $\oint \nabla\phi \cdot \overline{d\ell} = 0$ by vector identity.

While the magnetic flux density is not perturbed by the superposition of the gradient of a scalar field, the electric field strength is modified according to (5.2):

$$\overline{E} = -\nabla\Phi - \frac{\partial \overline{A}}{\partial t} - \frac{\partial \nabla\phi}{\partial t}. \tag{5.72}$$

However, for quasi-static analysis, the time-varying terms are neglected and $\overline{E}$ remains unchanged.

In summary, while inductance is unique, partial inductance can easily be shown to be not unique. The nonuniqueness is of no consequence since when combined, the partial inductances produce a unique inductance. The key to successful application of partial inductances is to ensure that they are only used in complete and closed loops.

5.10 Open Loop Modeling

To compute or measure inductance, closed loops are required. However, interconnect modeling requires open loops so that circuits can be connected. The case of a single loop is shown in Figure 5.18. Breaking the loop creates a two-wire interconnect, but the loop still has only one inductance describing it. The full inductance can be assigned to the top wire or the bottom wire, or it can be split between the two with a mutual inductance. The assignment is arbitrary, nonunique, and totally at

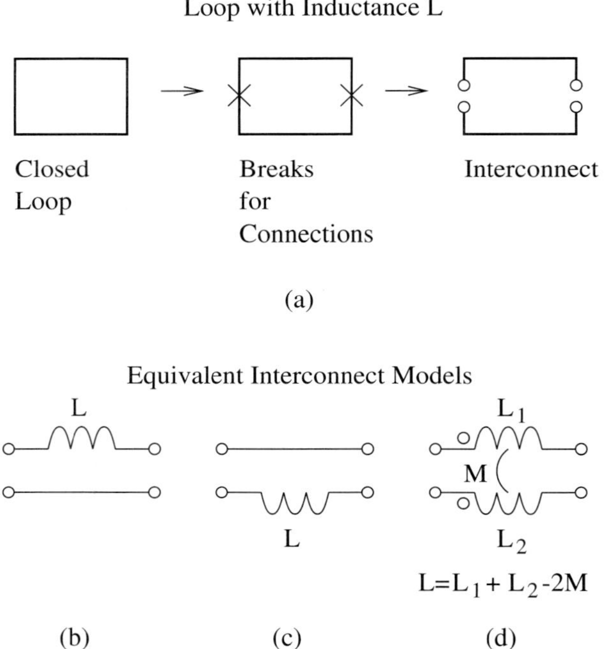

Figure 5.18. Interconnects require the opening of the closed loop to make connections. (a) The loop is broken in two places to create two ports, resulting in a two-wire interconnect. (b) The inductance is assigned to the top wire. (c) The inductance is assigned to the bottom wire. (d) The inductance is split between both wires.

the modeler's convenience; the only requirement is that the closed loop must have the proper inductance.

In Figure 5.18b and Figure 5.18c, one of the wires has no assigned inductance. This lead is often referred to as *perfect* or as the *reference* lead. The lead with the assigned inductance is then *referenced* to the reference lead. This terminology has little use since the assignments are arbitrary. It is important to remember that wires do not have inductance—closed loops have inductance.

For two loops, the interconnect is constructed as shown in Figure 5.19. Each loop generates a reference lead, but since the reference leads have no inductance and are always grounded, they can be combined into one reference lead, as shown in the final circuit model. Greater numbers of loops are handled in the same way.

Very often, the loops will share a common return path as shown for two loops in Figure 5.20. The return path is often a ground or power plane. The common leg increases the mutual inductance between the two loops, and this fact will be represented in the circuit model. Assuming that the inductances are assigned to the wires connecting points ac and bd, then the ground plane becomes the reference lead. The fact that the ground plane has no assigned inductance in the circuit model does not mean that planes have low inductance—loops have inductance; wires (even fat ones, like planes) do not.

The number of inductors and mutual inductors in the circuit model varies depending on whether a reference lead is included in the circuit model. If a reference lead is included, then N loops lead to N inductors and $N(N-1)/2$ mutual inductors. The component values are equal to the actual loop inductances and the mutual inductances between them. If a reference lead is not included, then the N loops lead to $N+1$ inductors and $N(N+1)/2$ mutual inductors, and the component values are arbitrary as long as the loop inductances and mutual inductances between loops of the circuit match the actual measurement or simulation of the real structure.

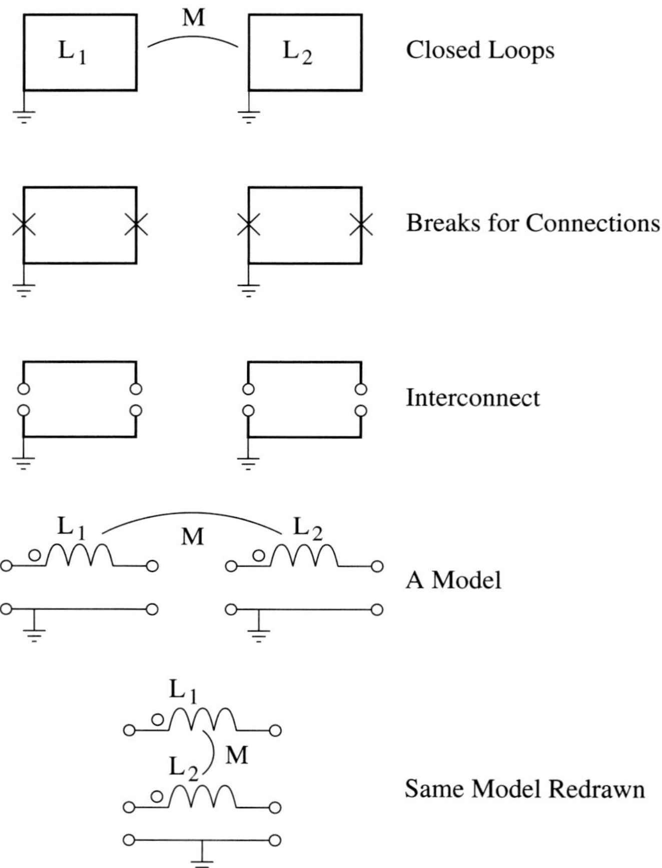

Figure 5.19. Example interconnect model for two closed loops. The loop inductances are assigned to the top wire for both loops.

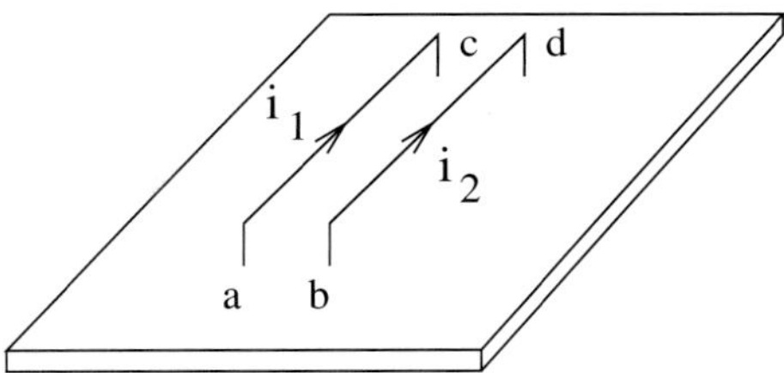

Figure 5.20. Two wire loops using a common leg serving as a return path.

Common Ground

A significant advantage of models with a reference lead is that circuits on both sides of the interconnect can utilize a common ground connection. For example, in Figure 5.21, the identical circuit is implemented in (a) and (b), except an extra ground is located in (b) on the other (driver) side of the interconnect. The connection through common ground shorts out the partial inductance of the ground connection in the interconnect. In (c), the interconnect model utilizes a reference lead that is conveniently chosen so that the ground lead has no assigned inductance. Therefore, common ground can be placed on both sides of the interconnect.

5.11 Manipulating the Reference Lead

It is very common for circuit simulations to be set up with multiple global grounds. In SPICE, nodes 0 and gnd are automatically global. To avoid shorting out the partial inductance of the ground connection in interconnect models (as in Figure 5.21b), the models should reference the ground lead (as in Figure 5.21c). Some methods and programs produce interconnect models without reference leads, so in these cases the model needs to be manipulated to create a reference lead.

Consider an N-lead interconnect with a common return path. If inductance is assigned to all N-leads plus the common return, then the interconnect is characterized by an $(N + 1) \times (N + 1)$ matrix, and the model is shown in Figure 5.22.

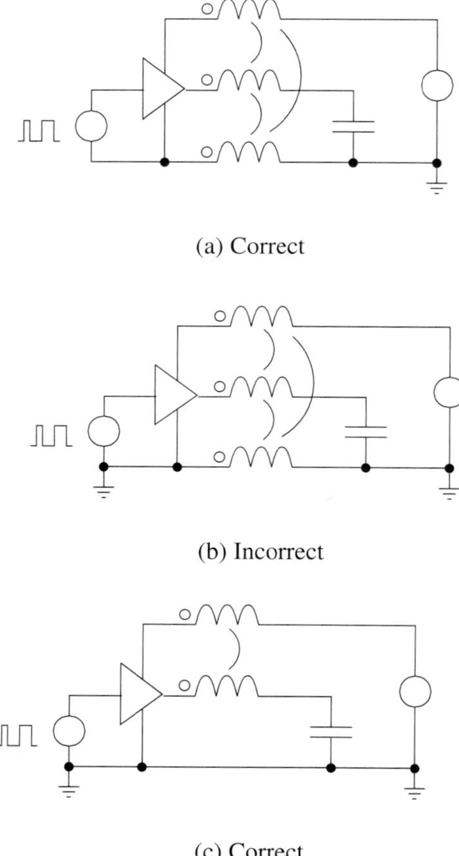

(a) Correct

(b) Incorrect

(c) Correct

Figure 5.21. Use of global grounds can be affected by the interconnect modeling scheme.

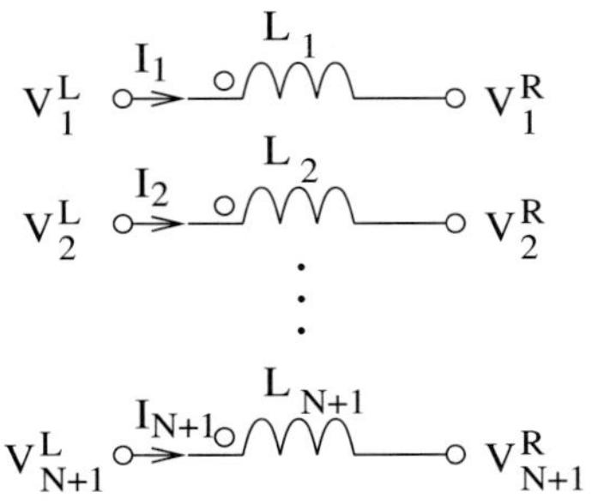

V_1^L I_1 L_1 V_1^R
V_2^L I_2 L_2 V_2^R
V_{N+1}^L I_{N+1} L_{N+1} V_{N+1}^R

Figure 5.22. Schematic for an N-lead interconnect with common return path and no reference lead.

Solving the circuit, Kirchhoff's voltage law requires

$$\overline{V}_L - \overline{V}_R = (\overline{\overline{R}} + j\omega\overline{\overline{L}})\overline{I}, \tag{5.73}$$

where the series resistance of each conductor is included for completeness; $\overline{V}_L$, $\overline{V}_R$ and $\overline{I}$ are vectors of length $N+1$; and, $\overline{\overline{R}}$ and $\overline{\overline{L}}$ are $(N+1)\times(N+1)$ matrices. Since all current passes through the interconnect, Kirchhoff's current law requires

$$I_i = -\sum_{\substack{k=1 \\ k\neq i}}^{N} I_k. \tag{5.74}$$

Substituting (5.74) into (5.73) yields

$$\overline{V}_L - \overline{V}_R = (\overline{\overline{R}} + j\omega\overline{\overline{L}})\overline{\overline{P}}\,\overline{I}_r, \tag{5.75}$$

where $\overline{I}_r = [\ I_1 \quad \cdots \quad I_{i-1} \quad I_{i+1} \quad \cdots \quad I_{N+1}\ ]^T$ and $\overline{\overline{P}}$ is an $(N+1)\times N$ matrix constructed from an $N\times N$ identity matrix with -1 inserted at each column in row i. For example, for four conductors with no reference lead,

$$\overline{\overline{P}} = \begin{bmatrix} 1 & 0 & 0 \\ 0 & 1 & 0 \\ -1 & -1 & -1 \\ 0 & 0 & 1 \end{bmatrix}.$$

The reference potential can be moved to lead i by subtracting its potential from those of the other leads. This operation can be performed by multiplying through (5.75) by $\overline{\overline{P}}^T$, yielding

$$\overline{\overline{P}}^T \overline{V}_L - \overline{\overline{P}}^T \overline{V}_R = \overline{\overline{P}}^T \overline{\overline{R}}\,\overline{\overline{P}}\,\overline{I}_r + j\omega \overline{\overline{P}}^T \overline{\overline{L}}\,\overline{\overline{P}}\,\overline{I}_r. \tag{5.76}$$

Equation (5.76) suggests the definitions

$$\begin{aligned} \overline{\overline{R}}' &= \overline{\overline{P}}^T \overline{\overline{R}}\,\overline{\overline{P}} \\ \overline{\overline{L}}' &= \overline{\overline{P}}^T \overline{\overline{L}}\,\overline{\overline{P}} \\ \overline{V}'_L &= \overline{\overline{P}}^T \overline{V}_L \\ \overline{V}'_R &= \overline{\overline{P}}^T \overline{V}_R. \end{aligned} \tag{5.77}$$

Both $\overline{\overline{R}}'$ and $\overline{\overline{L}}'$ are $N \times N$ matrices, while $\overline{V}'_L$ and $\overline{V}'_R$ are vectors with length N. The original voltage-current relationship in (5.73) is preserved with the reference potential set or moved to conductor i. With the voltage-current relationship unchanged, the new primed interconnect based on (5.77) produces exactly the same terminal characteristics as the original interconnect in (5.73).

Using (5.77), the inductance and resistance matrices of any interconnect of the form in Figure 5.22 can be manipulated to create a reference lead on any of the $N+1$ leads. If no reference is present, it can be created. If one is already present, it can be moved.

An interesting application for rereferencing of interconnect models is in comparing measurements and/or simulations from different sources. Different setups can produce different models because of the way the return path is set up. Models will normally be referenced to the return path, making comparison between two results difficult. If the models are rereferenced to one of the leads, the results can then be directly compared, except for the values which are known to be different for the return path. This technique is basically a way to set up the loops in a similar fashion so that the two models can be fairly compared.

Example with No Reference

An interconnect constructed using four identical wires is described by a partial inductance matrix given by

$$\overline{\overline{L}} = \begin{bmatrix} 10 & 6 & 4 & 3 \\ 6 & 10 & 6 & 4 \\ 4 & 6 & 10 & 6 \\ 3 & 4 & 6 & 10 \end{bmatrix}.$$

To reference the matrix to lead 4, the matrix

$$\overline{\overline{P}} = \begin{bmatrix} 1 & 0 & 0 \\ 0 & 1 & 0 \\ 0 & 0 & 1 \\ -1 & -1 & -1 \end{bmatrix}$$

is used in (5.77) to construct the equivalent 3×3 inductance matrix

$$\overline{\overline{L}}' = \begin{bmatrix} 14 & 9 & 5 \\ 9 & 12 & 6 \\ 5 & 6 & 8 \end{bmatrix}$$

with lead 4 as reference. The matrices $\overline{\overline{L}}$ and $\overline{\overline{L}}'$ can be used interchangeably, keeping in mind that all connections must pass through $\overline{\overline{L}}$, while $\overline{\overline{L}}'$ is referenced to lead 4, which does support current flow.

Hand calculations can verify the equivalence of the matrices. The only consideration is that all loops through the interconnect must provide the same inductance. Consider the loop out through lead 1 and back through lead 3. From $\overline{\overline{L}}$, the loop inductance is $10+10-2*4 = 12$, while for $\overline{\overline{L}}'$, the result is $14+8-2*5 = 12$, the same answer. For the loop out lead 3 and back on lead 4, the results are $10+10-2*6 = 8$ and 8 (read straight from the matrix since lead 4 has no partial inductance), so again the results are the same. Every closed loop will also yield identical closed loop inductances.

Example with Reference

An interconnect constructed using four identical wires referenced to a common return path is described by a partial inductance matrix given by

$$\overline{\overline{L}} = \begin{bmatrix} 10 & 6 & 4 & 3 \\ 6 & 10 & 6 & 4 \\ 4 & 6 & 10 & 6 \\ 3 & 4 & 6 & 10 \end{bmatrix}.$$

The common return path represents a fifth lead with no partial inductance. To move the reference to another lead, the fifth lead must be added to the matrix, yielding

$$\overline{\overline{L}} = \begin{bmatrix} 10 & 6 & 4 & 3 & 0 \\ 6 & 10 & 6 & 4 & 0 \\ 4 & 6 & 10 & 6 & 0 \\ 3 & 4 & 6 & 10 & 0 \\ 0 & 0 & 0 & 0 & 0 \end{bmatrix}.$$

To reference the matrix to lead 2, use

$$\overline{\overline{P}} = \begin{bmatrix} 1 & 0 & 0 & 0 \\ -1 & -1 & -1 & -1 \\ 0 & 1 & 0 & 0 \\ 0 & 0 & 1 & 0 \\ 0 & 0 & 0 & 1 \end{bmatrix}$$

to obtain

$$\overline{\overline{L}}' = \begin{bmatrix} 8 & 2 & 3 & 4 \\ 2 & 8 & 6 & 4 \\ 3 & 6 & 12 & 6 \\ 4 & 4 & 6 & 10 \end{bmatrix}.$$

Note that this 4×4 matrix represents leads 1, 3, 4, and 5 (the former reference), and now lead 2 is the reference. For example, the inductance of the loop out lead 1 and back lead 3 is $10 + 10 - 2 * 4 = 12$ and $8 + 8 - 2 * 2 = 12$.

5.12 Model Reduction

Using the algorithms in chapter 4, section 4.5, partial inductance matrices can be reduced in size to incorporate known information. The principal cases are when model leads are open-circuited, short-circuited, and paralleled together.

If a lead is open-circuited, then no current flows through the lead, and it does not affect the circuit. If a lead is shorted to ground on both ends, a loop is formed that can support the flow of current. While no voltage source directly drives a current in the loop, the mutual inductances linking it to other driven circuits will generate current flow, called *eddy currents*, which affect the operation of the circuit by reducing inductance and increasing resistance in other leads. Leads that are paralleled together also create closed loops that can support eddy currents.

5.13 Exercises

1. Two circular loops of thin wire are located as shown in Figure 5.23.

   (a) Using Neumann's mutual inductance formula with numerical integration, compute the mutual inductance between the two loops for b=5mm and a=1mm, 2mm, 4mm, 8mm, 16mm, 32mm, 64mm, and 128mm.

   (b) For wire radii of 0.01mm, $\mu_r = 1$, and $\sigma = 5.8 \times 10^7$S/m, compute the inductance and resistance of each loop. Assume one loop is closed while

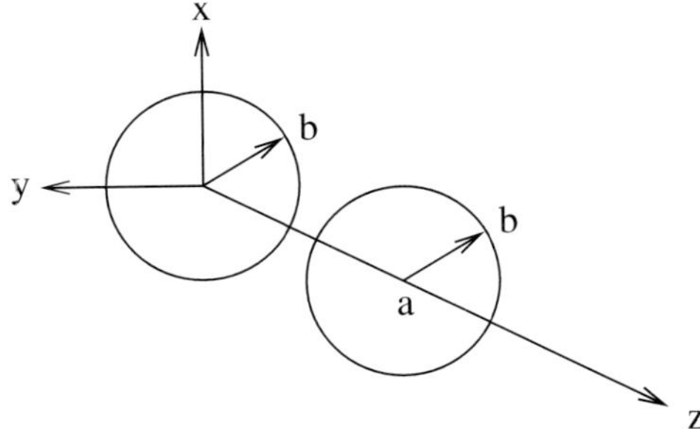

Figure 5.23. Two circular loops on the z-axis separated by distance a.

a small gap is introduced in the other to allow the insertion of a small sinusoidal voltage source of negligible source impedance. Compute the inductance and resistance of the resulting circuit for the spacings above at a frequency of 10MHz. Plot the results.

(c) Explain the trends in the plots.

2. Show that the partial mutual inductance between two orthogonal straight thin wires is zero.

3. Show that the partial mutual inductance between the two equal-length thin wires shown in Figure 5.24 is

$$M_p = \frac{\mu a \cos\theta}{2\pi} \ln\left[1 + \sqrt{\frac{2}{1-\cos\theta}}\right].$$

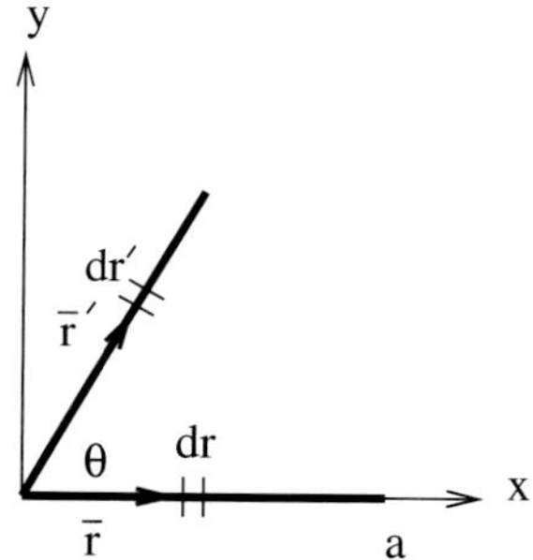

Figure 5.24. Setup for partial mutual inductance calculation for equal-length thin wires almost touching at one end.

4. Show that the inductance of a loop of thin wire of radius a formed into the shape of an equilateral triangle is

$$L = \frac{3\mu b}{2\pi}\left[\ln\left(\frac{2b}{a}\right) - \frac{3}{4} - \ln 3\right].$$

5. Show that the inductance of a loop of thin wire of radius a formed into a square with side lengths of b is

$$L = \frac{2\mu b}{\pi}\left[\ln\left(\frac{2b}{a}\right) - \ln(1+\sqrt{2}) + \sqrt{2} - \frac{7}{4}\right].$$

6. Show that the partial mutual inductance between a square loop of thin wire with side lengths a and a thin straight wire of length a at a distance b away, as shown in Figure 5.25, is given by

$$\begin{aligned} M_p &= \frac{\mu a}{2\pi}\left[\ln\left(\frac{a}{a+b} + \sqrt{1+\left(\frac{a}{a+b}\right)^2}\right) - \sqrt{1+\left(\frac{a+b}{a}\right)^2} + \frac{a+b}{a}\right] \\ &- \frac{\mu a}{2\pi}\left[\ln\left(\frac{a}{b} + \sqrt{1+\left(\frac{a}{b}\right)^2}\right) - \sqrt{1+\left(\frac{b}{a}\right)^2} + \frac{b}{a}\right] \end{aligned}$$

if the current directions are as shown.

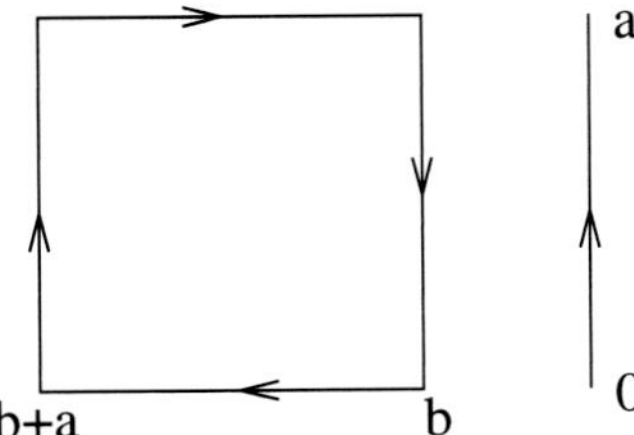

Figure 5.25. Setup for partial mutual inductance calculation between a square loop of thin wire and a straight thin wire.

7. Show that the external inductance per-unit-length for the coaxial line shown in Figure 5.26 is given by

$$\frac{L}{\ell} = \frac{\mu}{2\pi}\ln\left(\frac{b}{a}\right).$$

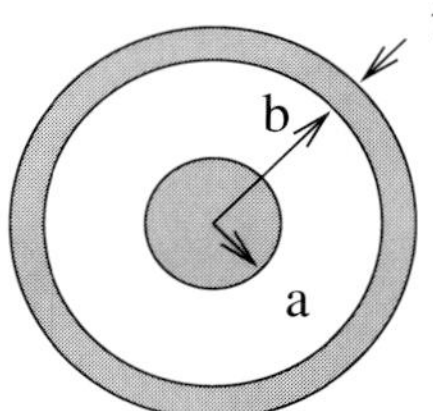

Figure 5.26. Cross section of a coaxial transmission line with outer conductor thickness of t.

8. Show that the internal inductance per-unit-length for the outer shell of the coaxial conductor in Figure 5.26 is given by

$$\frac{L}{\ell} = \frac{\mu}{2\pi}\left[\frac{(b+t)^4}{t^2(2b+t)^2}\ln\left(\frac{b+t}{b}\right) - \frac{5}{8} - \frac{b}{4t} - \frac{t}{8(2b+t)}\right],$$

assuming a uniform DC current distribution.

9. Show that the internal inductance per-unit-length for the conductive tube in Figure 5.27 is given by

$$\frac{L}{\ell} = \frac{\mu}{2\pi}\left[\frac{3}{8} - \frac{a}{4t} - \frac{t}{8(2a+t)} + \frac{\ln\left(\frac{a+t}{a}\right)}{\left(\frac{t}{a}\right)^2\left(2+\frac{t}{a}\right)^2}\right]$$

for uniform DC current distribution.

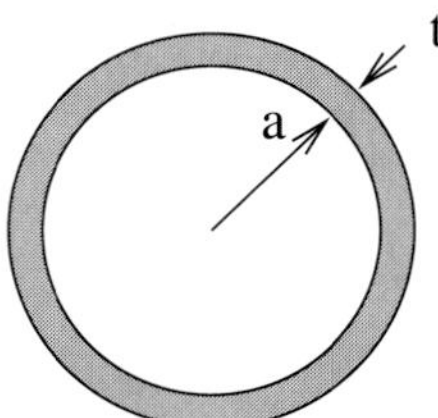

Figure 5.27. Cross section of conductive tube with wall thickness t.

10. Show that the inductance per-unit-length of the coaxial line in Figure 5.26 is

$$\frac{L}{\ell} = \frac{\mu}{2\pi}\left[\ln\left(\frac{b}{a}\right) + \frac{1}{4} + \frac{(b+t)^4}{t^2(2b+t)^2}\ln\left(\frac{b+t}{b}\right) - \frac{5}{8} - \frac{b}{4t} - \frac{t}{8(2b+t)}\right]$$

for uniform DC current distribution.

11. Show that the inductance per-unit-length of the annular coaxial line, formed by replacing the center conductor in Figure 5.26 with the ring in Figure 5.27, is

$$\frac{L}{\ell} = \frac{\mu}{2\pi}\left[\ln\left(\frac{b}{a+t}\right) + \frac{3}{8} - \frac{a}{4t} - \frac{t}{8(2a+t)} + \frac{\ln\left(\frac{a+t}{a}\right)}{\left(\frac{t}{a}\right)^2\left(2+\frac{t}{a}\right)^2} + \frac{(b+t)^4}{t^2(2b+t)^2}\ln\left(\frac{b+t}{b}\right) - \frac{5}{8} - \frac{b}{4t} - \frac{t}{8(2b+t)}\right]$$

for uniform DC current distribution.

12. If the inductance per-unit-length for the annular coax in problem 11 is provided the functional notation $L(a,b,t)$, where a is the inner radius, b is the outer radius, and t is the conductor thicknesses, show that a set of partial inductances for the annular coax is

$$L_i = L(a,c,t)$$

for the inner conductor,

$$L_o = L(b,c,t)$$

for the outer conductor, and

$$M = \frac{L(a,c,t) + L(b,c,t) - L(a,b,t)}{2} = M(a,b,c,t)$$

for the partial mutual. Note that $c > b+t$.

13. Show which of the two partial inductance matrices below represents an interconnect with lower inductance.

$$\overline{\overline{L}}_1 = \begin{bmatrix} 10 & 6 & 3 \\ 6 & 9 & 5 \\ 3 & 5 & 8 \end{bmatrix}$$

$$\overline{\overline{L}}_2 = \begin{bmatrix} 12 & 8 & 4 \\ 8 & 10 & 6 \\ 4 & 6 & 8 \end{bmatrix}$$

14. Using image theory, show that the inductance of a long round wire with radius a and length b located a distance d, measured from the center, from a perfectly conducting plane serving as the return path is

$$L = \frac{\mu b}{4\pi}\left(\frac{1}{2} + 2\ln\left[\frac{2d}{a}\right]\right).$$

15. Plot the input resistance and inductance for the circuit in Figure 5.28 for M=0 and M=4nH. Explain any differences in the results.

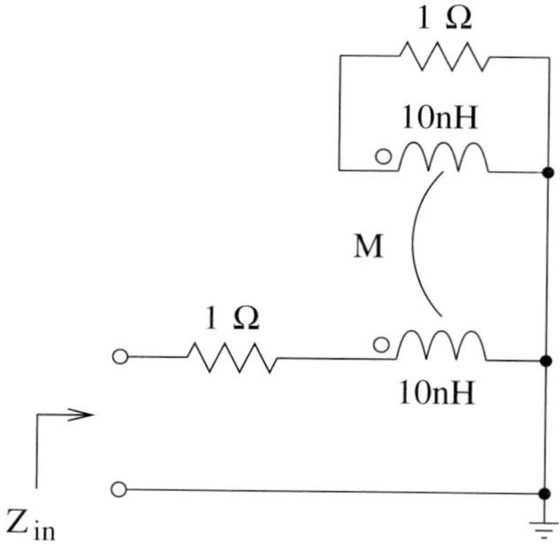

Figure 5.28. Circuit for problem 15.

16. Show that the partial inductances of the loop shown in Figure 5.29 can be replaced by a single inductance given by

$$L = L_1 + L_2 - 2M.$$

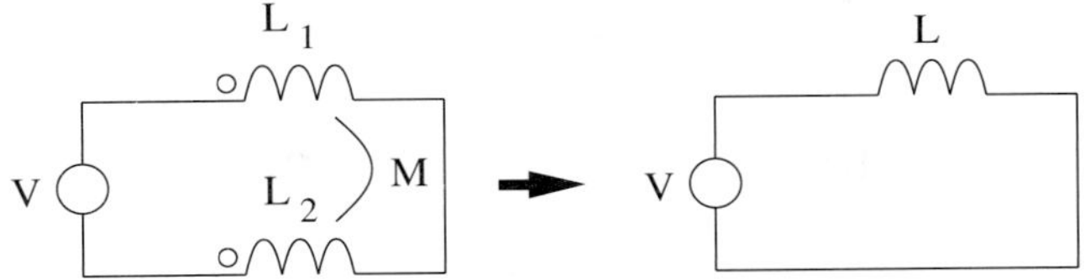

Figure 5.29. Equivalent partial inductance and inductance circuits.

17. Show that the partial inductances of the loop shown in Figure 5.30 can be replaced by a single inductance given by

$$L = \frac{L_1 L_2 - M^2}{L_1 + L_2 - 2M}.$$

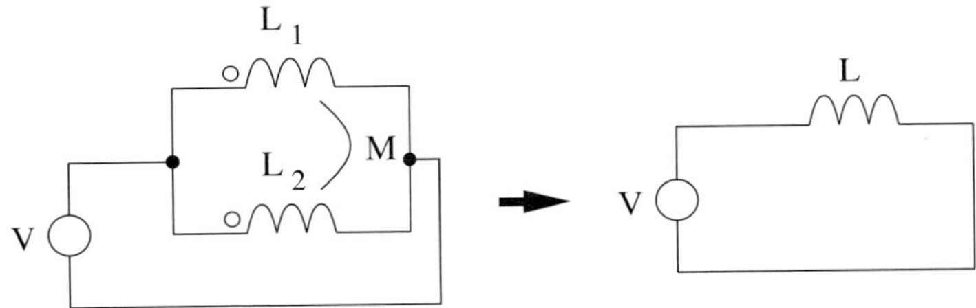

Figure 5.30. Another illustration of equivalent partial inductance and inductance circuits.

18. Show that (5.58) is true.

19. Show that (5.59) is true.

20. An inductance matrix is given by

$$\overline{\overline{L}} = \begin{bmatrix} L & M & M/\sqrt{3} \\ M & L & M \\ M/\sqrt{3} & M & L \end{bmatrix}.$$

(a) Find $\overline{\overline{T}}$ and $\overline{\overline{\lambda}}$ for a modal decomposition.

(b) Sketch a circuit that implements the inductance matrix using the modal decomposition.

(c) Derive conditions for the inductance matrix to remain passive.

21. A partial inductance matrix in H is

$$\overline{\overline{L}} = \begin{bmatrix} 1 \times 10^{-8} & 6 \times 10^{-9} & 4 \times 10^{-9} \\ 6 \times 10^{-9} & 8 \times 10^{-9} & 5 \times 10^{-9} \\ 4 \times 10^{-9} & 5 \times 10^{-9} & 7 \times 10^{-9} \end{bmatrix}.$$

The orthonormalized eigenvectors are computed to be

$$\overline{\overline{T}} = \begin{bmatrix} 0.643779089 & -0.323388434 & -0.693518857 \\ 0.595009865 & 0.781432736 & 0.187952496 \\ 0.481156675 & -0.533650448 & 0.695489363 \end{bmatrix},$$

while the corresponding eigenvalues in H are

$$\overline{\lambda} = \begin{bmatrix} 1.85350487 \times 10^{-8} & 2.10239342 \times 10^{-9} & 4.36255787 \times 10^{-9} \end{bmatrix}.$$

Simulate the circuit in Figure 5.31 using the partial inductance matrix directly and by using a modal decomposition implemented completely within SPICE, using voltage and current sources. Very small resistances may be needed in series with the partial self-inductors due to the loop formed by L_1 and L_2.

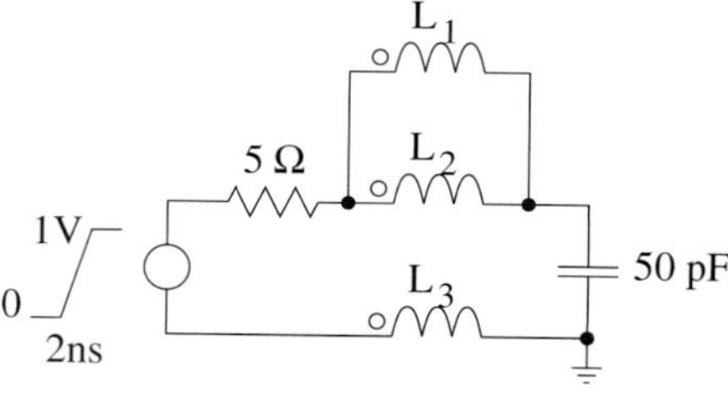

Figure 5.31. Schematic for testing the modal decomposition of a partial inductance matrix in problem 21.

22. Show that the inductance matrix can be simplified for an open-circuit loop by deleting the row and column in the matrix for that loop.

23. A partial inductance matrix with no reference lead is given by

$$\overline{\overline{L}} = \begin{bmatrix} 10 & 4 & 2 \\ 4 & 6 & 3 \\ 2 & 3 & 4 \end{bmatrix}.$$

Sketch three distinct but equivalent circuit models with no reference lead. Sketch another three distinct but equivalent circuit models with a reference lead.

24. Assume that each lead of an interconnect has some resistance, then the voltage drop across the interconnect is

$$\overline{v} = \overline{\overline{R}}\,\overline{i} + \overline{\overline{L}}\frac{d\overline{i}}{dt},$$

where $\overline{\overline{R}}$ is a diagonal matrix. Find the modal decomposition for this problem. Is there any simplification?

25. Show that the diagonal resistance matrix is passive when all entries are positive.

Chapter 6

CAPACITANCE

Complete modeling requires consideration of both electric and magnetic fields. Inductance relates magnetic flux to current for a given geometry. Similarly, capacitance relates the electric field as represented by the voltage to charge. In this chapter, the simple case of capacitance between two conducting objects is examined and then expanded for many conductors. Interconnects for digital systems typically have a great many conductors, with counts easily reaching into the hundreds, so a general multiconductor formulation of capacitance is essential. The concept of the capacitance matrix and its negative off-diagonal terms is fundamental.

6.1 Definition of Capacitance

Consider two initially uncharged conductors as shown in Figure 6.1. With no charge, the electric field is zero everywhere, so the voltage between the two conductors is also

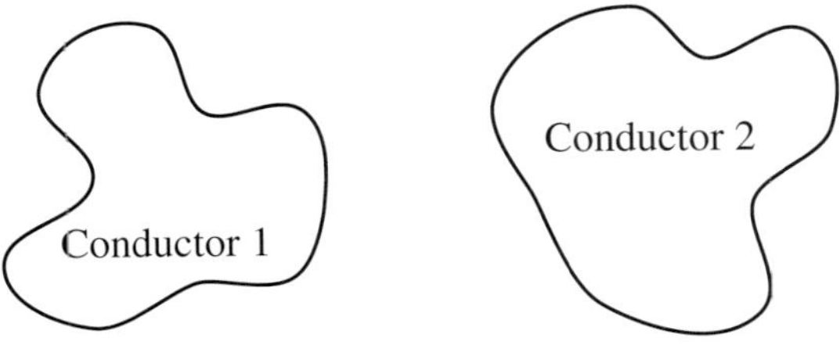

Figure 6.1. Two uncharged conductors.

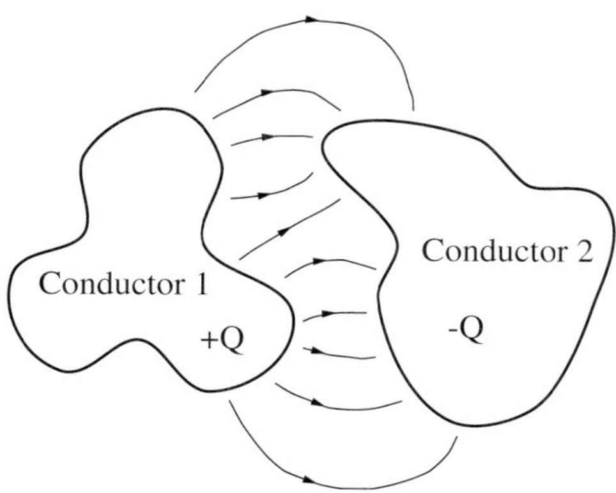

Figure 6.2. Two charged conductors with resulting electric field.

zero. Now transfer a small amount of charge from conductor 1 to conductor 2. The two conductors are now charged and produce an electric field as shown in Figure 6.2. The field represents a gradient in the electric potential, and a potential difference, or voltage, exists between the two conductors that is related to the quantity of charge transferred. In general,

$$v = f(Q).$$

The function f is unknown at this point, but the fact that it is linear can be deduced.

Expanding f in a Maclaurin series, then

$$v = a_0 + a_1 Q + a_2 Q^2 + a_3 Q^3 + \cdots . \tag{6.1}$$

When $Q = 0$ then $v = 0$, so a_0 must also equal be zero. The higher-order expansion coefficients can be investigated through examination of Gauss's law.

In integral form, Gauss's law relates electric flux and charge according to

$$\oint\!\!\oint_S \overline{D} \cdot \overline{ds} = \iiint_V \rho \, dV,$$

where $\overline{D}$ is the electric flux density and ρ is the charge density. $\oint\!\oint_S \overline{D} \cdot \overline{ds}$ sums the normal component of the electric flux density over a closed surface, and $\iiint_V \rho \, dV$

finds the total charge enclosed within that surface. For simple materials, $\overline{D} = \epsilon\overline{E}$, and with the total charge, $Q = \iiint_V \rho\, dV$, then Gauss's law applied to conductor 1 provides

$$\oint\!\!\oint_{S_1} \epsilon\overline{E} \cdot \overline{ds} = Q, \tag{6.2}$$

where S_1 is a closed surface around conductor 1 only. Changes in the value of Q do not affect either S_1 or ϵ, so any change in Q must be reflected in the electric field strength, $\overline{E}$. Because (6.2) is linear, a doubling of Q must be accompanied by a doubling of $\overline{E}$ since S_1 is arbitrarily set. From its definition in (5.4), voltage is linear with respect to the electric field, so doubling the charge also doubles the voltage. Since the voltage is linearly dependent on the total charge, all of the nonlinear terms in (6.1) must be zero. The general relationship between voltage and charge must be

$$v = a_1 Q,$$

which is conventionally written as $Q = v/a_1$, with $1/a_1$ named the capacitance and given the symbol C:

$$Q = Cv.$$

Capacitance relates the charge present on conductors to the voltage, or potential difference, between them.

If the direction of the problem is reversed by taking the conductor with charge $-Q$ as the basis, then the sign on the voltage is reversed. Then $-Q = -Cv$, and the original definition of capacitance is recovered. Therefore, the capacitance between two conductors is the same whether looking from one conductor or the other, so capacitance is reciprocal.

Example

Capacitance serves the role of a geometry factor to encapsulate the relationship between charge and voltage for a particular configuration of conductors. Consider the ideal parallel-plate capacitor shown in Figure 6.3. Assuming no fringing, then Gauss's law applied to the surface described by the dotted line yields

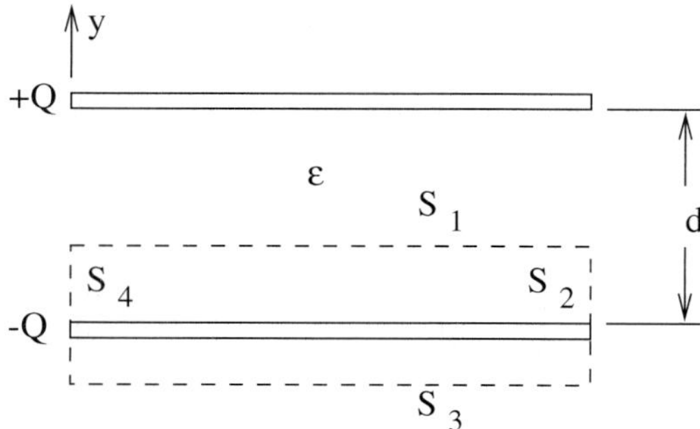

Figure 6.3. Capacitance calculation setup for the parallel-plate capacitor.

$$\begin{aligned}\oint\!\!\oint_S \overline{D}\cdot\overline{ds} &= \iint_{S_1} D_y dx\, dz \\ &= D_y \iint_{S_1} dx\, dz \\ &= \epsilon E_y A = -Q. \end{aligned} \tag{6.3}$$

The voltage between the plates is

$$\begin{aligned} v &= -\int \overline{E}\cdot\overline{d\ell} \\ &= -\int_0^d E_y dy \\ &= \frac{Q}{\epsilon A}\int_0^d dy \\ &= \frac{Qd}{\epsilon A}. \end{aligned} \tag{6.4}$$

Solving for Q, then

$$Q = \frac{\epsilon A}{d} v,$$

so the capacitance can be identified as $C = \frac{\epsilon A}{d}$. The capacitance captures the physical properties of the structure and encapsulates them into a single constant relating the voltage to the charge.

6.2 Capacitance between Several Conductors

When more than two conductors are present, the simple relationship $Q = Cv$ can be repeatedly applied to obtain the capacitance relationships for the whole group. Consider N conductors as illustrated in Figure 6.4, where each conductor is charged and so has a potential difference with the grounded reference conductor. While the quasi-static approximation holds, the potential difference between any two conductors is the difference in their voltages.

Between conductors i and j, a voltage of $v_i - v_j$ exists that is supported by the charge $Q_{ij} = C_{ij}(v_i - v_j)$ on conductor i and by $-Q_{ij}$ on conductor j. The total charge on conductor i is the sum of these charges from each pairing between conductor i and every other conductor. Explicitly writing out the sum yields

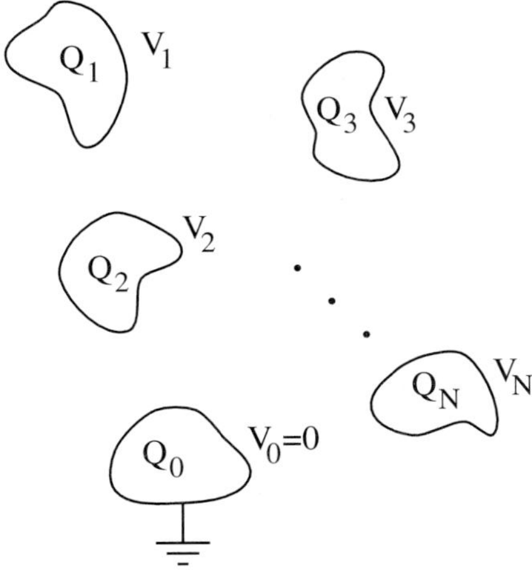

Figure 6.4. N-conductor setup for examination of capacitance between several conductors.

$$
\begin{aligned}
Q_1 &= C_{10}v_1 + C_{12}(v_1 - v_2) + \cdots + C_{1N}(v_1 - v_N) \\
Q_2 &= C_{21}(v_2 - v_1) + C_{20}v_2 + \cdots + C_{2N}(v_2 - v_N) \\
&\vdots \\
Q_N &= C_{N1}(v_N - v_1) + C_{N2}(v_N - v_2) + \cdots + C_{N0}v_N,
\end{aligned}
$$

where the subscript 0 means electrostatic ground at zero potential. Collecting terms on the voltages yields

$$
\begin{aligned}
Q_1 &= (C_{10} + C_{12} + \cdots + C_{1N})v_1 - C_{12}v_2 - \cdots - C_{1N}v_N \\
Q_2 &= -C_{21}v_1 + (C_{21} + C_{20} + \cdots + C_{2N})v_2 - \cdots - C_{2N}v_N \\
&\vdots \\
Q_N &= -C_{N1}v_1 - C_{N2}v_2 - \cdots + (C_{N1} + C_{N2} + \cdots + C_{N0})v_N,
\end{aligned}
$$

which can be succinctly stated in vector form as

$$\overline{Q} = \overline{\overline{C}}\,\overline{v}, \tag{6.5}$$

where $\overline{\overline{C}}$ is the capacitance matrix (or short-circuit capacitance matrix) with the general form

$$\overline{\overline{C}} = \begin{bmatrix} C_{10}+C_{12}+\cdots+C_{1N} & -C_{12} & \cdots & -C_{1N} \\ -C_{21} & C_{20}+C_{21}+\cdots+C_{2N} & \cdots & -C_{2N} \\ \vdots & \vdots & \ddots & \vdots \\ -C_{N1} & -C_{N2} & \cdots & C_{N0}+C_{N1}+C_{N2}+\cdots \end{bmatrix}. \tag{6.6}$$

For discussion, the entries in the capacitance matrix can be labeled as

$$\overline{\overline{C}} = \begin{bmatrix} \tilde{C}_{11} & \tilde{C}_{12} & \cdots & \tilde{C}_{1N} \\ \tilde{C}_{21} & \tilde{C}_{22} & \cdots & \tilde{C}_{2N} \\ \vdots & \vdots & \ddots & \vdots \\ \tilde{C}_{N1} & \tilde{C}_{N2} & \cdots & \tilde{C}_{NN} \end{bmatrix}.$$

Note how the diagonal terms of the capacitance matrix are not the capacitances to ground for the conductors. In other words, $C_{i0} \neq \tilde{C}_{ii}$. To find the capacitance to ground for conductor i, sum across either row or column i. Mathematically stated, then

$$C_{i0} = \sum_{j=1}^{N} C_{ij} = \sum_{j=1}^{N} C_{ji}.$$

The diagonal term by itself gives the capacitance to ground when all other leads are grounded. This case represents a somewhat worst-case capacitive loading for the conductor, called the *loading capacitance*. The diagonal terms are also called the *coefficients of capacitance*.

The off-diagonal terms of the capacitance matrix are always negative, indicating that a positive voltage differential between conductors i and j induces a negative charge on j. The capacitance between conductors is easily obtained from the capacitance matrix as

$$C_{ij} = -\tilde{C}_{ij}, \quad i \neq j.$$

The off-diagonal terms are often referred to as *mutual capacitances* and as *coefficients of induction*.

Because of the reciprocity of capacitance, capacitance matrices are also reciprocal, so $\tilde{C}_{ij} = \tilde{C}_{ji}$, and the matrix is symmetric. The special form of the capacitance matrix has implications for its passivity, which is examined in section 6.6.

With these definitions, a circuit model implementing a capacitance matrix can be constructed by inspection. The model elements are often referred to as *two-terminal capacitances* to distinguish the model values from the matrix entries. A simple example in Figure 6.5 demonstrates the construction of a circuit model from a capacitance matrix.

Occasionally, a matrix will be encountered in which all of the matrix entries are positive. While not a capacitance matrix, it still may be referred to as one. Such a matrix will typically be a compilation of (two-terminal) capacitor values for use in a circuit model; principally, the diagonal elements are the capacitances to ground and the off-diagonal elements are the capacitances between conductors. In this case,

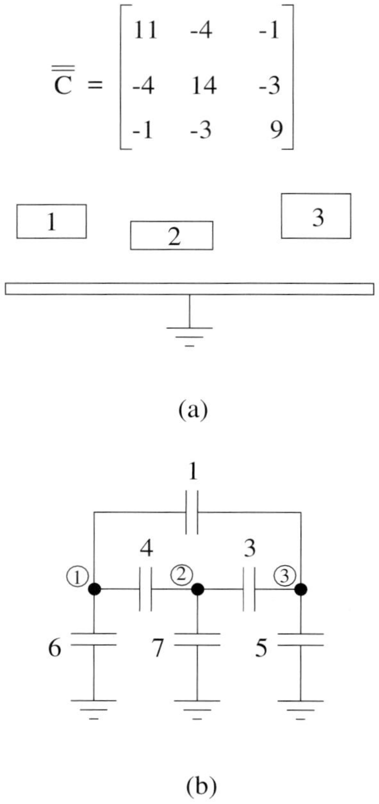

$$\overline{\overline{C}} = \begin{bmatrix} 11 & -4 & -1 \\ -4 & 14 & -3 \\ -1 & -3 & 9 \end{bmatrix}$$

(a)

(b)

Figure 6.5. Example capacitance matrix and its corresponding circuit model.

the matrix would be of the form

$$\overline{\overline{C}} = \begin{bmatrix} C_{10} & C_{12} & \cdots & C_{1N} \\ C_{21} & C_{20} & \cdots & C_{2N} \\ \vdots & \vdots & \ddots & \vdots \\ C_{N1} & C_{N2} & \cdots & C_{N0} \end{bmatrix}, \tag{6.7}$$

which carries the same information as the capacitance matrix in (6.6) but requires less effort to translate to a circuit model. It is important to note that (6.7) is a compilation for convenience only and is not suitable for calculations of voltage or charge vectors using (6.5) or any other formulation for a capacitance matrix.

6.3 Energy Definition of Capacitance

Since voltage is path-dependent, general-purpose electromagnetic simulators often use an energy-based definition for capacitance for easier setup. From circuit theory, the energy stored in a capacitor is

$$\varepsilon = \frac{1}{2}Cv^2,$$

while the result from electromagnetic field theory is

$$\varepsilon = \frac{1}{2}\iiint_V \overline{D}\cdot\overline{E}\,dV.$$

Equating these two definitions and solving for C yields the energy definition for capacitance as

$$C = \frac{1}{v^2}\iiint_V \overline{D}\cdot\overline{E}\,dV. \tag{6.8}$$

A numerical simulation would normally set $v = 1$ for convenience.

Generalizing (6.8) for the case of multiple conductors leads to

$$C_{ij} = \frac{1}{v_i v_j}\iiint_V \overline{D}_i\cdot\overline{E}_j\,dV. \tag{6.9}$$

The two field quantities ($\overline{D}_i$ and $\overline{E}_j$) must be found from two separate computations; otherwise, there is no way to separate the two. A practical approach is to ground all conductors except i to compute $\tilde{C}_{ii}$, a diagonal element of the capacitance matrix, and then store the field. A second calculation with all conductors grounded except j, yielding $\tilde{C}_{jj}$, provides the second field quantity. Forming the dot product of these two solutions and applying (6.9) yields the capacitance between the two conductors, C_{ij}, the negative of an off-diagonal entry in the capacitance matrix. The full capacitance matrix can be filled out by repeating the process for every pair of conductors.

6.4 Frequency Dependence

For materials typically used in interconnects, such as metals, plastics, glass, fiberglass, and ceramics, capacitance between conductors is remarkably constant with respect to frequency changes. The definition of capacitance suggests the mechanisms for this constancy. Since $C = Q/v$, capacitance will change along with changes in the charge and/or voltage. While DC leakage is negligible, and while time-varying fields deposit negligible charge (a restatement of the quasi-static assumption), the charge remains fixed. Also, under quasi-static conditions, the electric field remains constant while frequency is varied, so the voltage remains constant as well. Therefore, under these assumptions, the capacitance is constant with respect to frequency.

The frequency-independence of capacitance (while the lumped approximation holds) can be exploited to measure the inductance of the loop formed by a single interconnect and its return path. With both ends open-circuited, a low-frequency capacitance measurement can establish C. The interconnect is then shorted at one end and a high-frequency sine-wave source is applied at the other, so a closed loop is established. A frequency sweep by the source can be used to find the resonant frequency of the circuit, ω_o. Since C is constant, then the inductance at the resonant frequency is then solvable, knowing that $\omega_o = 1/\sqrt{LC}$.

Example

If a vector network analyzer (VNA) is used as the source, a plot of the reflection coefficient on a Smith chart (see chapter 8, section 8.3.5) allows a simple identification of ω_o as the frequency where the reflection coefficient is real. For example, suppose a capacitance measurement of an open-circuit package lead yields 5pF. One end is shorted and the other is driven with a VNA, resulting in the Smith chart in Figure 6.6. The frequency where it crosses the real axis is 1.54GHz, so solving for L provides 2.14nH. The Smith chart also shows that at low frequencies, the normalized impedance is 0.1 and real, so the resistance of the circuit at low frequencies is $0.1 \times 50\Omega = 5\Omega$, assuming the Smith chart is normalized to 50Ω. The data used

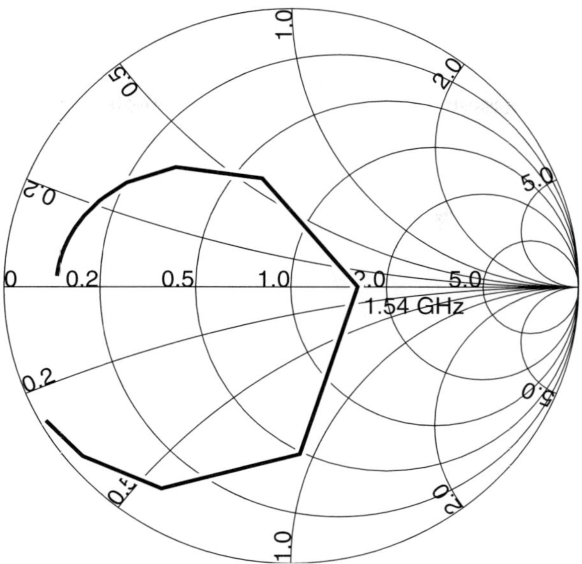

Figure 6.6. Smith chart of the reflection coefficient of an RLC circuit.

to plot the Smith chart use a 5pF capacitor in parallel to a 5Ω resistor and a 2nH inductor in series. The high losses of this example pull the frequency sufficiently to introduce a 7% error in the inductance extraction.

6.5 Circuit Equations with Capacitance

Capacitance can be folded into general circuit analysis through the continuity equation, which in integral form is

$$\oint\!\!\oint_S \overline{J} \cdot \overline{ds} = -\frac{d}{dt} \int\!\!\int\!\!\int_V \rho \, dV. \tag{6.10}$$

$\oint\!\oint_S \overline{J} \cdot \overline{ds}$ sums the total current flowing out of a volume and $\int\!\int\!\int_V \rho \, dV$ sums the total charge within the volume. Therefore, the continuity equation states that net current flow out of a volume must be matched by a time rate of loss of charge within the volume.

The closed surface in (6.10) can be broken into several pieces as

$$\oint\!\!\oint_S \overline{J} \cdot \overline{ds} = \sum_{k=1}^{N} \int\!\!\int_{S_i} \overline{J} \cdot \overline{ds},$$

and since current is defined as $i = \int\!\int_S \overline{J} \cdot \overline{ds}$, then

$$\oint\!\!\oint_S \overline{J} \cdot \overline{ds} = \sum_{k=1}^{N} i_k. \tag{6.11}$$

Replacing the total charge, $Q = \int\!\int\!\int_V \rho dV$, in (6.10) with $Q = Cv$ and incorporating (6.11) yields

$$\sum_{k=1}^{N} i_k = -\frac{d}{dt}(Cv).$$

As discussed in section 6.4, the capacitance is frequency-independent, and therefore time-independent, so it can be brought out from under the time-derivative. Defining $i = -\sum_{k=1}^{N} i_k$ as the total current entering (note the change in current direction) the volume from all sources, then the circuit relationship for capacitance is

$$i = C\frac{dv}{dt}.$$

The circuit relationship for a single capacitor can be generalized for the presence of multiple conductors. For a closed surface around conductor k, the total current entering the enclosed volume is

$$\begin{aligned} i_k &= \frac{d}{dt}Q_k \\ &= \frac{d}{dt}(-C_{k1}v_1 - C_{k2}v_2 - \cdots \\ &\quad +(C_{k0} + C_{k1} + \cdots)v_k - \cdots). \end{aligned} \tag{6.12}$$

Distributing the time derivative to the voltage terms and stating in vector notation, then

$$\overline{i} = \overline{\overline{C}}\frac{d\overline{v}}{dt}. \tag{6.13}$$

6.6 Modal Decomposition and Passivity

A modal decomposition exists for capacitance, but its utility is less than that for inductance, where the modal decomposition eliminates the great many mutual inductances. Because conductors shield each other, most capacitances between conductors within interconnects are negligible. With few mutual capacitances, there is little incentive to implement a modal decomposition. However, modal decomposition can prove that capacitance matrices are always passive.

6.6.1 Modal Decomposition

Paralleling the development in chapter 5, section 5.8.1, the real and symmetric $N \times N$ capacitance matrix can be diagonalized by finding the eigenvalues and eigenvectors of

$$\overline{\overline{C}}\,\overline{v}_k = \lambda_k \overline{v}_k, \qquad k = 1, 2, \ldots, N.$$

After orthonormalizing the eigenvectors, then

$$\overline{\overline{T}}^T \overline{\overline{C}}\,\overline{\overline{T}} = \begin{bmatrix} \lambda_1 & 0 & \cdots & 0 \\ 0 & \lambda_2 & \cdots & 0 \\ \vdots & \vdots & \ddots & \vdots \\ 0 & 0 & \cdots & \lambda_N \end{bmatrix}, \tag{6.14}$$

where

$$\overline{\overline{T}} = \begin{bmatrix} \overline{v}_1 & \overline{v}_2 & \cdots & \overline{v}_N \end{bmatrix}.$$

Inverting (6.14) and substituting for $\overline{\overline{C}}$ in (6.13) yields

$$\overline{i} = \overline{\overline{T}}\,\overline{\overline{\lambda}}\,\overline{\overline{T}}^T \frac{d\overline{v}}{dt}.$$

Moving $\overline{\overline{T}}^T$ under the time derivative and defining modal voltages and currents as in equation (5.63), then

$$\overline{i}_m = \overline{\overline{\lambda}} \frac{d\overline{v}_m}{dt}. \tag{6.15}$$

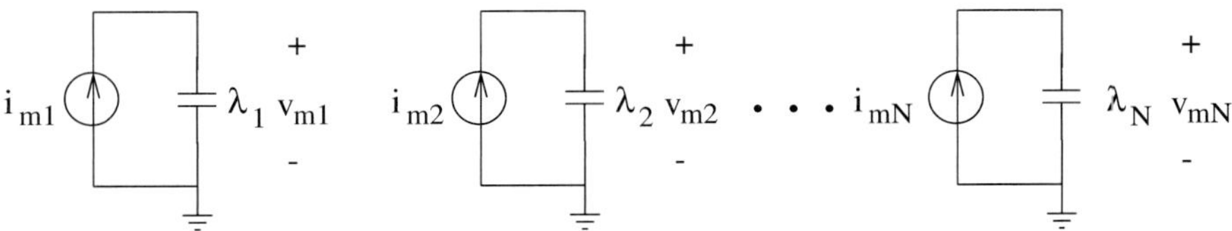

Figure 6.7. Decoupled circuits for computing modal voltages from modal currents to implement a capacitance matrix with a modal decomposition.

The circuit relationships embodied in (6.15) for the modal voltages and currents can be implemented with decoupled circuits as shown in Figure 6.7, which would ideally be coded into the simulator but could be implemented manually.

For the interconnect shown in Figure 6.8a, the modal currents are $\bar{i}_m = \overline{\overline{T}}^T(\bar{i}^L - \bar{i}^R)$. The modal voltages computed in the N decoupled circuits of Figure 6.7 are recombined according to $\bar{v} = \overline{\overline{T}}\,\bar{v}_m$ to obtain the values for the voltage sources in Figure 6.8b.

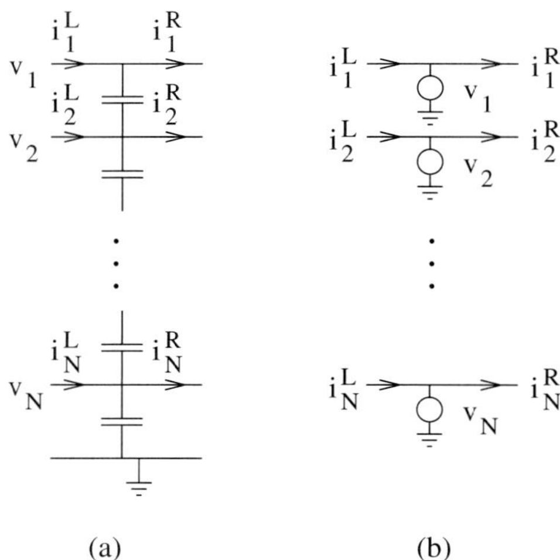

Figure 6.8. The capacitive interconnect model in (a) can be replaced by the model in (b) based on a modal decomposition of the capacitance matrix.

6.6.2 Passivity

For the capacitance matrix to be passive, each subcircuit in Figure 6.7 must be passive. These circuits will be passive as long as the modal capacitances given by the eigenvalues are nonnegative (see problem 2). Therefore, the capacitance matrix is passive when

$$\lambda_k \geq 0 \ \forall \ k.$$

In practice, positive eigenvalues are required since a zero eigenvalue represents an open-circuited modal current.

Capacitance matrices have a special form in which the off-diagonals are negative, the diagonal elements are positive, and the sum of any row or column is greater than zero (with some capacitance to ground assumed). Mathematically,

$$\tilde{C}_{ii} > - \sum_{\substack{j=1 \\ j \neq i}}^{N} \tilde{C}_{ij}. \tag{6.16}$$

For the standard eigenvalue problem, $\overline{\overline{A}}\overline{x} = \lambda\overline{x}$, Gersgorin's circle theorem[1] states that every eigenvalue of $\overline{\overline{A}}$ lies in at least one of the circles $C_1, C_2, \cdots, C_N$, where C_i is centered at A_{ii} with a radius

$$r_i = \sum_{\substack{j=1 \\ j \neq i}}^{N} |A_{ij}|.$$

Gersgorin's circle theorem is graphically illustrated in Figure 6.9.

Replacing the general matrix elements A_{ij} with their equivalents from the capacitance matrix, then

1. Richard Bellman, *Introduction to Matrix Analysis*. Philadelphia: SIAM, 1995, p. 107.

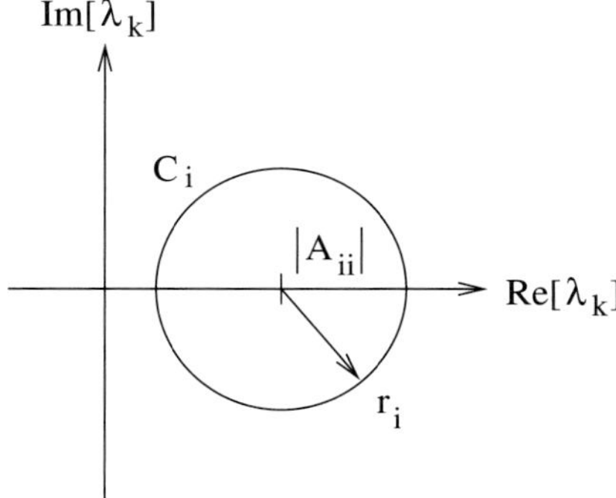

Figure 6.9. Graphical representation of Gersgorin's circle theorem.

$$r_i = -\sum_{\substack{j=1 \\ j \neq i}}^{N} \tilde{C}_{ij}$$

Finally, the property in (6.16) ensures that

$$r_i < \tilde{C}_{ii}.$$

Since circle C_i is centered at $\tilde{C}_{ii}$, the radius of the circle is small enough so that the circles stay entirely in the right-hand plane. Since the eigenvalues always lie within at least one circle, then the eigenvalues are always positive.

Unlike inductance matrices, where active matrices are possible and must be checked for and avoided, capacitance matrices are always passive. Practically, this means that capacitance matrices can be assembled or simplified, such as by deleting small terms, without worrying about creating an active matrix.

6.7 Reference and Capacitance

Compared to inductance, construction of capacitive interconnect models is straightforward since each conductor already has open-circuited ends available for connections. Each conductor has capacitance to ground and to every other conductor. The values for these capacitances can be found directly from the capacitance matrix. Since ground is defined when setting up the capacitance solution, its effects are built into all of the capacitance values, which can be said to be referenced to the particular ground selection.

Recall that for inductance, a reference lead is naturally created when a closed loop is opened to permit connections to a circuit. The reference lead can take any partial inductance value, and zero is a common choice. Whatever value is chosen, the remaining partial inductances can be said to reference this lead.

Reference can then take two meanings in an interconnect model. For capacitance, it is the conductor assigned to zero potential (ground), while for inductance, it is a conductor assigned an arbitrary partial inductance value, with zero being common, but not necessary, due to the nonuniqueness of partial inductance. These assignments do not need to be made to the same conductor. Usually, a large plane of conductor is present within the circuit, and this conductor will be assigned to be both the ground reference for the capacitance and the zero-valued partial inductance reference for the inductance.

Ground is often described as being "perfect" with no inductance. Descriptions along these lines confuse the two meanings of reference. First, ground is an electrostatic concept that has no place in inductance descriptions, which are magnetostatic and therefore decoupled from electrostatics. Second, ground is a conductor like any other, so it can have partial inductance, but most models choose to assign its partial inductance value to be zero.

6.8 Model Reduction

Using the algorithms in chapter 4, section 4.5, capacitance matrices can be reduced in size to incorporate known information. The principal cases are when model leads are grounded, floated, and paralleled together.

The reduction for grounded conductors can be found by setting the voltage for each grounded conductor to zero in (6.5) and then retaining the charges for the nongrounded conductors. The net result is that the row and column in the capacitance matrix for each grounded conductor is deleted from the matrix.

For floating conductors, the charge is zero. The charge on a conductor can be set to zero by deleting the appropriate row and column in $\overline{\overline{C}}^{-1}$. The reduced capacitance matrix for floating conductors is then given by

$$\overline{\overline{C}}' = \left(\overline{\overline{A}}^T \overline{\overline{C}}^{-1} \overline{\overline{A}}\right)^{-1},$$

using (4.16).

For paralleled conductors, the admittance relationship between voltage and current can be used with the formulation in chapter 4, section 4.5.2 to parallel conductors together.

6.9 Exercises

1. Assume that a 3×3 capacitance matrix is known for a three-conductor interconnect. If line 1 is grounded while line 3 has 1V applied, what is the voltage on line 2, assuming it is floating? What are the charges on all three conductors?

2. Show that a capacitance matrix is passive when all of its eigenvalues are greater than or equal to zero.

3. What does it mean if the sum across a row or column of a capacitance matrix is zero? How can this be physically realized?

4. Compute the capacitance of a parallel-plate capacitor using the energy definition of capacitance. Neglect the fringing fields.

5. A capacitance matrix is given by

$$\overline{\overline{C}} = \begin{bmatrix} C & -C_m & -C_m/\sqrt{3} \\ -C_m & C & -C_m \\ -C_m/\sqrt{3} & -C_m & C \end{bmatrix}.$$

(a) Find $\overline{\overline{T}}$ and $\overline{\overline{\lambda}}$ for a modal decomposition.

(b) Sketch a circuit that implements the capacitance matrix using the modal decomposition.

(c) Derive conditions for the capacitance matrix to remain passive.

6. Given the capacitance matrix for a three-conductor interconnect as

$$\overline{\overline{C}} = \begin{bmatrix} 10 & -3 & -1 \\ -3 & 5 & -2 \\ -1 & -2 & 5 \end{bmatrix},$$

(a) Find the capacitance matrix for the two-conductor interconnect when lead 2 is grounded. Sketch a schematic.

(b) Find the capacitance matrix for the two-conductor interconnect when lead 2 is floating. Sketch a schematic.

Chapter 7

RESISTANCE

Interaction between resistance and inductance creates two effects important to interconnect modeling: the *skin effect* and *current crowding*, also referred to as the *proximity effect*. In both cases, currents distribute differently at low and high frequencies to minimize the overall reactance of the circuit. At low frequencies, inductive reactance is negligible, so the currents spread out to minimize resistance. A side effect of this optimization is that the inductance increases. At high frequencies, inductive reactance dominates resistance, so the currents crowd together to minimize inductance, and resistance is increased. Redistribution of current towards the outside surface of a conductor is called the *skin effect*. Redistribution of current parallel to the surface is called *current crowding*. Both types are depicted in Figure 7.1, where at low frequencies in Figure 7.1a, the current is spread across the cross section of the wire and the width of the ground plane. At high frequencies in Figure 7.1b, the currents redistribute to the surface of the wire (skin effect) and to the area under the wire (current crowding and skin effect).

A widely discussed generic example of current crowding is shown in Figure 7.2a, where an interconnect uses a plane to return the current, but a slot in the ground plane interferes with the current path. The components of the current supporting the high-frequency content of the signal (i.e., the parts creating sharp edges in high-speed signaling) crowd underneath the signal line to maximize the mutual inductance between the line and the plane. The slot forces the current away from

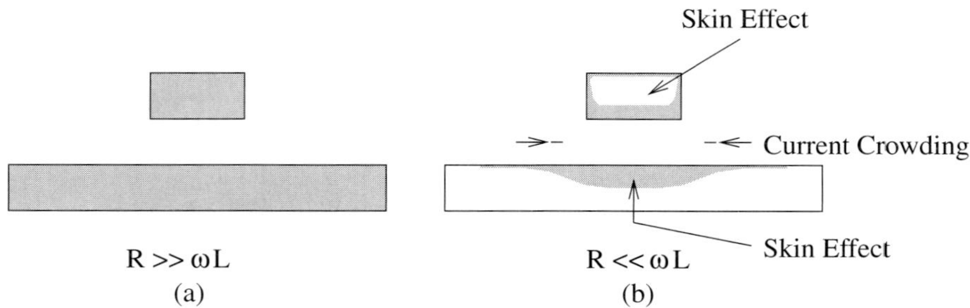

Figure 7.1. Current crowding in interconnects: (a) low-frequency current distribution, (b) high-frequency current distribution.

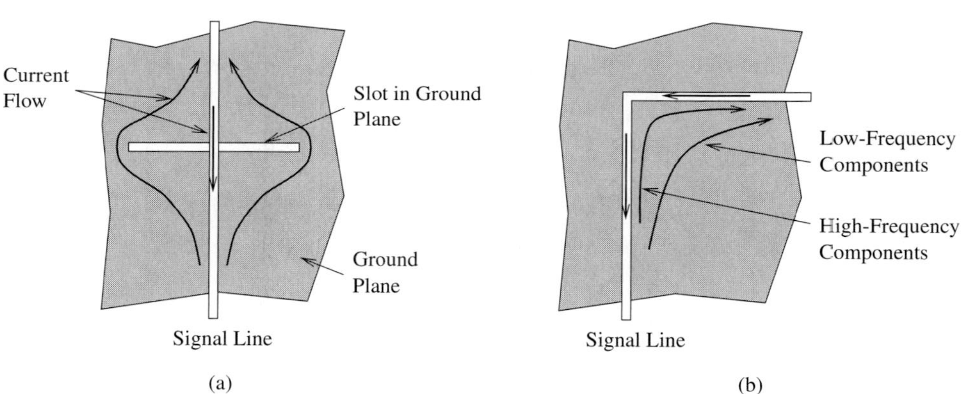

Figure 7.2. Two examples of current crowding in interconnects: (a) return currents on a plane must route around a slot, (b) low-frequency components can cut the corner of a bend.

the signal line, thereby defeating the beneficial current crowding and causing the inductance to increase. A secondary effect is increased crosstalk if the slot causes the ground currents to pass near other signal lines. Good high-frequency design requires the minimization or elimination of slots in ground planes.

A second example of current crowding is sketched in Figure 7.2b. The high-frequency components of the current on the ground plane track under the signal line to minimize inductance, while the low-frequency components cut the corner to minimize resistance. Unlike slots, where all frequency components are rerouted, current crowding at bends has little performance impact.

Skin effects and current crowding lead to frequency-dependent resistance that affects circuits by inducing conductor-loss dispersion. Low-frequency components of the signal are barely affected, while high-frequency components can see significant attenuation. By knocking down the amplitudes of only the high-frequency components, conductor-loss dispersion causes edge rate degradation and rounding of sharp corners. While signal degradation of any kind is normally regarded as undesirable, too much high-frequency content can lead to excessive crosstalk. In circuits with the potential for high levels of crosstalk, inclusion of skin effects can be important to avoid pessimistic simulations of noise levels.

Analytical approaches to the skin effect and current crowding do not yield very much in the way of useful formulas. For the skin effect, the round wire far from its return path can be exactly solved. Current crowding is far too dependent on geometrical and material considerations for the existence of a canonical problem. To clearly demonstrate current crowding and the skin effect, several problems are set up and solved in an intuitive circuit simulation style.

The methodology employed is a manual implementation of a technique called Partial Equivalent Electric Circuit (PEEC), which is itself a straightforward application of circuit theory. The basic approach is to subdivide all conductors and then assume that the current is uniform in each piece. The DC resistance and inductance of each piece is computed based on a library of formulas, then a circuit is formed with these elements. Finally, the circuit is solved with a circuit simulator such as SPICE. The interaction between resistance and inductance naturally leads to skin effects and current crowding. Accuracy improves with finer-grained subdivision of the original geometry.

7.1 Skin Effect

The skin effect is simply current crowding taking place within the cross section of a single conductor. The effect can be demonstrated by subdividing a square wire into three equal subsections, as shown in Figure 7.3a. With this subsectioning scheme, skin effects can only develop in the vertical direction. To enable skin effects in the horizonal direction, a nine-subsection or greater scheme is required, as suggested in Figure 7.5. Each subsection has resistance, partial self-inductance, and partial mutual inductance to the other two pieces. Assuming that the DC values are available, then the circuit can be described as shown in Figure 7.3b. Note that $M_1 > M_2$ due to proximity.

The voltage drop across the resistor/inductor pairs is given by

$$\overline{V} = \overline{\overline{Z}}\,\overline{I}.$$

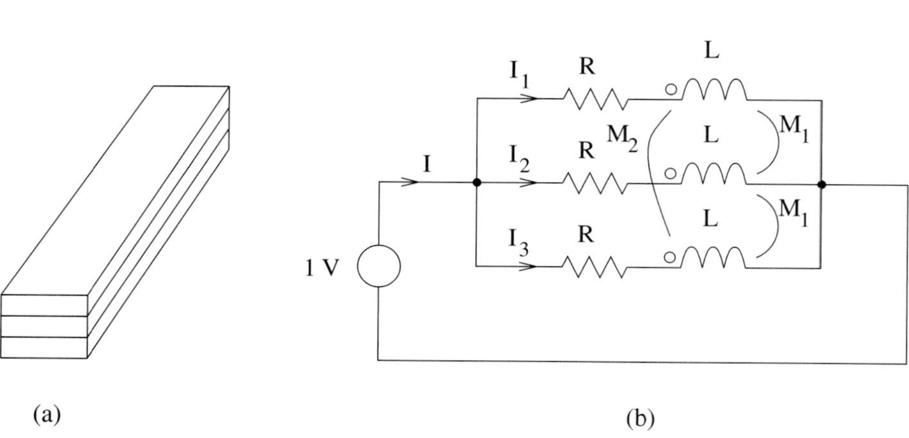

Figure 7.3. A rectangular conductor subdivided into three sections.

The parallel connections force all voltages to be equal to one, so for this circuit

$$\begin{bmatrix} 1 \\ 1 \\ 1 \end{bmatrix} = \begin{bmatrix} R + j\omega L & j\omega M_1 & j\omega M_2 \\ j\omega M_1 & R + j\omega L & j\omega M_1 \\ j\omega M_2 & j\omega M_1 & R + j\omega L \end{bmatrix} \begin{bmatrix} I_1 \\ I_2 \\ I_3 \end{bmatrix}.$$

Inverting the matrix enables a direct solution for the currents, then $I = I_1 + I_2 + I_3$. Since $V = 1$, then the equivalent impedance for the original conductor is $Z = 1/I$. The equivalent resistance is $\text{Re}[Z]$, while the equivalent inductance is $\text{Im}[Z]/\omega$.

To see the skin effect, the magnitudes of the currents can be compared. Taking the ratio of the current in the outer section[1] to that in the inner yields

$$\left|\frac{I_1}{I_2}\right|^2 = 1 + \frac{(M_1 - M_2)(2L - 3M_1 + M_2)\omega^2}{R^2 + (L - 2M_1 + M_2)^2\omega^2}.$$

In terms of frequency, this equation is of the form

$$f(\omega) = 1 + \frac{a\omega^2}{1 + b\omega^2}, \tag{7.1}$$

which increases monotonically from a DC value to the asymptotic limit of $1 + a/b$. The current in the wire is then evenly distributed at DC and crowds to the outside of the wire as frequency increases. With only three subsections to the wire, there is a limit on how much crowding occurs. Note that it is assumed that $2L - 3M_1 + M_2 > 0$, which is normally true since $L > M_1 > M_2$.

Redistribution of current due to the skin effect causes the resistance and inductance of the wire to change. The equivalent resistance is

$$R_{\text{eq}} = \frac{R}{3} + \frac{2}{3}\frac{(M_1 - M_2)^2 R\omega^2}{9R^2 + (3L - 4M_1 + M_2)^2\omega^2}.$$

Note that the basic form of the frequency dependence tracks that of the current distribution. At DC, the resistance is the expected value of $R/3$. As frequency increases, the area utilized by the current decreases, so R monotonically increases. The important and general result here is that the skin effect always increases resistance with frequency.

1. By symmetry, $I_1 = I_3$.

The equivalent inductance is

$$L_{\mathrm{eq}} = \frac{1}{9}(3L + 4M_1 + 2M_2) - \frac{2}{9}\frac{(M_1 - M_2)^2(3L - 4M_1 + M_2)\omega^2}{9R^2 + (3L - 4M_1 + M_2)^2\omega^2}.$$

As for the resistance, the basic form follows that of the current distribution. As the currents vacate the center of the wire, internal inductance is reduced, and the inductance decreases monotonically from its DC value to an asymptotic limit. The skin effect always decreases inductance with frequency. Just as for the current, under normal conditions $3L - 4M_1 + M_2 > 0$.

The increase in resistance as the currents shift to the skin of the conductor is the defining characteristic of the skin effect. With increasing frequency, more of the current crowds to the outside of the wire, causing increased resistance and decreased inductance. The skin depth is defined as the distance from the surface to where the current drops to $1/e$ of its value at the surface. Therefore, the skin depth is frequency-dependent and decreases with increasing frequency.

For any given interconnect, the importance of the skin effect can be examined by considering the ratio of the skin depth to the conductor thickness (δ/t), and there are three regimes to consider. For $\delta/t \gg 1$, the current has the DC distribution, resistance is low, inductance is high, and there is little frequency dependence. For $\delta/t \approx 1$, resistance begins increasing while inductance begins decreasing. This transition region is characterized by a knee in the frequency dependence of both resistance and inductance, and can be seen in the example in Figure 7.7, where the knee frequency is approximately 0.1MHz. For $\delta/t \ll 1$, the currents reside on the outer surface of the conductor. Resistance increases strongly with frequency as less and less area is available. Inductance is constant because the internal inductance is shielded.

At frequencies above the knee frequency, the resistance has a square root dependence on frequency. For conductors with no sharp corners and at frequencies where the conductor is several skin depths thick, the classic skin depth formula is

$$\delta = \sqrt{\frac{2}{\omega\mu\sigma}},$$

where $\mu = \mu_r \mu_o$ and $\mu_o = 4\pi \times 10^{-7}$H/m is the permeability of free space. For the vast majority of interconnect materials, the relative permeability is $\mu_r = 1$.

Magnetic materials with $\mu_r \gg 1$ are used in some packages. Nickel is often used in very thin layers as an adhesion promoter, but it is generally so thin that it has insignificant impact. Leadframes are often fabricated from Alloy 42, especially in ceramic packages such as CERDIP and CQFP, because of the excellent match of the coefficients of thermal expansion (CTE) between Alloy 42 and the glasses used to seal together the two ceramic halves. Alloy 42 has a high relative permeability near 200, which leads to well-developed skin effects and high losses at remarkably low frequencies. Plating with nonmagnetic metals can reduce or eliminate the detrimental effects of Alloy 42 by providing a parallel current path through metal with low permeability.

7.2 Current Crowding

A simple structure that can be easily analyzed to demonstrate current crowding is shown in Figure 7.4a, where conductor 1 represents a signal line with a return

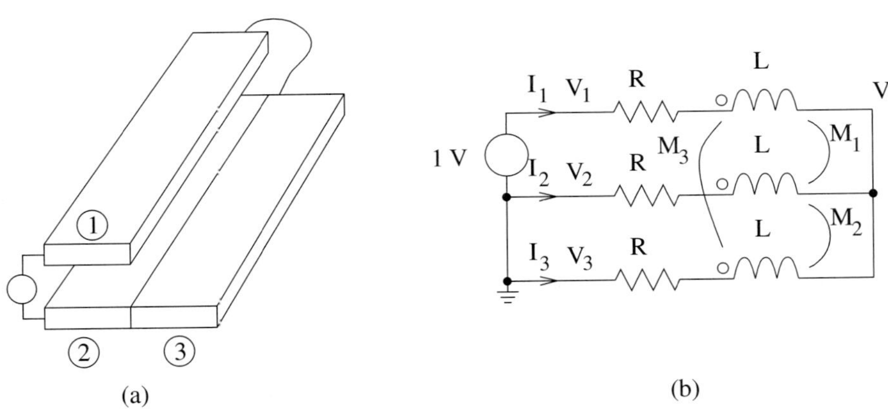

Figure 7.4. A loop formed by a single conductor and a return path with two parallel conductors.

path formed by conductors 2 and 3. The subsectioning scheme enables current crowding on the return path. The proximity of the various subsections indicates that $M_1 > M_3$. Skin effects are not possible but could be included by subdividing the conductors in thickness. Each subsection has resistance, partial self-inductance, and partial mutual inductance to the other two pieces. Assuming that the DC values are available, then the circuit can be described as shown in Figure 7.4b.

The circuit has the constraints that

$$\begin{aligned} I_1 + I_2 + I_3 &= 0 \\ V_1 &= 1 \\ V_2 &= 0 \\ V_3 &= 0. \end{aligned}$$

With these, the voltage drop across the resistor/inductor pairs is given by

$$\begin{bmatrix} 1-V \\ -V \\ -V \end{bmatrix} = \begin{bmatrix} R+j\omega L & j\omega M_1 & j\omega M_3 \\ j\omega M_1 & R+j\omega L & j\omega M_2 \\ j\omega M_3 & j\omega M_2 & R+j\omega L \end{bmatrix} \begin{bmatrix} I_1 \\ I_2 \\ -I_1-I_2 \end{bmatrix},$$

which provides three equations for the three unknowns V, I_1, and I_2.

To see current crowding, the magnitudes of the currents can be compared. Taking the ratio of the currents in the two conductors of the return path yields

$$\left|\frac{I_2}{I_3}\right|^2 = 1 + \frac{4(L-M_2)(M_1-M_3)\omega^2}{R^2 + (L-M_1-M_2+M_3)^2\omega^2}.$$

In terms of frequency, this equation is of the form in (7.1) that monotonically increases from a DC value to a high-frequency asymptotic limit. At DC, the currents are evenly distributed on the return path, but as frequency increases, the currents crowd underneath the signal line.

Redistribution of current due to crowding causes the resistance and inductance of the loop to change. The equivalent impedance is found by solving for I_1 and then inverting. The equivalent resistance is

$$R_{\text{eq}} = \frac{3}{2}R + \frac{1}{2}\frac{(M_1-M_3)^2 R\omega^2}{R^2 + (L-M_2)^2\omega^2}.$$

Note that the basic form of the frequency dependence tracks that of the current distribution. At DC, the resistance is the expected value of $3R/2$. As frequency increases, the area utilized by the current decreases, so R monotonically increases. The important and general result here is that current crowding always increases resistance with frequency.

The equivalent inductance is

$$L_{\text{eq}} = \frac{3}{2}L - M_1 + \frac{1}{2}M_2 - M_3 - \frac{1}{2}\frac{(L - M_2)(M_1 - M_3)^2\omega^2}{R^2 + (L - M_2)^2\omega^2}.$$

As for the resistance, the basic form follows that of the current distribution. As the currents crowd under the signal line, mutual inductance is enhanced so loop inductance drops. Current crowding always decreases inductance with increasing frequency.

7.3 PEEC Method

The Partial Equivalent Electric Circuit (PEEC) method builds up models of complex interconnect structures from simple primitive elements, such as rectangular bars, to find frequency-dependent resistance and inductance. Although many approaches are possible, the easiest is to compute the DC resistance and static partial self-inductance for each primitive element, plus the partial mutual inductance between each pair of elements. The resistances and inductances are then assembled into a complete circuit and solved with a circuit simulator. Accuracy improves with finer-grained subdivision of the original geometry. The interaction between resistance and inductance naturally leads to frequency-dependent skin effects and current crowding.

The previous two sections use a manual implementation of the PEEC method to derive frequency-dependent impedances for either the skin effect or current crowding. However, all crowding effects can be accommodated simultaneously if the conductors are sufficiently subdivided. For example, Figure 7.5 shows an interconnect with two signals and a plane return path. The conductors are subdivided to capture

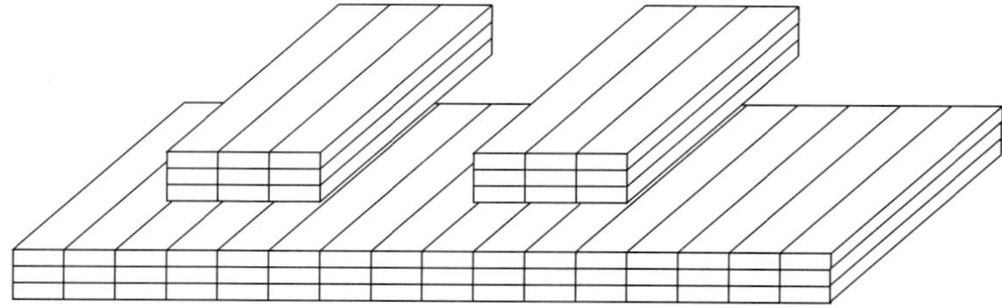

Figure 7.5. Fine subdivision for capturing all current crowding effects using the PEEC method.

skin effects, current crowding on the return path, and current crowding between the signals. If the signals bend, then further subdivision of the conductors is needed in the direction of current flow.

Since the PEEC method leads to a conventional circuit, any circuit simulation technique is applicable. For example, a full PEEC-based model of a socket can be included with driver and receiver models in a SPICE simulation. The main advantage of including the full PEEC model is that the interconnect's frequency-dependent behavior is completely included. Unfortunately, this approach becomes impractical for complex PEEC models that can quickly overwhelm any circuit simulator.

To limit the circuit size for practical circuit simulation, the size of the PEEC model must usually be reduced. The PEEC model by itself is a linear circuit, while general-purpose circuit simulators must handle nonlinear circuit elements. A good strategy is to precompute a simplified PEEC model using a linear solver, then the simplified model can be used in the nonlinear circuit simulator. The primary disadvantage of this methodology is that the circuit elements in the reduced PEEC model are frequency-dependent, but time-domain simulators often do not accommodate frequency-dependent elements well, if at all.

7.3.1 General Formulation

The problem setup is the same whether a general-purpose or dedicated solver is used. All of the conductors in the problem must first be subdivided into N canonical primitive structures, such as rectangular bars, for which formulas for resistance, partial self-inductance, and partial mutual inductance are known. These small, generically shaped conductors are called *branches*. Many implementations are possible, but the most straightforward approach assumes that the current is uniform across the cross section of the branches. Then relatively simple DC resistance and static inductance formulas are applicable.

The resistance and partial self-inductance of each branch is computed along with the partial mutual inductance between each branch. The results can be assembled into a diagonal $N \times N$ resistance matrix and a dense $N \times N$ partial inductance matrix. The voltage drop across each branch is

$$\begin{aligned} \overline{V}_b &= (\overline{\overline{R}} + j\omega\overline{\overline{L}})\overline{I}_b \\ &= \overline{\overline{Z}}_p\overline{I}_b, \end{aligned} \tag{7.2}$$

where $\overline{I_b}$ are the branch currents. It is assumed that the structure is subdivided in an appropriate way to enable all necessary current flow.

To complete the model of the interconnect, the branches are tied together at a number of nodes, M, where the number of nodes depends on the subdivision used. For example, the breakdown in Figure 7.4 uses three nodes connecting three branches. Because voltages are potential differences, one node must serve as the reference with a voltage of zero. This node, called the *datum node*, is not included in the problem setup—it is assumed. The same situation occurs in general-purpose circuit simulators such as SPICE, where the datum node is labeled as "0" and forms the ground reference.

At this point, the interconnect has been broken into its component parts and modeled with a circuit consisting of resistances and partial self- and mutual inductances. The circuit must be solved with either a general-purpose or dedicated simulator.

7.3.2 Dedicated Solver

A dedicated solver[2] for the PEEC method can be constructed by satisfying Kirchhoff's voltage and current laws for the branch voltages and currents. The goal is to eliminate the branch voltages and currents in favor of node voltages and currents. Then the nodes internal to the circuit can be eliminated to obtain a simplified model relating just the external nodes, or ports. The model is used by connecting the ports to other system components, such as drivers and receivers.

The branch voltages are related to the node voltages by differences: the difference of the two node voltages at the ends of a branch is the *branch voltage.* This relationship for the entire problem can be stated as

$$\overline{V}_b = \overline{\overline{A}}\,\overline{V}_n, \tag{7.3}$$

where $\overline{\overline{A}}$, called the *incidence matrix*, stores all of the connection information in an $N \times M$ matrix. It can be constructed by setting $A_{bi} = 1$ and $A_{bj} = -1$ when the current flows from node i to node j through branch b. For example, the incidence matrix for the network in Figure 7.6 is

$$\overline{\overline{A}} = \begin{bmatrix} -1 & 0 & 0 & 0 & 1 \\ 0 & -1 & 0 & 0 & 1 \\ 0 & 0 & -1 & 1 & 0 \\ 0 & 0 & 0 & 1 & -1 \end{bmatrix},$$

showing how the node voltages subtract to obtain the branch voltages.

Continuing the formulation, the total current into the nodes is given by

$$\overline{I}_n = \overline{\overline{A}}^T \overline{I}_b, \tag{7.4}$$

where $\overline{I}_b$ are the branch currents driven by the branch voltages, and $\overline{I}_n$ are currents driven into the nodes by external sources. For the example in Figure 7.6,

2. R. B. Wu, C. N. Kuo, and K. K. Chang, "Inductance and resistance computations for three-dimensional multiconductor interconnection structures," *IEEE Trans. Microwave Theory Tech.*, vol. 40, no. 2, Feb. 1992, pp. 263–71.

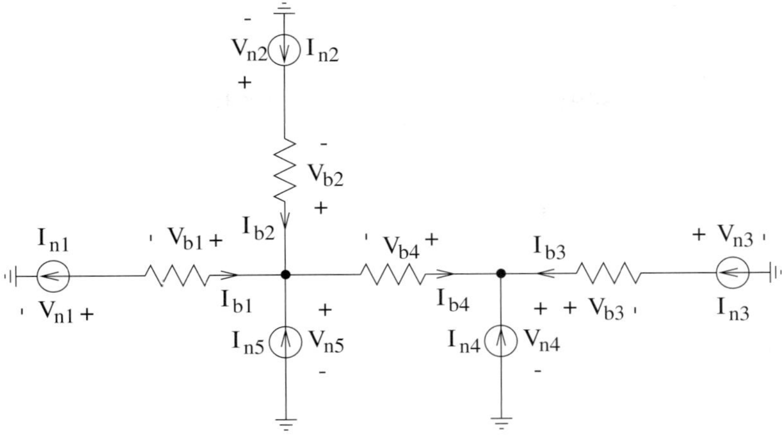

Figure 7.6. Node and branch notation for a four-branch/five-node network.

$$\overline{\overline{A}}^T = \begin{bmatrix} -1 & 0 & 0 & 0 \\ 0 & -1 & 0 & 0 \\ 0 & 0 & -1 & 0 \\ 0 & 0 & 1 & 1 \\ 1 & 1 & 0 & -1 \end{bmatrix},$$

showing how the node currents are summed from the branch currents. In general, most of the $\overline{I}_n$ are zero since current is externally supplied only at the ports.

The branch voltages and currents can be eliminated by substituting (7.2) and (7.3) into (7.4) to get

$$\overline{I}_n = \overline{\overline{A}}^T \overline{\overline{Z}}_p^{-1} \overline{\overline{A}}\, \overline{V}_n. \tag{7.5}$$

The currents and voltages at the vast majority of the nodes are of no interest. The nodes of interest are for external connections at the ports, so (7.5) needs to be reduced. For all of the internal nodes, $I_{n_i} = 0$, so the reduction can be performed by inverting (7.5) and then deleting the rows and columns representing the internal nodes. The remaining matrix is then the desired equivalent impedance matrix relating the node voltages at the ports to the external node currents entering the ports. Row and column deletion can be stated mathematically as shown in chapter 4, equation (4.16), leading to a succinct formula for the equivalent impedance as

$$\overline{\overline{Z}}_{eq} = \overline{\overline{B}}^T \left(\overline{\overline{A}}^T \overline{\overline{Z}}_p^{-1} \overline{\overline{A}} \right)^{-1} \overline{\overline{B}} \tag{7.6}$$

since

$$\overline{V}_n = \overline{\overline{Z}}_{eq} \overline{I}_n \tag{7.7}$$

for the ports only.

While the formulation of the PEEC method represented by (7.6) is quite simple, the solution requires two matrix inversions plus four matrix multiplications. Matrix inversion is an expensive numerical operation, so PEEC methods can be slow for large numbers of branches. Fortunately, several efficiency enhancements[3] are available that enable the PEEC method to solve very large problems.

Skin effects and current crowding can be investigated if the branch currents are known. Combining (7.2) with (7.3) and (7.7) yields

$$\overline{I}_b = \overline{\overline{Z}}_p^{-1} \overline{\overline{A}} \, \overline{\overline{Z}}_{eq} \overline{I}_n,$$

where $\overline{I}_n$ are the port currents. The branch currents can be computed for any given excitation at the ports.

7.3.3 In-Circuit Solution

In a typical application, (7.6) is used to obtain a reduced model at a given frequency. If the frequency dependence is strong over the bandwidth of interest, then a table of frequency-dependent data can be generated and then modeled with a wideband technique, such as a ladder network or a rational function polynomial.

For cases where $\overline{\overline{Z}}_p$ is sufficiently small, the reduction in (7.6) can be skipped and $\overline{\overline{Z}}_p$ can be included directly into the full circuit simulation. The frequency

3. M. Kamon, M. Tsuk, and J. White, "FASTHENRY: A multipole-accelerated 3-D inductance extraction program," *IEEE Trans. Microwave Theory Tech.*, vol. 42, no. 9, Sept. 1994, pp. 1750–58.

dependence of the structure is then fully captured in the circuit simulation without resorting to a wideband modeling technique. Only when the simulation time is too slow should $\overline{\overline{Z}}_p$ be reduced.

In other words, $\overline{\overline{Z}}_{eq}$ represents a precharacterization step that reduces runtime in the circuit simulation while freezing the model at a single frequency. Once precharacterization is performed, frequency dependence is recovered only through additional modeling.

7.3.4 Practical Issues

The PEEC method can solve very large resistance and inductance problems with high accuracy while capturing skin effects and current crowding. A fundamental limit is that the lumped approximation must hold. In practice, the geometry cannot be broken down into arbitrarily small primitives, so the user typically has considerable influence on the solution through the particular subdivision scheme selected. Therefore, use of PEEC methods can require some electromagnetics expertise from the user to obtain good results. Finally, the PEEC modeling scheme is not extendable to capacitance calculations, so another method, such as the boundary element method, is required with new modeling requirements. In other words, the practitioner may need to build two separate models of the same structure to obtain a complete lumped model.

7.3.5 Example: PEEC Computation of Coaxial Inductance

The per-unit-length inductance of coaxial transmission line can be computed using the PEEC method by breaking the conductors into annular rings, as shown in Figure 7.8. The PEEC method requires formulas for the partial self-inductance of each ring, the partial mutual inductance between any pair of rings, plus the DC resistance of each ring. A suitable set of equations for the partial self- and mutual inductances is available from chapter 5, problem 12. The DC resistance is easily computed from the ring areas.

The PEEC method for this calculation is implemented in a C program listed in Appendix B. Sample computed results are shown in Figure 7.7. Note the convergence in the inductance calculation for increasing numbers of thinner rings: accuracy at higher frequencies is a function of granularity of the decomposition. For resistance, thinner rings increase the upper frequency for accurate resistance calculation, but the exact profile is not obtainable because PEEC models always plateau at the resistance of the outer ring when all of the currents have crowded there.

An interesting aspect of this problem is the use of a reference ring located outside the coax to allow the use of inductance formulas for partial inductance calculations. Except for numerical stability issues raised by far placements of the reference ring, the solution is independent of the reference ring location.

7.4 Ladder Networks

PEEC methods can handle arbitrary conductor cross sections while accurately computing frequency-dependent resistance and inductance due to the skin effect and current crowding. However, as a modeling tool, a PEEC breakdown inconveniently produces many elements since every branch is coupled to every other branch by partial mutual inductance. For round wires, a simple ladder network can model the skin effect while using no partial mutual inductances. Ladders can also be used to approximate wires with other cross sections.

The PEEC method is an integral solution to Maxwell's equations, so to develop a different model, a differential equation relating the magnetic field and current is derived and solved with a finite-element solution that leads directly to a ladder network. The approach does not incorporate the return path and so produces a model of the frequency-dependent partial self-inductance of the wire.

In good conductors, Maxwell's equations can be specialized to a simpler form. Conduction currents dominate displacement currents, so $\overline{J} \gg \frac{\partial \overline{D}}{\partial t}$ and displacement currents can be neglected. Ohm's law holds, so $\overline{J} = \sigma \overline{E}$, where σ is the conductivity. Finally, the constitutive relation $\overline{B} = \mu \overline{H}$ holds for regular metals. With these modifications, Faraday's and Ampere's laws in differential form are converted from

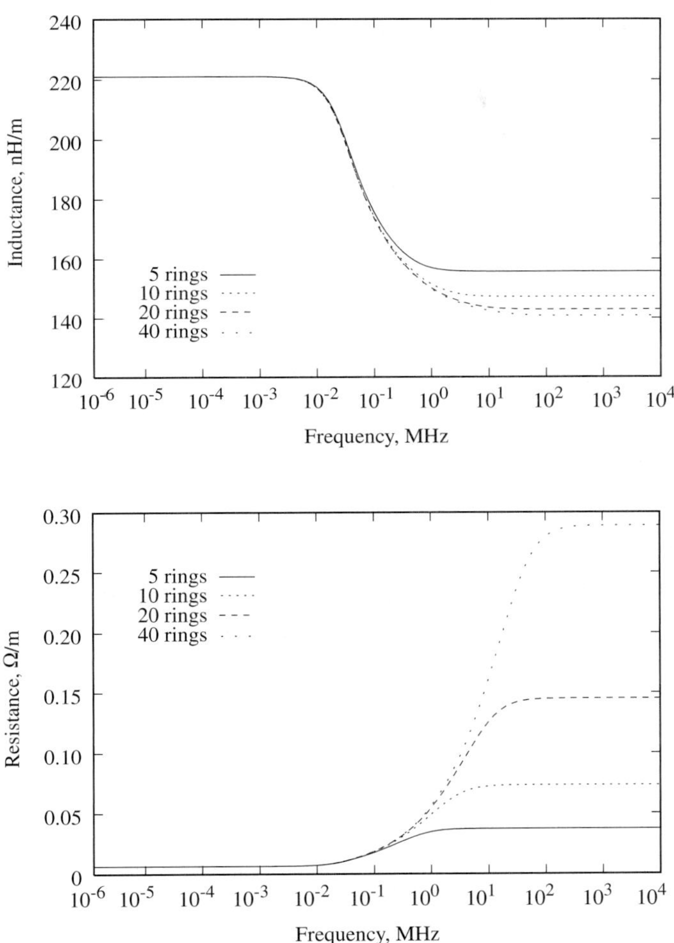

Figure 7.7. Inductance and resistance of copper coax with $a = 0.001\text{m}$ $b = 0.002\text{m}$ with an outer conductor thickness of just under 0.001m for several levels of decomposition.

$$\nabla \times \overline{E} = -\frac{\partial \overline{B}}{\partial t}$$

$$\nabla \times \overline{H} = \overline{J} + \frac{\partial \overline{D}}{\partial t}$$

to

$$\nabla \times \overline{J} = -\mu\sigma \frac{\partial \overline{H}}{\partial t} \tag{7.8}$$

$$\nabla \times \overline{H} = \overline{J}. \tag{7.9}$$

Rotational symmetry holds for round wires. It is convenient in these cases to use cylindrical coordinates where

$$\nabla \times \overline{A} = \frac{1}{r} \begin{vmatrix} \hat{r} & r\hat{\phi} & \hat{z} \\ \frac{\partial}{\partial r} & \frac{\partial}{\partial \phi} & \frac{\partial}{\partial z} \\ A_r & rA_\phi & A_z \end{vmatrix} \tag{7.10}$$

for any vector $\overline{A}$. With symmetry and the lumped approximation, then $\frac{\partial}{\partial \phi} = \frac{\partial}{\partial z} = 0$ and (7.10) simplifies to

$$\nabla \times \overline{A} = -\frac{\partial A_z}{\partial r}\hat{\phi} + \frac{1}{r}\frac{\partial}{\partial r}(rA_\phi)\hat{z}. \tag{7.11}$$

From chapter 5, problem 7, the magnetic field strength external to a round wire is

$$\overline{H} = \frac{i}{2\pi r}\hat{\phi}, \tag{7.12}$$

where i is the total current enclosed within the radius r. Substituting this into (7.9) and using (7.11) yields

$$\overline{J} = \frac{1}{2\pi r}\frac{\partial i}{\partial r}\hat{z}, \tag{7.13}$$

which shows that $\overline{J}$ is directed only in the $\hat{z}$ direction. (This is not much of a surprise since this fact is assumed to derive (7.12).) With only a $\hat{z}$ component, the subscript z can be dropped.

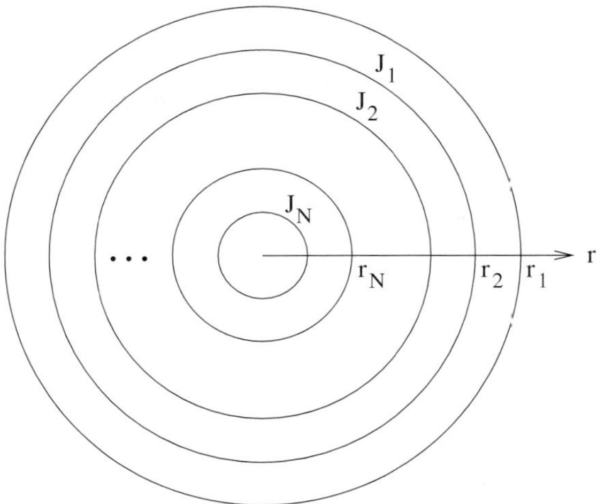

Figure 7.8. Subdivision of a round wire into concentric rings for finite-element solution.

Using a finite-element method,[4] the cross section of the round wire can be subdivided into concentric conductive rings as shown in Figure 7.8. The rings are numbered from the outside in. To match up to (7.12), the variable i_k is defined to be the sum of the currents in ring k and all the rings it encloses. The total current in the wire is then i_1.

The current density in ring k is J_k and is assumed to be constant with respect to r (i.e., the DC current distribution). All of the J_k flow in the same direction. The current in ring k is $i_k - i_{k+1}$ and can be found by integrating (7.13) from the ring's inner radius to its outer radius to get

$$\begin{aligned} i_k - i_{k+1} &= \pi(r_k^2 - r_{k+1}^2)J_k \\ &= A_k J_k, \end{aligned} \tag{7.14}$$

where A_k is just the area of ring k.

4. C. H. Yen, Z. Fazarinc, and R. L. Wheeler, "Time-domain skin-effect model for transient analysis of lossy transmission lines," *Proc. IEEE*, vol. 70, no. 7, July 1982, pp. 750–757.

Combining (7.8) and (7.12) with (7.11) and the fact that $\overline{J}$ has only a z component yields

$$\frac{\partial J}{\partial r} = \mu\sigma \frac{1}{2\pi r}\frac{\partial i}{\partial t}.$$

Converting the spatial derivative into a difference equation and using (7.14) to eliminate J yields

$$\frac{(i_{k-1} - i_k)/A_{k-1} - (i_k - i_{k+1})/A_k}{r_{k-1} - r_k} = \mu\sigma \frac{1}{2\pi r_k}\frac{di_k}{dt}. \tag{7.15}$$

This equation can be converted into a circuit equation by defining

$$R_k = \frac{1}{\sigma A_k}$$

as the resistance per-unit-length of ring k, and

$$L_k = \frac{\mu}{2\pi}\frac{r_{k-1} - r_k}{r_k}$$

as the partial self-inductance per-unit-length of the ring. Combining these definitions with (7.15) produces

$$R_{k-1}(i_{k-1} - i_k) - R_k(i_k - i_{k+1}) = L_k \frac{di_k}{dt}. \tag{7.16}$$

The difference equation in (7.16) is implemented for each ring with a small subcircuit as shown in Figure 7.9a. The subcircuits are joined together at the common resistors to obtain the complete ladder network in Figure 7.9b, shown for a four-step ladder. The ladder produces the resistance and partial self-inductance for a round wire. A physical model still requires the closure of the loop.

The ladder does not need to be very deep to capture all frequency dependence from DC to the breakdown of the lumped approximation. Depending on the closeness of fit, ladders with depths of three to five are sufficient. Accuracy and depth can be optimized through selection of the ring areas. For example, a constant resistance ratio R_k/R_{k+1} has advantages.[5]

5. C. H. Yen, Z. Fazarinc, and R. L. Wheeler, "Time-domain skin-effect model for transient analysis of lossy transmission lines," *Proc. IEEE*, vol. 70, no. 7, July 1982, pp. 750–757.

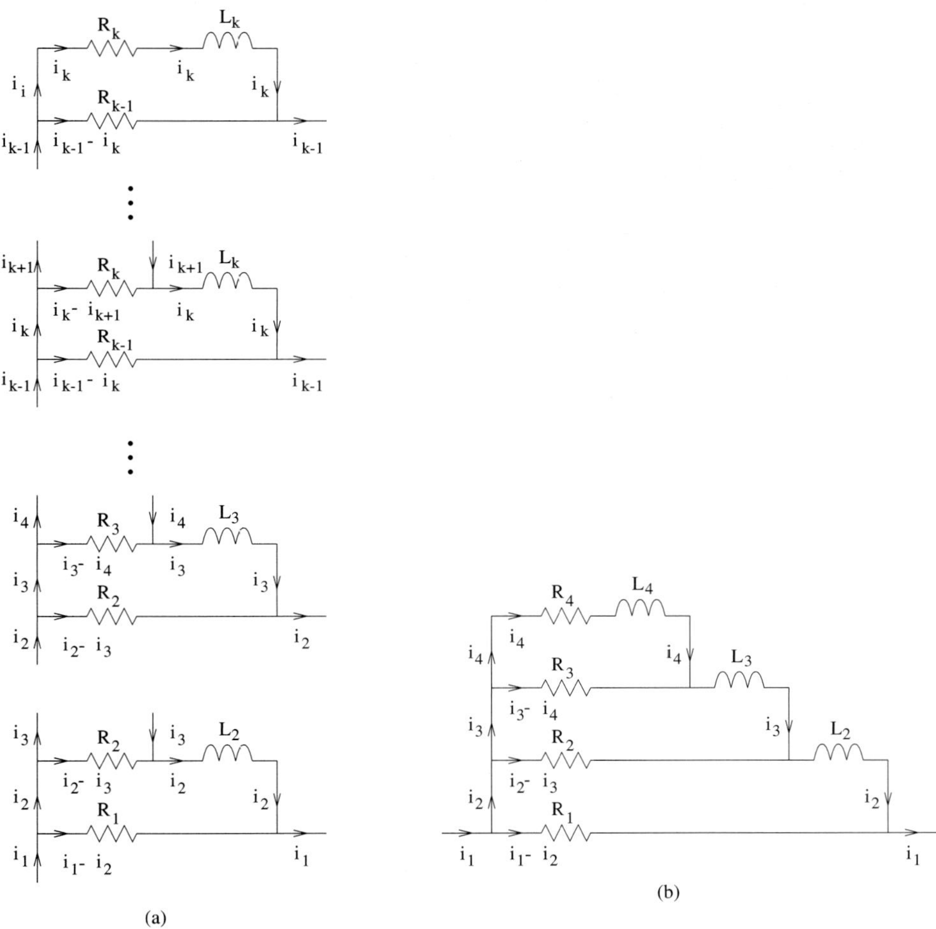

Figure 7.9. Ladder networks can compute the frequency-dependent inductance and resistance of round wires: (a) circuits representing each difference equation in (7.16), (b) complete ladder combining the subcircuits in (a) for a four-step ladder.

For general wire cross sections, a ladder can be fitted to existing data through optimization with a circuit simulator or through a continued fraction expansion from a rational function approximation. Rational function approximation is discussed in chapter 10, section 10.4.1, and an example of ladder construction with continued fractions is included.

7.5 Transresistance

Nonzero off-diagonal terms, called *transresistances*, often appear in resistance matrices, and they represent resistance in shared sections of interconnects, such as a ground plane serving as a common return path. For example, the two-port model in Figure 7.10 has the resistance matrix

$$\overline{\overline{R}} = \begin{bmatrix} R_1 + R_2 & R_2 \\ R_2 & R_2 + R_3 \end{bmatrix},$$

using the result from problem 14 in chapter 4.

SPICE Implementation

Given a resistance matrix with off-diagonal terms, it can be very difficult to synthesize a network to implement the shared resistance. Unfortunately, SPICE does not provide a transresistance element as it does the mutual inductor for implementing "transinductance." The transresistance element must be synthesized using a current-controlled voltage source.

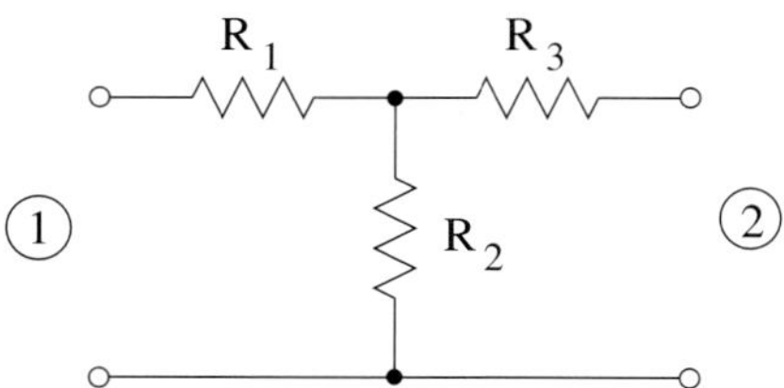

Figure 7.10. A two-port circuit with common resistance leading to off-diagonal entries in the resistance matrix.

The transresistance element relates the voltage on line m to the current on line n according to

$$v_m = R_{mn} i_n.$$

The current-controlled voltage source allows a direct implementation of this formula. The current is detected in line n using any component (although particular SPICE versions may require the use of a 0-volt voltage source), and then a voltage with weight $R_{mn} i_n$ is inserted into line m. For example, the 2×2 resistance matrix given by

$$\overline{\overline{R}} = \begin{bmatrix} R_{11} & R_{12} \\ R_{12} & R_{22} \end{bmatrix}$$

can be implemented in SPICE as

```
Ra n1 n2 R11
Rb n4 n5 R22
Ha n2 n3 Rb R12
Hb n5 n6 Ra R12.
```

The input and output terminals for lead 1 are `n1` and `n3`, while for lead 2 they are `n4` and `n6`.

7.6 Exercises

1. Using the PEEC formalism in (7.6), set up the skin effect problem in Figure 7.3.

2. Using the PEEC formalism in (7.6), set up the current crowding problem in Figure 7.4.

3. (a) Using the PEEC method, compute and plot the frequency-dependent resistance and inductance of the braided twin-lead transmission line in Figure 7.11. The braids consist of seven packed strands of 22-gauge copper ($\sigma = 5.8 \times 10^7$S/m) wire. Cover the frequency range from DC to the high-frequency asymptotic limit for the inductance.

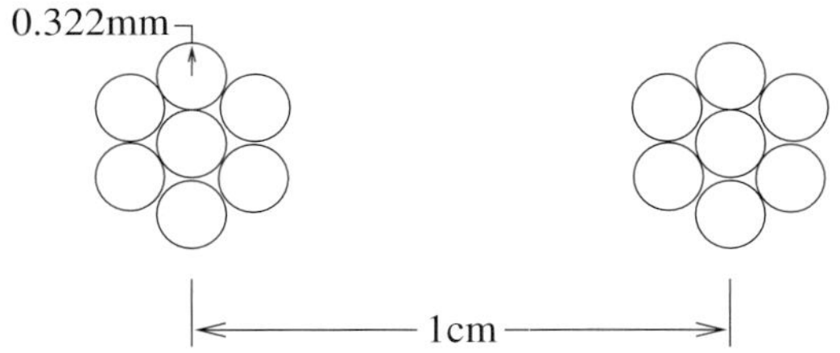

Figure 7.11. Twin-lead transmission line with braided conductors.

(b) For one of the braided leads, compute and plot the currents in each strand as a function of frequency for a total lead current of 1A.

(c) Assuming the transmission line is embedded in a homogeneous material with a dielectric constant of 4, estimate the capacitance of the transmission line and compute the frequency-dependent characteristic impedance and phase velocity.

(d) Discuss any need to apply the PEEC method to the individual strands of wire.

4. For a trace on a PCB using 1oz copper, what is the approximate maximum clock frequency for which skin effect losses are negligible? Assume that the edge rate is 20% of the clock period and that the trace width is much greater than its thickness. Let $\sigma_{\mathrm{Cu}} = 5.8 \times 10^7\mathrm{S/m}$, and 1oz copper is 35.6$\mu$m thick.

5. Using the technique of section 7.4, derive a ladder network to model the skin effect in a wide rectangular slab of conductor using a subdivision into horizontal bars.

6. A two-port network is described by the impedance matrix

$$\overline{\overline{Z}} = \begin{bmatrix} 10 & 5 \\ 5 & 8 \end{bmatrix} + \jmath\omega \begin{bmatrix} 3 \times 10^{-9} & 2 \times 10^{-9} \\ 2 \times 10^{-9} & 5 \times 10^{-9} \end{bmatrix}.$$

Implement this matrix in SPICE with and without current-controlled voltage sources. Using SPICE's AC capability, compute the network's 2×2 impedance matrix and show that the two implementations give the same result for a frequency sweep from DC to 10GHz.

Chapter 8

MEASUREMENT OF PARASITICS

Interconnections in digital systems are intentionally made to be very good to minimize degradation in signal quality as signals move from driver to receivers. As a result, the parasitic capacitances and inductances in the interconnects are quite small and require careful characterization if accurate values are to be obtained. For example, package lead inductances can easily range down to 1nH to 15nH with capacitances of 0.1pF to 2pF. For the reactance of a 1nH loop to reach a measurable 1Ω, the characterization frequency must be as high as 160MHz.

Interconnect parasitics are inconveniently configured in a mechanical sense. A wirebond cannot be plugged into an LCR meter to find its inductance. Special fixturing and/or probes are typically required along with careful consideration of the path for the return currents. The measured inductance is critically dependent on the area of the loop formed by the signal and return path. Similarly, capacitance is strongly affected by the distance to nearby conductors.

Typical equipment in digital laboratories includes source generators, oscilloscopes, logic analyzers, and low-frequency probes. The bandwidths on these instruments are geared towards digital needs and are typically too low for parasitic characterization.

Parasitic extraction usually requires dedicated instruments to achieve the required accuracy. The three main choices are the impedance analyzer (IA), vector network analyzer (VNA), and time-domain reflectometer (TDR). While a source generator and an oscilloscope can be used together to create a TDR, dedicated TDRs are available as a single instrument that achieves much higher performance. The IA and VNA produce swept frequency results, while the TDR's output is a time-domain waveform. Through the Fourier transform, time-domain and frequency-domain results are equivalent, so the choice of instrument is to a certain extent a matter of preference. However, bandwidth limitations and/or time windowing can limit the opportunities for conversion between domains.

8.1 Measurement Counts

Interconnects typically contain many terminals, so the number of measurements can become quite large. With a typical one- or two-port laboratory instrument, one measurement finds one unknown, so filling out an impedance, admittance, or scattering matrix requires as many measurements as there are matrix entries. For a nonreciprocal matrix with N terminals, the total number of required measurements is N^2. For a reciprocal matrix, $N(N-1)/2$ measurements are required for the off-diagonal terms, while N are required for the diagonal, so the total is $N(N+1)/2$. In some techniques, the diagonal terms are found automatically when measuring the off-diagonal terms, so the number of measurements for these cases is $N(N-1)/2$.

Because of the N^2 dependency for all cases, the number of required measurements can quickly become prohibitive. For example, a 100-lead QFP has 200 ports: 100 outside connections plus 100 inside of the package. A complete characterization to produce a 200×200 S-parameter (or impedance, or admittance) matrix requires 200(201)/2=20,100 measurements. An autoprober can help complete the measurements in reasonable time.

8.2 Impedance Analyzer

A schematic of an impedance analyzer (IA) is shown in Figure 8.1. A sinusoidal

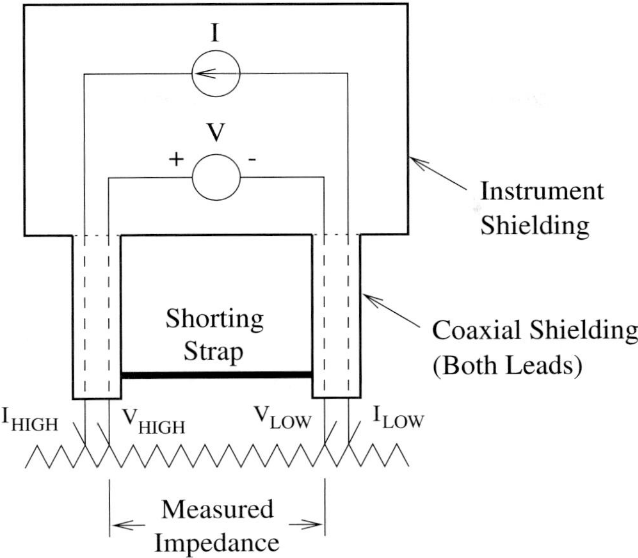

Figure 8.1. Schematic of an impedance analyzer.

constant-current source I at frequency ω drives current through the device under test (DUT), then a high-input-impedance vector voltmeter measures the voltage drop. A vector voltmeter measures both magnitude and phase at a given frequency, so the impedance is $Z_{\text{meas}} = V/I$. The parasitics are then $R = \text{Re}[Z_{\text{meas}}]$ and $L = \text{Im}[Z_{\text{meas}}]/\omega$, with similar expressions using admittance for cases involving conductance and capacitance.

One of the principal advantages of the IA is its use of a four-probe (or Kelvin probe) connection to the DUT to eliminate contact resistance. If the current probes poorly contact the DUT, the added series resistance does not affect the current because a constant-current source is used. If the voltage probes poorly contact the DUT, the extra series resistance adds insignificant voltage to the measured voltage because negligible current flows into the high-impedance voltmeter. Therefore, the measurement is unaffected by the quality of the contacts, and four-probe setups are essential for measuring small resistances.

The upper frequency limit on typical impedance analyzers falls in a range around 1MHz to 15MHz. For the very small parasitics often seen in packaging, sockets, and connectors, high characterization frequencies are required to obtain sufficient reactance that can be measured with good accuracy and repeatability. With its relatively low maximum frequency, the IA often must give way to higher-frequency instruments such as the VNA or TDR.

Characterization of the impedance matrix of a multiconductor interconnect with an IA requires two sets of measurements. The diagonal terms can be found by applying the two leads of the IA between port i and ground with all other ports left open-circuited. From chapter 4, equation (4.1), the voltage at port i is then

$$V_i = Z_{ii} I_i,$$

so the measured impedance is Z_{ii}. The diagonal terms of an $N \times N$ impedance matrix can be found by N one-port impedance measurements.

For the second set of measurements, the two leads of an IA are applied to ports i and j, with all other ports left open-circuited. Care must be taken not to accidentally ground the ports. The voltages at these two ports are

$$V_i = Z_{ii} I_i + Z_{ij} I_j$$

and

$$V_j = Z_{ji} I_i + Z_{jj} I_j.$$

Because the IA acts as a current source, $I_i = -I_j = I_{\text{meas}}$. Furthermore, the measured voltage is $V_{\text{meas}} = V_i - V_j$, leading to the conclusion that

$$V_{\text{meas}} = (Z_{ii} - 2Z_{ij} + Z_{jj}) I_{\text{meas}},$$

where reciprocity is used to set $Z_{ji} = Z_{ij}$. The IA measures the impedance $Z_{\text{meas}} = V_{\text{meas}}/I_{\text{meas}}$, so the off-diagonal impedance matrix element can be found as

$$Z_{ij} = \frac{1}{2}\left(Z_{ii} + Z_{jj} - Z_{\text{meas}}\right).$$

Since Z_{ii} and Z_{jj} are known from the first set of diagonal measurements, the off-diagonal elements can be found at the rate of one per measurement.

If an admittance matrix is required, an impedance matrix can be measured and then inverted. Experimentally, open circuits are far easier to deal with than are the short circuits that are required when directly measuring multiconductor admittance matrices.

8.3 Vector Network Analyzer

As frequency increases, the concept of voltage breaks down as uniqueness is lost. Another concept must be used for characterization of circuits, and traveling waves work quite well. The vector network analyzer (VNA) is a multiport instrument operating on the principle of directional traveling waves.

The sinusoidal voltage on a transmission line is a superposition of traveling waves moving in opposite directions. From chapter 2, equation (2.27), the voltage is

$$V(z,\omega) = V^{+}(\omega)e^{-\gamma(\omega)z} + V^{-}(\omega)e^{+\gamma(\omega)z}.$$

Similarly from (2.28), the current on the line is

$$I(z,\omega) = \frac{1}{Z_o(\omega)}\left[V^{+}(\omega)e^{-\gamma(\omega)z} - V^{-}(\omega)e^{+\gamma(\omega)z}\right].$$

Consider a single-port VNA with an applied load Z_L at $z = 0$, then

$$\begin{aligned} Z_L(\omega) &= \frac{V(0,\omega)}{I(0,\omega)} \\ &= Z_o(\omega)\frac{V^{+}(\omega) + V^{-}(\omega)}{V^{+}(\omega) - V^{-}(\omega)} \\ &= Z_o(\omega)\frac{1+\Gamma(\omega)}{1-\Gamma(\omega)}, \end{aligned} \tag{8.1}$$

where $\Gamma(\omega) = V^{-}(\omega)/V^{+}(\omega)$ is the frequency-dependent reflection coefficient at the load. Therefore, the load impedance can be determined if the incident and reflected wave voltage amplitudes can be found. At sufficiently high frequencies, the directional waves can be separated with a directional coupler. The performance of directional couplers strongly impacts the lower frequency limit and total bandwidth of network analyzers.

The definition of the reflection coefficient relates the reflected wave to the incident wave as $V^- = \Gamma V^+$. For a two-port instrument, this can be generalized to include both ports as

$$\begin{bmatrix} V_1^- \\ V_2^- \end{bmatrix} = \begin{bmatrix} S_{11} & S_{12} \\ S_{21} & S_{22} \end{bmatrix} \begin{bmatrix} V_1^+ \\ V_2^+ \end{bmatrix}. \tag{8.2}$$

The reflected waves are written in terms of the incident waves using the scattering matrix.

S-parameters can be measured two at a time with a two-port vector network analyzer. V_2^+ can be forced to zero by matching port 2 internally to the VNA, then (8.2) simplifies to

$$V_1^- = S_{11} V_1^+$$

and

$$V_2^- = S_{21} V_1^+.$$

By measuring V_1^-, V_2^- and V_1^+, the S-parameters S_{11} and S_{21} can be found. A simple schematic of a VNA is shown in Figure 8.2, where directional couplers separate the directional waves. Repeating the measurement by driving port 2 while matching port 1 yields the remaining two S-parameters. The VNA may have internal switches to accomplish the alternate driving and matching of ports, but if not, the DUT can simply be reversed.

Scalar network analyzers do not detect the phase of the S-parameters and are useful for scalar measurements such as the gain of an amplifier. By avoiding phase measurements, a substantially less expensive instrument can be produced. For parasitic measurement, it is necessary to measure phase to separate the real and imaginary parts of impedance (to obtain R and L) and admittance (to obtain G and C), so scalar network analyzers are not suited to parasitic extraction.

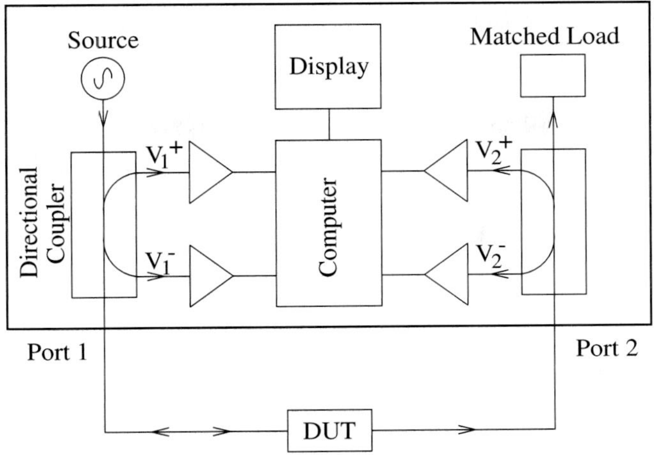

Figure 8.2. Schematic of a two-port VNA configured to measure S_{11} and S_{21}.

The VNA measures all quantities with respect to its built-in system impedance, Z_o, which is typically 50Ω. When the magnitudes of the S-parameters stray too far from the system impedance, through low reflection or coupling, the VNA cannot accurately resolve the quantities. How far is too far? It depends on the care taken in setup and calibration. Reasonable efforts should yield accurate measurements to about $Z_o/1000$ to $Z_o * 1000$, and very careful setups could yield substantially more. Note that this range exceeds typical VNA guaranteed specifications by a wide margin.

The advantages of the VNA include high-frequency operation to bring the reactances of small parasitics into a measurable range, precision calibration with explicit control of ground reference points, and high signal-to-noise ratio (60-100dB).

The principal disadvantages of the VNA are two-point contact, so contact resistance limits the ability to measure small resistances, and the need for postprocessing of S-parameter data to yield circuit models consisting of resistors, inductors, capacitors, and transmission lines.

Coupling terms are measured with the IA by bridging two DUT terminals with the single port of the instrument. This technique is not possible with the VNA because the ground lead of the probe would short a terminal to ground. Measurements of coupling terms with the VNA requires two ports on the instrument to enable the separate probing of each terminal. Fortunately, two-port VNAs are readily constructed. Note that this discussion applies in equal measure to the time-domain reflectometer, which is discussed in section 8.4.

8.3.1 Calibration

Calibration plays a key role in the use of the VNA and is required at every use to account for variations in phase and power. Every new setup requires recalibration so that phase changes induced by cables and fixtures can be subtracted out. This situation is in sharp contrast to other instruments, such as voltmeters, oscilloscopes, impedance analyzers, and time-domain reflectometers, which require only periodic calibration by a calibration service. Success with a VNA requires careful and proper calibration before every use and as changing setups and conditions warrant.

There are a wide variety of calibration methods, including SOLT (short-open-load-through), TRL (through-reflect-load), and LRM (load-reflect-match). The acronym hints at the standards that must be measured to achieve calibration. Depending on the DUT, some calibration methods yield higher accuracy than others, and while the SOLT method is generally considered to have the lowest accuracy, its accuracy is still quite adequate for interconnect characterization. Without extensive efforts, the SOLT is also the easiest to apply. A brief explanation of the SOLT method follows.

Calibration for a one-port measurement requires the measurement of three standards: a short circuit, an open circuit, and a matched load. The quality of the standards and their applications directly determines the accuracy of the measurement. All standards have some parasitic values associated with them, and network analyzers allow the incorporation of the values, if known, to improve accuracy. Vendors of standards will typically provide the parasitic values, and it is important to use the correct set for a given set of standards.

Calibration for a two-port measurement is a two-step process. First, each port undergoes a one-port calibration. Second, the two ports are connected together with a through connection to complete the calibration. Additional calibration for leakage between the ports is also possible but not generally needed for interconnect work.

After calibration, the S-parameters produced by a VNA reference the points in the setup at which the calibration standards were applied. Any additional path or discontinuity to the DUT must be removed by other means. For example, calibration with coaxial standards calibrate the VNA up to the coaxial connectors, while calibration with a reference probe card calibrates up to the probe tips.

8.3.2 Single Lumped Parasitic Extraction

Capacitance and inductance are straightforward quantities to measure with a VNA, although a small amount of numerical processing is required. To measure series resistance and inductance, the DUT is shorted to ground at one end and measured at the other. Assuming that port 1 is used on a VNA, then from (8.1) the impedance is

$$R + \jmath\omega L = Z_o \frac{1 + S_{11}}{1 - S_{11}},$$

where Z_o is the system impedance of the VNA. Equating real and imaginary parts yields R and L. A similar process yields G and C, using admittance and an open-circuit sample.

For the measurement of small inductances, a lower frequency limit is implied by the magnitude of the reactance compared to Z_o. Assuming that the system impedance is 50Ω and that the VNA is accurate down to $50\Omega/1000 = 0.05\Omega$, then an inductance of 0.1nH can be accurately resolved for frequencies above 80MHz (neglecting losses). Better resolution requires higher frequencies, but an upper frequency limit exists where the lumped approximation breaks down. Inductance decreases with frequency due to the skin effect and current crowding, and the appearance of an increasing inductance signals the breakdown of the lumped approximation.

8.3.3 Measurements of Multiport Networks with VNAs

The following three sections provide detail on measuring Z-, Y-, and S-parameter matrices using a two-port VNA. Two formulations are possible—the diagonal terms can be measured separately or in conjunction with the off-diagonal terms. Separately measuring the diagonal terms allows confirmation of good contact of the first port before landing the second to measure the off-diagonal term.

Impedance Matrix

The impedance matrix relates current to voltage according to $\overline{V} = \overline{\overline{Z}}\,\overline{I}$, which can be expanded as

$$\begin{bmatrix} V_1 \\ V_2 \\ \vdots \\ V_N \end{bmatrix} = \begin{bmatrix} Z_{11} & Z_{12} & \cdots & Z_{1N} \\ Z_{21} & Z_{22} & \cdots & Z_{2N} \\ \vdots & \vdots & \ddots & \vdots \\ Z_{N1} & Z_{N2} & \cdots & Z_{NN} \end{bmatrix} \begin{bmatrix} I_1 \\ I_2 \\ \vdots \\ I_N \end{bmatrix}. \tag{8.3}$$

For a one-port measurement, if port i is driven with the VNA and all other ports are left unconnected, then $I_j = 0\ \forall j \neq i$, and the voltage measured at port i is given simply as $V_i = Z_{ii} I_i$. Assuming that port 1 of the VNA is performing the measurement and that its impedance is Z_o, then from (4.6) and (4.7)

$$\sqrt{Z_o}(a_1 + b_1) = Z_{ii}\frac{1}{\sqrt{Z_o}}(a_1 - b_1).$$

Since $S_{11} = \frac{b_1}{a_1}$, then

$$Z_{ii} = Z_o \frac{1 + S_{11}}{1 - S_{11}}. \tag{8.4}$$

For a two-port measurement, assume that port 1 of the VNA drives port i of the N-port network, that port 2 drives port j, and that both ports have a characteristic impedance Z_o. When all other ports are open-circuited, their currents vanish, and from (8.3)

$$\begin{aligned} V_1 &= Z_{ii} I_1 + Z_{ij} I_2 \\ V_2 &= Z_{ji} I_1 + Z_{jj} I_2. \end{aligned}$$

Applying chapter 4, equation (4.8) to equate this 2×2 impedance matrix to the measured two-port S-parameters yields

$$
\begin{aligned}
Z_{ii} &= Z_o \frac{(1+S_{11})(1-S_{22})+S_{12}S_{21}}{(1-S_{11})(1-S_{22})-S_{12}S_{21}} \\
Z_{ij} &= Z_o \frac{2S_{12}}{(1-S_{11})(1-S_{22})-S_{12}S_{21}} \\
Z_{ji} &= Z_o \frac{2S_{21}}{(1-S_{11})(1-S_{22})-S_{12}S_{21}} \\
Z_{jj} &= Z_o \frac{(1-S_{11})(1+S_{22})+S_{12}S_{21}}{(1-S_{11})(1-S_{22})-S_{12}S_{21}}.
\end{aligned} \tag{8.5}
$$

Equation (8.5) can be used alone, so that one two-port measurement provides all four impedance matrix entries, or just the off-diagonal terms can be used if the diagonal self terms have already been determined from one-port measurements using (8.4).

Impedance matrices are very convenient to measure because the unused ports remain open-circuited. As the two ports of the VNA are moved to map out the full matrix in a pair-wise sequence, nothing needs to be done to the unused ports before the measurement is taken. The technique works as long as the small capacitance and radiation at the open-circuited ports are negligible. Typically, the technique will be adequate for lower frequencies where data for lumped modeling is taken.

Y-Parameters

The admittance matrix relates voltage to current according to $\overline{I} = \overline{\overline{Y}}\,\overline{V}$, which can be expanded as

$$
\begin{bmatrix} I_1 \\ I_2 \\ \vdots \\ I_N \end{bmatrix} = \begin{bmatrix} Y_{11} & Y_{12} & \cdots & Y_{1N} \\ Y_{21} & Y_{22} & \cdots & Y_{2N} \\ \vdots & \vdots & \ddots & \vdots \\ Y_{N1} & Y_{N2} & \cdots & Y_{NN} \end{bmatrix} \begin{bmatrix} V_1 \\ V_2 \\ \vdots \\ V_N \end{bmatrix}. \tag{8.6}
$$

For a one-port measurement, if port i is driven with the VNA and all other ports are shorted to ground, then $V_j = 0\ \forall j \neq i$, and the current measured at port

i is given simply as $I_i = Y_{ii}V_i$. Assuming that port 1 of the VNA is performing the measurement and that its impedance is Z_o, then from chapter 4, equations (4.6) and (4.7)

$$\frac{1}{\sqrt{Z_o}}(a_1 - b_1) = Y_{ii}\sqrt{Z_o}(a_1 + b_1).$$

Since $S_{11} = \frac{b_1}{a_1}$, then

$$Y_{ii} = Y_o \frac{1 - S_{11}}{1 + S_{11}},$$

where $Y_o = 1/Z_o$ is the system admittance of the VNA.

For a two-port measurement, assume that port 1 of the VNA drives port i of the N-port network, that port 2 drives port j, and that both ports have a characteristic impedance of Z_o. When all other ports are shorted to ground, their voltages vanish and from (8.6)

$$\begin{aligned} I_1 &= Y_{ii}V_1 + Y_{ij}V_2 \\ I_2 &= Y_{ji}V_1 + Y_{jj}V_2. \end{aligned}$$

Applying (4.8), converted for admittance, to equate this 2×2 admittance matrix to the measured two-port S-parameters yields

$$\begin{aligned} Y_{ii} &= Y_o \frac{(1 - S_{11})(1 + S_{22}) + S_{12}S_{21}}{(1 + S_{11})(1 + S_{22}) - S_{12}S_{21}} \\ Y_{ij} &= Y_o \frac{-2S_{12}}{(1 + S_{11})(1 + S_{22}) - S_{12}S_{21}} \\ Y_{ji} &= Y_o \frac{-2S_{21}}{(1 + S_{11})(1 + S_{22}) - S_{12}S_{21}} \\ Y_{jj} &= Y_o \frac{(1 + S_{11})(1 - S_{22}) + S_{12}S_{21}}{(1 + S_{11})(1 + S_{22}) - S_{12}S_{21}}. \end{aligned}$$

The technique is valid while the inductance and resistance of the short circuits are negligible compared to the values within the network. However, the requirement that all ports of the N-port network not connected to the VNA must be shorted to ground presents practical drawbacks when implemented experimentally. As the VNA measurement is moved from one port to another on the network, the short circuits must be repeatedly made and then removed, a process that can be very time-consuming and can potentially create a wear-out problem if large numbers of ports are to be characterized.

S-Parameters

Measurement of S-parameters is straightforward mathematically when using a VNA because the measured quantities are already S-parameters. However, the measurement of multiport S-parameters, two ports at a time, requires the tedious mounting and removal of matched terminations at the extra ports. The S-matrix relates incident and reflected traveling waves according to $\overline{b} = \overline{\overline{S}}\,\overline{a}$, which can be expanded as

$$\begin{bmatrix} b_1 \\ b_2 \\ \vdots \\ b_N \end{bmatrix} = \begin{bmatrix} S_{11} & S_{12} & \cdots & S_{1N} \\ S_{21} & S_{22} & \cdots & S_{2N} \\ \vdots & \vdots & \ddots & \vdots \\ S_{N1} & S_{N2} & \cdots & S_{NN} \end{bmatrix} \begin{bmatrix} a_1 \\ a_2 \\ \vdots \\ a_N \end{bmatrix}.$$

For a one-port measurement, if port i is driven with the VNA and all other ports are matched in the system impedance of the VNA, then $a_j = 0\ \forall j \neq i$, and VNA directly reports $S_{ii} = b_1/a_1$.

For a two-port measurement, assume that port 1 of the VNA drives port i of the N-port network, that port 2 drives port j, and that both ports have a characteristic impedance of Z_o. If all remaining are matched in the system impedance of the VNA, then $a_k = 0\ \ \forall k \neq i, j$ and

$$\begin{aligned} b_1 &= S_{ii}a_1 + S_{ij}a_2 \\ b_2 &= S_{ji}a_1 + S_{jj}a_2. \end{aligned} \tag{8.7}$$

The VNA produces these results directly.

As the measurement ports on the VNA are moved to map out the full S-matrix, the matched loads must also be moved. Practical issues involving the load size, wear-out from repeated mounting and demounting loads, and time make the measurement difficult for large numbers of ports.

8.3.4 Conversions between Types

Multiport circuits can be described with impedance, admittance, or scattering matrices (and others as well, such as hybrid, transmission, and ABCD parameters). Conversions between the types are mathematically exact, so in principle, one is not preferred over another. For example, multiport impedance matrices are easiest to measure because unused ports are terminated in open circuits, so it is tempting to measure $\overline{\overline{Z}}$ and then to derive $\overline{\overline{Y}} = \overline{\overline{Z}}^{-1}$ or $\overline{\overline{S}}$ with (4.9). However, finite precision in real data can prevent some conversions due to error accumulation.

Consider the conversion from $\overline{\overline{S}}$ to $\overline{\overline{Z}}$. An N-conductor interconnect plus a port numbering scheme is depicted in Figure 8.3. At low frequencies, the S-parameters tend towards

$$\overline{\overline{S}} = \begin{bmatrix} 0 & +1 & 0 & 0 & \cdots \\ +1 & 0 & 0 & 0 & \cdots \\ 0 & 0 & 0 & +1 & \cdots \\ 0 & 0 & +1 & 0 & \cdots \\ \vdots & \vdots & \vdots & \vdots & \ddots \end{bmatrix}.$$

In converting from $\overline{\overline{S}}$ to $\overline{\overline{Z}}$ using (4.14), the critical term is $(\overline{\overline{I}} - \overline{\overline{S}})^{-1}$, where the inverse must exist before the conversion can be performed. Checking $\overline{\overline{I}} - \overline{\overline{S}}$ yields

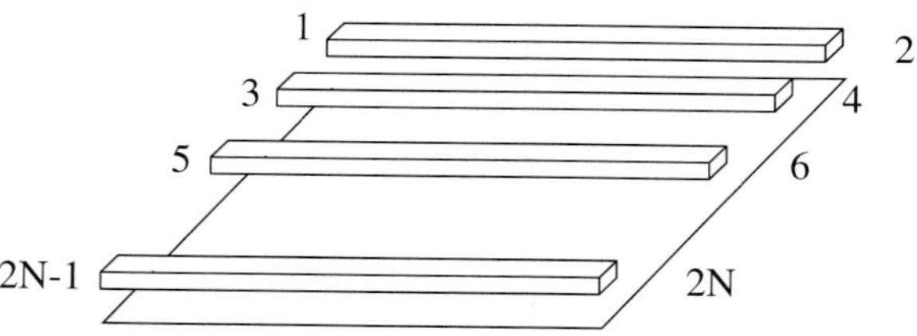

Figure 8.3. N-conductor interconnect with $2N$ ports.

$$\overline{\overline{I}} - \overline{\overline{S}} = \begin{bmatrix} +1 & -1 & 0 & 0 & \cdots \\ -1 & +1 & 0 & 0 & \cdots \\ 0 & 0 & +1 & -1 & \cdots \\ 0 & 0 & -1 & + & \cdots \\ \vdots & \vdots & \vdots & \vdots & \ddots \end{bmatrix},$$

but the inverse of this matrix does not exist. Similar problems arise if the target is $\overline{\overline{Y}}$ instead of $\overline{\overline{Z}}$. Therefore, small errors in $\overline{\overline{S}}$ can lead to large errors in $\overline{\overline{Z}}$ or $\overline{\overline{Y}}$.

8.3.5 Smith Chart

The Smith chart is a tool developed by microwave engineers to assist with calculations involving transmission lines, and much insight can be gained through its use. A rudimentary knowledge of the Smith chart is quite useful since it is used as a major display mode for VNAs. When calibrating and using a VNA, the Smith chart provides immediate feedback as to whether the DUT is inductive, capacitive, lossy, open, or short. Good quality measurements can be facilitated by monitoring the Smith chart display.

A printed Smith chart looks quite complex as a result of its use in pre-computer days as a functional computation device, but it is in fact a simple plot of the reflection coefficient Γ on the complex plane. A grid of Z_L values computed with (8.1) is superimposed onto the plot, enabling the direct graphical conversion between Γ and Z_L.

A complex plane is shown in Figure 8.4a with a superimposed circle of radius 1. Since $|\Gamma| > 1$ implies an active circuit (more power propagates from the load than towards it), the unit circle indicates a perfectly reflective, lossless, and passive load. All points inside the circle indicate passive, lossy, and imperfect reflections with some power imparted to the load.

In Figure 8.4b, a very simple Smith chart is shown, annotated with important special cases. The reflection coefficient for a lossy inductive load and real characteristic impedance is

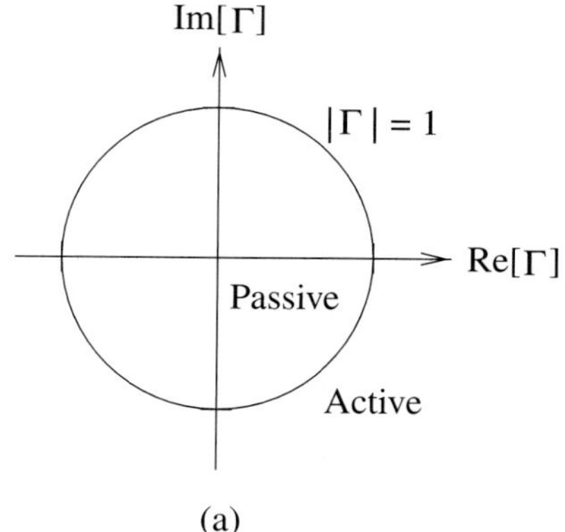

(a)

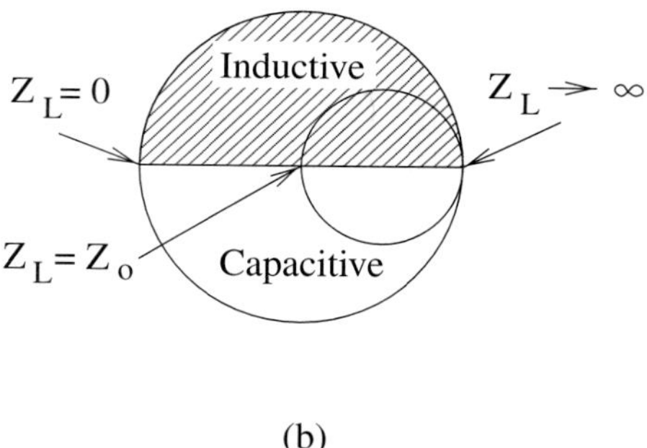

(b)

Figure 8.4. Reflection coefficients can be plotted on a complex plane: (a) annotated with rectangular coordinates, (b) overlayed with impedance to form a Smith chart.

$$
\begin{aligned}
\Gamma &= \frac{Z_L - Z_o}{Z_L + Z_o} \\
&= \frac{R + \jmath\omega L - Z_o}{R + \jmath\omega L + Z_o} \\
&= \frac{(R - Z_o)(R + Z_o) + (\omega L)^2}{(R + Z_o)^2 + (\omega L)^2} + \jmath\frac{2\omega L Z_o}{(R + Z_o)^2 + (\omega L)^2}.
\end{aligned}
$$

With a positive imaginary part and a real part that can be positive or negative, inductive loads always appear in the top half of the Smith chart. Conversely,

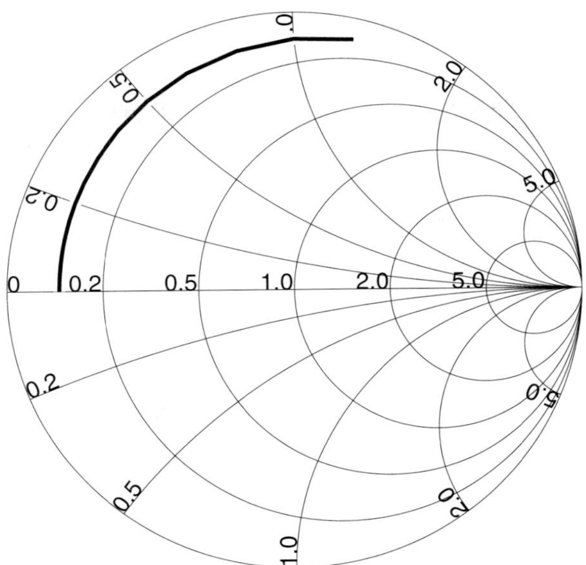

Figure 8.5. Impedance of a 5Ω resistor in series with a 10nH inductor plotted on a Smith chart, with $Z_o = 50\Omega$ over the frequency range from 100kHz to 1GHz.

capacitive loads always appear in the bottom half. For a purely resistive load, the reflection coefficient is always on the real axis.

As an example, the impedance of a load consisting of a 5Ω resistor in series with a 10nH inductor is plotted on a Smith chart in Figure 8.5 for frequencies ranging from 100kHz to 1GHz. The chart is overlayed with impedance values normalized to 50Ω. At low frequencies, the reactance of the inductor is negligible, so the plot starts on the real axis at $\Gamma = (5-50)/(5+50) = -0.8182$ or, equivalently, $Z_L/Z_o = 5/50 = 0.1$. As frequency increases, inductive reactance increases, pushing the trace farther into the upper half of the Smith chart.

For a resonant circuit, the trace on the Smith chart will cross over from one half to the other, as demonstrated in chapter 6, Figure 6.6 for a capacitor in parallel with a series resistor and inductor; at low frequencies the circuit looks inductive at its input terminals, but above 1.54GHz the circuit looks capacitive.

8.4 Time-Domain Reflectometer

The basic use of a time domain-reflectometer (TDR) is to isolate faults in long runs of inaccessible cable. The classic example is a buried phone line that has developed a short or open circuit. The location of the fault can be found by injecting a single voltage pulse on the line and capturing the reflection on an oscilloscope. The injected pulse propagates down the line, reflects off the discontinuity, and then returns to form a pulse on the oscilloscope. The flight time can then be measured from the trace, and knowing the propagation speed on the cable, the spot to dig can be computed.

A TDR can be constructed from a pulse generator and an oscilloscope. Purpose-built instruments offer a single, easier-to-use, and generally higher-performance package. A simple schematic of a TDR is shown in Figure 8.6. The signal generator injects a source waveform v_s with a source impedance matched to the system impedance to absorb any returning reflections. A high input-impedance oscilloscope measures the return voltage v_r (plus the injected waveform on its way out). The key to using a TDR is in interpreting the reflection result.

Assume that v_s is the unit step function and that the transmission line has a delay of τ. After the step is injected, it takes until time 2τ before any reflection can return to be measured at v_r. During this time the voltage evenly divides on the source impedance and transmission line impedance, so the measured voltage at v_r is 0.5V. After 2τ has elapsed, the DUT can begin to be distinguished. An example simulated TDR trace is shown in Figure 8.7 for a 10nH load. When the initial edge

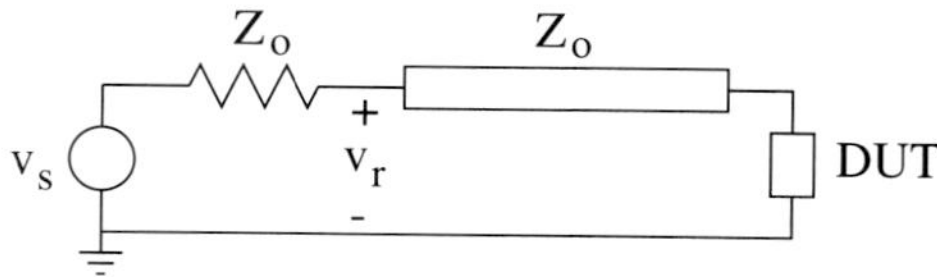

Figure 8.6. Simple schematic of a time-domain reflectometer (TDR).

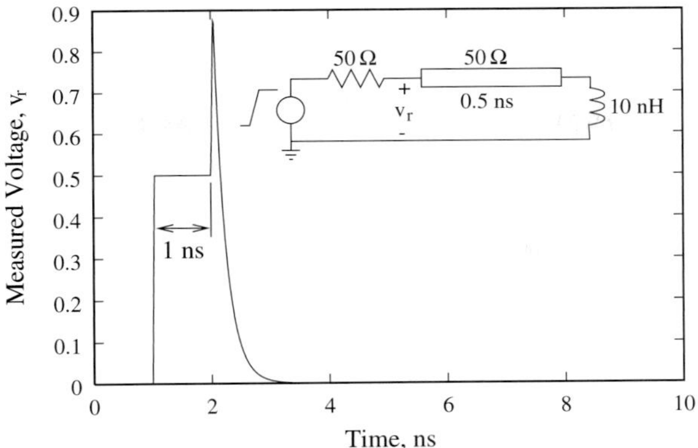

Figure 8.7. TDR trace using the simple model of Figure 8.6.

hits the inductor, it appears as an open circuit. Voltage doubles on an open circuit, so the trace moves up towards 1V. The inductor is a short to the low-frequency components, so as time advances, the inductor looks more and more like a short, so the voltage moves back towards zero. Note the 1ns delay before the TDR "sees" the inductor, due to the 0.5ns delay of the transmission line.

The advantages of the TDR include intuitive (for many) operation in the time-domain, multiple numbers of ports, and extraction of impedance profiles. The disadvantages include low signal-to-noise ratio ($\sim$ 20dB) compared to the VNA, post-processing for parasitic extraction (although the computation is built into many TDR instruments), and shadowing of impedance profile information behind large reflection coefficients.

8.4.1 Inductance Extraction

A basic setup for the extraction of inductance and mutual inductance using a TDR with two channels is shown in Figure 8.8. Port 1 drives line 1, while port 2 senses the voltage generated on line 2. All inductance associated with cables, connectors, and fixtures are absorbed into L_1 and L_2. The loop equations describing this setup

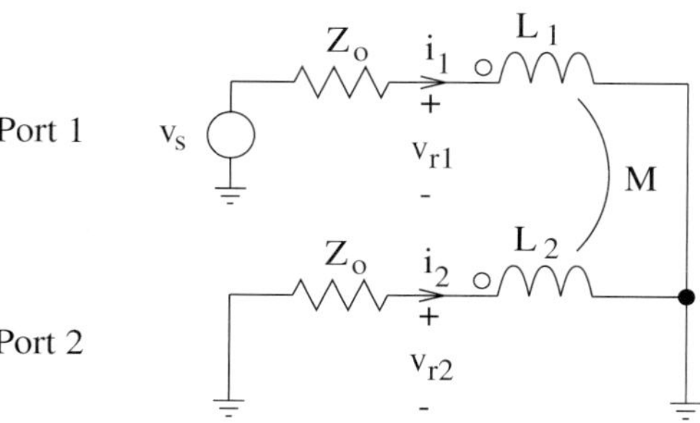

Figure 8.8. TDR setup for inductance and mutual inductance extraction.

are

$$v_{r1} = L_1 \frac{di_1}{dt} + M \frac{di_2}{dt} \tag{8.8}$$

and

$$v_{r2} = M \frac{di_1}{dt} + L_2 \frac{di_2}{dt}. \tag{8.9}$$

To find the inductance of line 1, integrate (8.8) to obtain

$$\int_{-\infty}^{t} v_{r1}(\tau) d\tau = L_1 i_1(t) + M i_2(t). \tag{8.10}$$

Choose t large enough so that all waveforms are completely settled, then DC conditions prevail and $i_1(t) = v_o/Z_o$, where v_o is the step voltage of the TDR channel, and $i_2(t) = 0$. Under these assumptions, (8.10) can be solved for L_1 giving

$$L_1 = \frac{Z_o}{v_o} \int_{-\infty}^{t} v_{r1}(\tau) d\tau, \tag{8.11}$$

for large t. The inductance of line 2 can similarly be found by driving line 2 rather than line 1.

To find the mutual inductance between lines 1 and 2, integrate (8.9) to obtain

$$\int_{-\infty}^{t} v_{r2}(\tau) d\tau = M i_1(t) + L_2 i_2(t).$$

For large t, DC conditions prevail, and $i_1(t) = v_o/Z_o$ while $i_2(t) = 0$, so solving for M yields

$$M = \frac{Z_o}{v_o} \int_{-\infty}^{t} v_{r2}(\tau) d\tau,$$

for large t.

To remove the inductances of cables, connectors, and fixtures, a set of measurements are made with a good short-circuit in place of the DUT. These inductances are subtracted from those found with the DUT in place to isolate the inductances of the DUT alone. Combining the two measurements into a single pair of formulas yields

$$L_1 = \frac{Z_o}{v_o} \int_{-\infty}^{t} \left(v_{r1}(\tau) - v_{r1_{\text{short}}}(\tau)\right) d\tau \tag{8.12}$$

and

$$M = \frac{Z_o}{v_o} \int_{-\infty}^{t} \left(v_{r2}(\tau) - v_{r2_{\text{short}}}(\tau)\right) d\tau, \tag{8.13}$$

where the subscript "short" means the measurement with the good short-circuit in place.

8.4.2 Capacitance Extraction

A basic setup for the extraction of capacitance and mutual capacitance using a TDR with two channels is shown in Figure 8.9. Port 1 drives line 1, while port 2 senses the voltage generated on line 2. All capacitance associated with cables, connectors, and fixtures are absorbed into C_1, C_2 and C_M. The nodal equations describing this setup are

$$\frac{v_{r1}}{Z_o} - \frac{v_s}{Z_o} + (C_1 + C_M)\frac{dv_{r1}}{dt} - C_M \frac{dv_{r2}}{dt} = 0 \tag{8.14}$$

and

$$\frac{v_{r2}}{Z_o} + (C_2 + C_M)\frac{dv_{r2}}{dt} - C_M \frac{dv_{r1}}{dt} = 0. \tag{8.15}$$

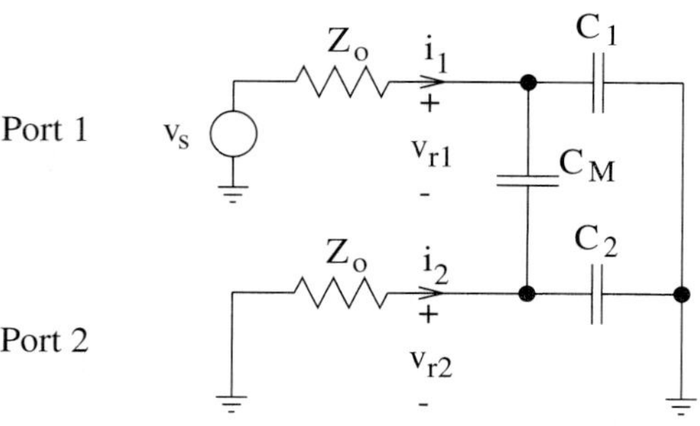

Figure 8.9. TDR setup for capacitance and mutual capacitance extraction.

Integrating (8.14) yields

$$\frac{1}{Z_o}\int_{-\infty}^{t}(v_{r1}(\tau)-v_s(\tau))\,d\tau+(C_1+C_M)\,v_{r1}(t)-C_M v_{r2}(t)=0. \tag{8.16}$$

For large t, such that the waveforms on both channels are settled, DC conditions prevail and $v_{r1}(t)=v_o$ (the step voltge of the TDR) and $v_{r2}(t)=0$. Under these conditions, (8.16) becomes

$$C_1+C_M=\frac{1}{Z_o v_o}\int_{-\infty}^{t}(v_s(\tau)-v_{r1}(\tau))\,d\tau. \tag{8.17}$$

Similarly, integrating (8.15) then evaluating for large t yields

$$C_M=\frac{1}{Z_o v_o}\int_{-\infty}^{t}v_{r2}(\tau)d\tau.$$

So the mutual capacitance is found first, then the capacitance to ground can be found. Note that, in general, for multiport interconnects, all mutual capacitance terms must be characterized before the capacitances to ground can be evaluated.

To remove the capacitances of cables, connectors, and fixtures, a set of open-circuit measurements are made with the DUT removed. These capacitances are subtracted from those found with the DUT in place to isolate the capacitances of the DUT alone. Combining the two measurements into a single pair of formulas yields

$$C_1+C_M=\frac{1}{Z_o v_o}\int_{-\infty}^{t}\left(v_{r1_{\text{open}}}(\tau)-v_{r1}(\tau)\right)d\tau$$

and

$$C_M = \frac{1}{Z_o v_o} \int_{-\infty}^{t} \left(v_{r2}(\tau) - v_{r2_{\mathrm{open}}}(\tau)\right) d\tau.$$

Note how the detailed time profile of the source, $v_s(t)$, drops out since it is assumed to be the same in both measurements.

8.4.3 Impedance Profile

A TDR can provide estimates of the characteristic impedances and propagation delays of cascades of transmission lines. Assume that the transmission line connecting the DUT to the TDR has zero length. (Its actual profile can be extracted along with the transmission lines in the DUT.) If the DUT presents an impedance Z_L, then

$$v_r = v_s \frac{Z_L}{Z_o + Z_L},$$

and solving for Z_L yields

$$Z_L = Z_o \frac{v_r}{v_s - v_r}. \tag{8.18}$$

The impedance profile is then found by computing Z_L from v_r at each time point and plotting Z_L vs. time.

v_s is needed and can be found in a number of ways. For example, for no load or transmission line, then $v_s = v_r$, or with a very long, matched transmission line, then $v_s = 2v_r$. Practical issues may favor one technique over another. Alternatively, v_s can be approximated by v_o, the step voltage magnitude, which approximates the open circuit voltage.

The simulation in Figure 8.10 demonstrates impedance profile extraction using $v_s = v_o$. The transmission lines are accurately characterized in terms of impedance and delay (note the factor of 2 involved due to the round-trip nature of TDR measurements); however, multiple reflections exist and create the impression that more structure exists after 4ns than actually exists in reality.

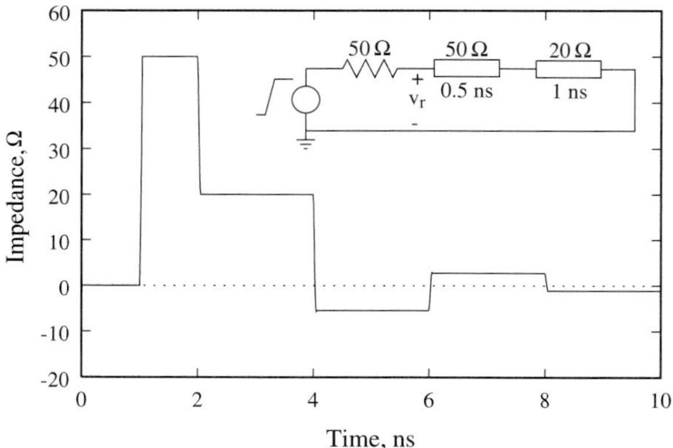

Figure 8.10. TDR trace using the simple model of Figure 8.6 and the impedance profile extraction formula (8.18).

Each change in characteristic impedance causes the TDR trace to bump up or down to a new impedance level. Increasing impedance implies increased inductance, decreased capacitance, or both. Conversely, decreasing impedance implies decreased inductance, increased capacitance, or both. For parasitic extraction, such as the discontinuity presented by an imperfect coaxial connector, bumps up imply an inductive parasitic while bumps down imply a capacitive parasitic. These concepts are summarized in Figure 8.11. Note that these relationships are derived in chapter 2, sections 2.4.3 and 2.4.4.

8.4.4 Layer Peeling

For complex structures, multiple reflections can make it difficult to identify the correct impedances and delays in an impedance profile. Transmission line theory can be used to derive a postprocessing extraction algorithm that removes the multiple reflections. Of the many techniques that have been developed, the one presented here is called *layer peeling.*

A lossless interconnect is assumed to consist of a cascade of equal-delay transmission lines with unknown characteristic impedances as represented in Figure 8.12a.

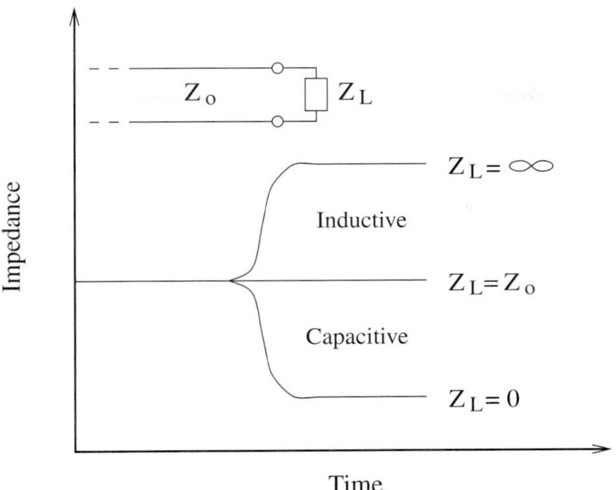

Figure 8.11. Basic relationships for impedance discontinuities in TDR traces.

A TDR is used to measure the reflection coefficient at time points 0, τ, 2τ, ..., $N\tau$. Since TDRs discern structures according to their round trip times, the delay of each section of transmission line is then $\tau/2$. Because samples are only available out to time $N\tau$, just N sections of transmission lines can be characterized because for those farther out, the signal does not have time to return to the TDR.

It is shown in chapter 2, equation (2.16) that the total voltage can be written as the sum of a forward- and backward-traveling voltage wave, so $v(t) = v^{+}(t) + v^{-}(t)$. For layer peeling, the traveling waves must be tracked at both the left and the right sides of each transmission line section. For this purpose, a superscript "L" indicates the left side, while an "R" indicates the right. The transmission lines in Figure 8.12a are annotated with their traveling waves.

In addition, the traveling waves are time-dependent. In Figure 8.12b, the traveling waves on each section of transmission line are plotted for each time point. Because it takes time for the TDR pulse to reach further into the interconnect, the traveling wave amplitudes are zero for earlier time steps at the greater depths. This

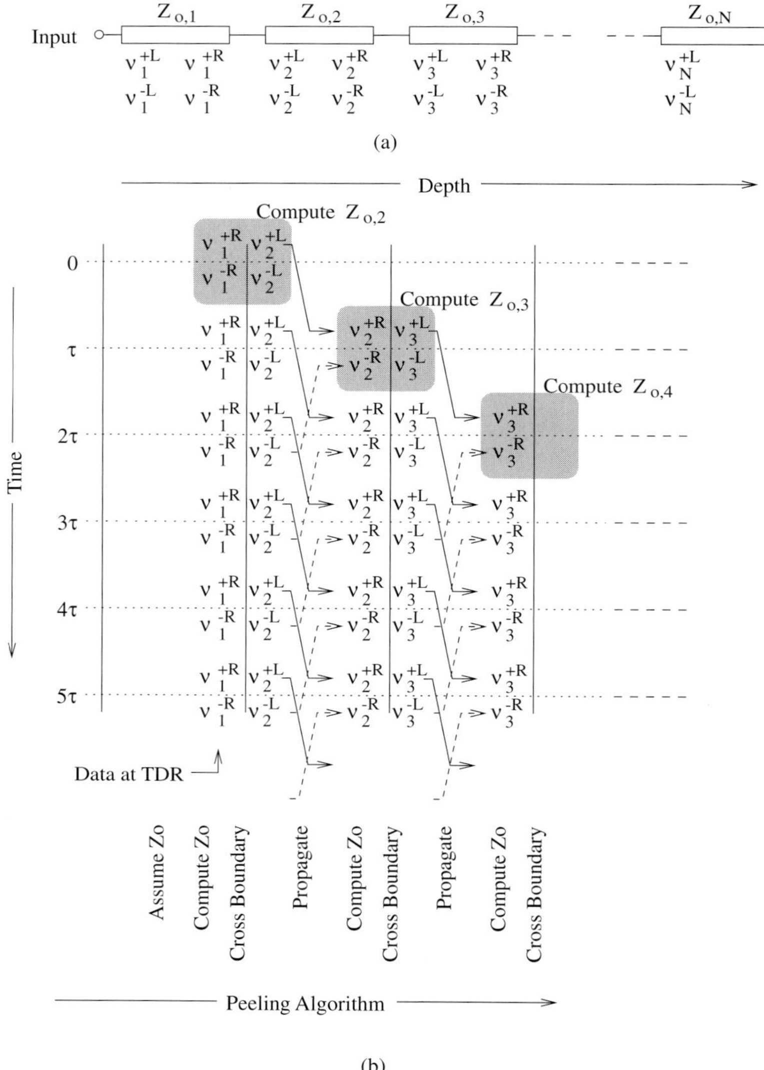

Figure 8.12. Layer peeling for removing multiple reflections from TDR-generated impedance profiles: (a) the interconnect is assumed to be a cascade of equal-delay transmission lines, (b) the traveling waves are moved at all time points from left to right into the interconnect, allowing the extraction of one characteristic impedance per step.

fact is exploited to extract the characteristic impedances of the transmission line segments.

The peeling algorithm is represented schematically in Figure 8.12b, and it consists of three steps. It is assumed that the left-side traveling waves on transmission line i are known at all time steps and that $Z_{o,i}$ is known. In step 1, the traveling waves are propagated across the section of transmission line. Positive-going waves (left-to-right) transfer their values to the next time step, while negative-going waves (right-to-left) transfer to the prior time step. In step 2, the characteristic impedance of the next transmission line segment is computed according to chapter 2, equation (2.40), which can be rearranged to yield

$$Z_{o,i+1} = Z_{o,i} \frac{1 + v_i^{-R}/v_i^{+R}}{1 - v_i^{-R}/v_i^{+R}}.$$

This formula must be used at a time step where there is no negative-going wave in section $i+1$ (i.e., $v_{i+1}^{-L} = 0$), and these positions are indicated in Figure 8.12b. In step 3, the traveling wave amplitudes on the left side of section $i+1$ are computed from those on the right side of section i according to (2.38) and (2.39). At this point, the algorithm is back to the starting point and can be repeated N times to extract the characteristic impedances of N transmission line segments. Since both forward and backward traveling waves are tracked at all times, multiple reflections are accounted for and do not show up in the final extracted impedance profile.

To kick off the peeling algorithm, v^{+L} and v^{-L} for each time step as measured at the TDR is required. Referring to Figure 8.6, assume that the TDR is set to report voltage (rather than reflection coefficient or impedance), so v_r at each time step is available. A calibration measurement is required to split the total voltage into its traveling wave components. A measurement with the DUT replaced by a precision matched load with $R = Z_o$ will result in no reflection, so $v^{-L} = 0$ and $v^{+L} = V_{r,\text{matched}}$. For subsequent measurements with the DUT in place,

$$v^{-L} = v_r - v_{r,\text{matched}}.$$

The characteristic impedance of the first section of transmission line to be de-embedded is assumed to be Z_o, and this is reasonable since there is a short section of transmission line between the sampling circuitry and the probe tips. The full length of this connection will be extracted along with the DUT.

Layer peeling uses the characteristic impedance of the current transmission line segment to compute the impedance of the next. Therefore, any error in the current section is propagated to the next, so errors accumulate as the extraction moves deeper into the interconnect. Algorithms of this type tend to be sensitive to noise, and layer peeling is no exception. Commercial-grade implementations of layer peeling must implement a noise control methodology. Robust and general schemes are not yet available, so ad hoc methods are required. For example, most applications have a limited range of possible impedance values, so the extraction code can cap the computed values. Also, many applications consist of cascades of constant impedance values, so the extraction code could quash attempts to make small impedance changes. With a good strategy, peeling can be sufficiently noise-tolerant to produce a useful extraction tool.

Examples

The following examples are computed using the sample code in Appendix E. Calibration is performed using a matched load. Because layer peeling takes time of flight into account, the extracted impedance profile reflects the true delay of the interconnect. In contrast, a simple TDR plot of impedance profiles, as discussed in section 8.4.3, incorporates a factor of 2 in the delay due to the round-trip nature of TDR measurements. The sample code adjusts the simple TDR extraction by a factor of 2 in time to line it up to the peeled response.

The first example is a repeat of the simple transition from 50Ω to 20Ω from Figure 8.10. The original and peeled results are shown in Figure 8.13. The multiple reflections have been completely removed. Note the change in time scales between Figures 8.13 and 8.10.

The second example involves a more complex profile that demonstrates the dif-

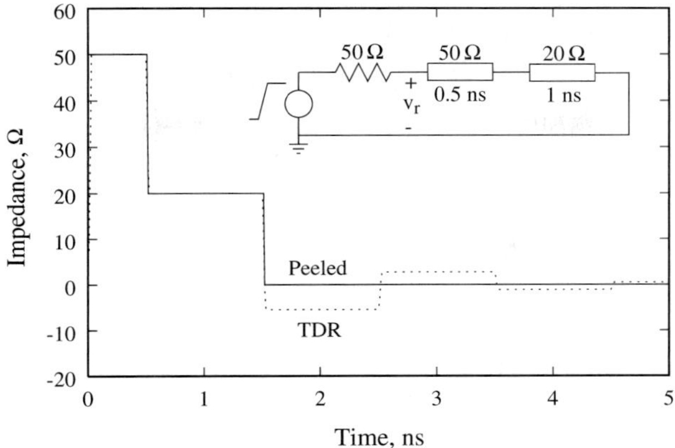

Figure 8.13. TDR-based and peeled impedance profiles for a simple profile.

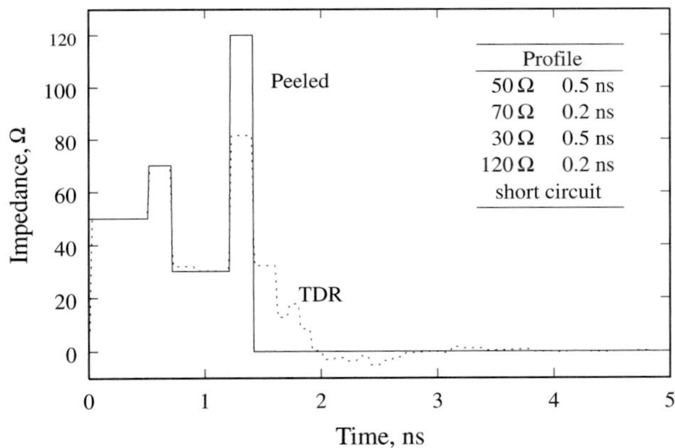

Figure 8.14. TDR-based and peeled impedance profiles for a complex profile.

ficulties that multiple reflections can cause. The profile and results are shown in Figure 8.14, where a simple interpretation of the TDR trace would result in a large underestimation of the impedance of one section of the profile.

The third example repeats the second with random noise added to the two TDR signals (matched load and profile extraction). Two results are shown. In Figure 8.15,

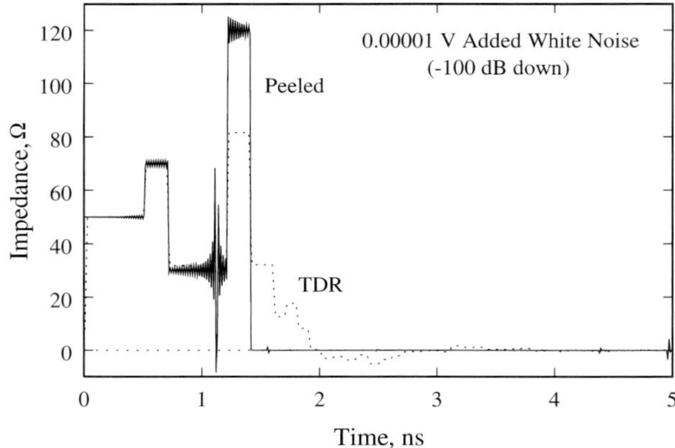

Figure 8.15. Effects of -100dB of noise on the results in Figure 8.14.

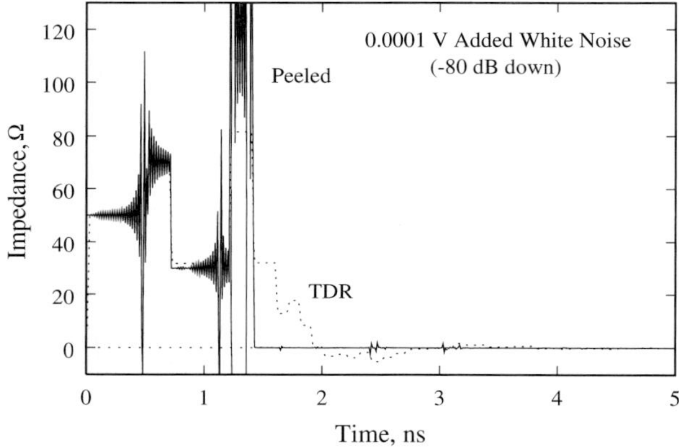

Figure 8.16. Effects of -80dB of noise on the results in Figure 8.14.

-100dB of white noise is added, while in Figure 8.16, -80dB of noise is added. Both of these noise levels are quite small, yet both strongly impact the extracted results. The -100dB results may still be useful, but the -80dB results are not. To be viable, the extraction must withstand as much as -20dB to -40dB of added noise.

A side benefit, somewhat visible in the examples above, is that layer peeling can sharpen the extracted impedance edges compared to a TDR trace. For a straight

TDR profile extraction, the edge rate of the excitation waveform sets the limit on the resolution of the extraction. In principle, application of the peeling algorithm results in impedance resolution down to the sampling rate of the TDR, so peeling would allow the extraction of fine detail from a very slow TDR edge. This could be very useful for extracting profiles of small structures. However, noise becomes more problematic as the edge rate slows, so significant sharpening may or may not occur, depending on the impedance profile.

8.4.5 Resolution

The ability of a TDR to resolve an impedance discontinuity is limited by the rise time of the TDR pulse that is injected into the DUT. Without peeling, a TDR can resolve discontinuities with delays equal to or greater than the rise time, a condition that can be stated mathematically as

$$t_r < TOF.$$

For longer rise times, the discontinuities smear together. For example, in Figure 8.17 two 0.1ns discontinuities spaced just 0.1ns apart fail to resolve for rise times greater than 0.1ns. Also observable in this plot is the error in the impedance of the second discontinuity due to multiple reflections. With peeling, a TDR can often resolve discontinuities with delays down to about one-third the rise time.

8.4.6 Multiport TDR Measurements

Since commercial TDR instruments often support several channels with high bandwidth, they are often used to perform crosstalk measurements. A four-channel TDR can be set up as shown in Figure 8.18 to simultaneously measure impedance on channel 1, delay on channel 4 (with time-domain transmission (TDT)), near-end noise on channel 2, and far-end noise on channel 3. Since typical TDR channels present a 50Ω input impedance, the setup may not match real systems having unterminated nets. In this case, the TDR measurements are useful for validating circuit models, which can then be used to estimate the response of the real system. TDRs pro-

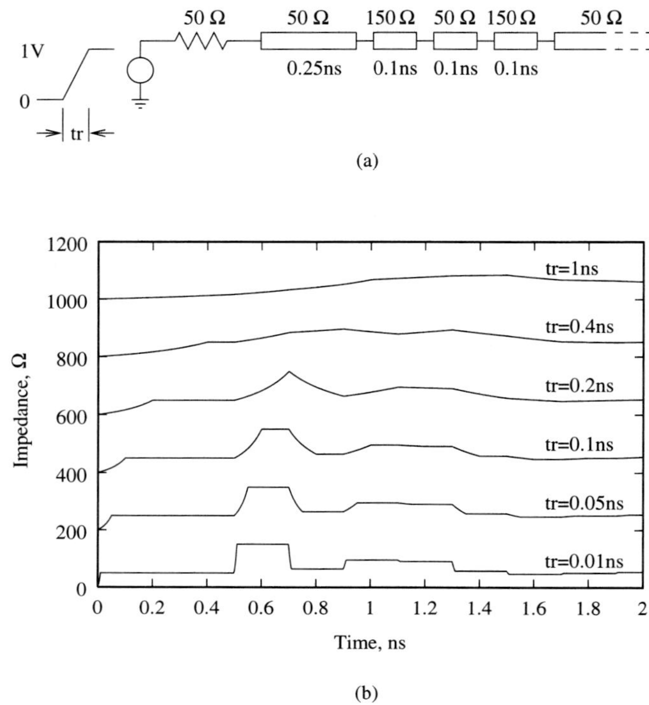

Figure 8.17. SPICE demonstration of the effect of rise time on TDR resolution: (a) circuit with two short and closely spaced impedance discontinuities, (b) TDR impedance profiles for several rise times. These unpeeled waveforms are separated by 200Ω offsets.

duce very fast rise times that are much faster than most systems, so the measured crosstalk is greater than that seen in slower systems.

Two-port TDR systems can replicate the VNA function with the addition of suitable data processing. Time-domain waveforms can be collected and then transformed to the frequency domain with Fourier or fast Fourier transforms. Calibration data, such as for SOLT calibration, can be collected to enable error correction. Setups in this mode are referred to as *time-domain network analyzers*, and a carefully executed system can achieve S-parameter measurements almost as accurate as mea-

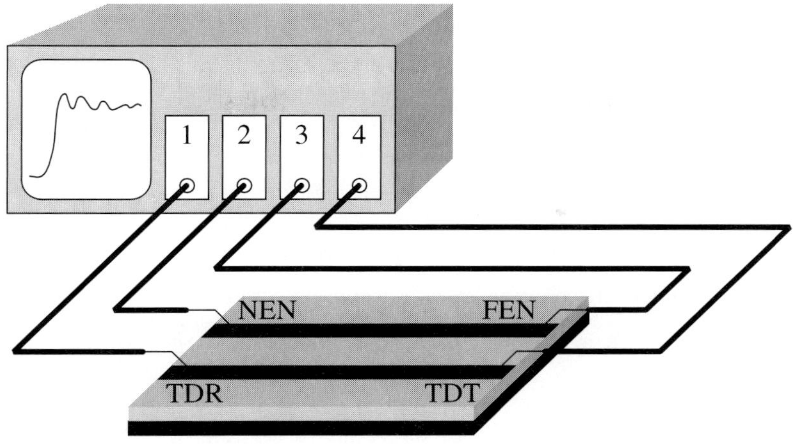

Figure 8.18. Multichannel TDR setup for characterizing crosstalk.

surements acquired with an actual VNA.

Two-port TDR systems can also be used to measure even- and odd-mode impedances and delays. If two TDR channels are identically excited in phase, then only the even mode on a symmetric pair of transmission lines will be excited. Similarly, the odd mode can be excited if the two TDR channels are identically excited out of phase.

8.5 Tradeoffs

Since the frequency and time domains are related by the Fourier transform, instrument selection is somewhat a matter of personal preference. However, domain equivalence requires sufficient bandwidth or time span in the measured data, and practical considerations may limit the breadth of the data and hamper the ability to transform domains. Once the data is in the domain in which modeling will take place, a few basic tradeoffs exist.

The IA and VNA produce frequency-dependent results. Unless the circuit simulator accepts frequency-dependent component values, and SPICE does not unless enhanced to perform convolution, then the modeler is faced with two choices. First, the data can be processed into a model that the circuit simulator will accept, such as a ladder network or a rational polynomial function. Second, the data can be approximated as frequency-independent so that the value at just one frequency can be used. For example, the skin effect produces frequency-dependent resistance and inductance. For highest accuracy, a model covering a specified bandwidth can be constructed. However, the change in inductance is often small, so if the resistance is negligible, as it often is in interconnects, a good approximation is a frequency-independent inductance.

A TDR produces parasitic values using an integrated waveform of a specific shape and edge rate, so the resulting value is shape-dependent. Modeling difficulty arises when the TDR's edge rate and the application's edge rate do not match. Since the TDR's edge is almost always much faster than the application's, then the high-frequency behavior is emphasized in the TDR-derived model. If additional accuracy is needed, TDR measurements should be taken with the edge slowed to match that of the application.

A basic tradeoff is established. Frequency-domain instruments produce data good for any edge rate (limited by the bandwidth of the data), but modeling may be quite difficult. Time-domain instruments produce data good for one edge rate, but modeling is trivial. The required accuracy in the models determines the hardness of these limits.

A second tradeoff involves bandwidth and modeling depth. Since an IA or a VNA sets up standing waves, the entire structure is excited and can, in principle, be resolved into a model. For the TDR, high losses or a high reflection coefficient can effectively shield following structure and limit the depth of the model into the interconnect. In both cases, the ability to resolve deep and/or hidden structure is dependent on the signal-to-noise ratio of the instrument and setup. Frequency-domain instruments typically achieve much higher signal-to-noise ratios than time-domain instruments and can be more effective in some cases in resolving structure deep into an interconnect.

8.6 Exercises

1. Develop and describe a technique using the impedance analyzer to measure general $N \times N$ admittance matrices.

2. Demonstrate that the method in section 8.2 produces the correct 2×2 capacitance matrix by plugging in the expected values an impedance analyzer would measure.

3. As suggested in Figure 8.19, a one-port VNA can be modeled as a perfect VNA that characterizes loads through a two-port interconnect. Using measurements of short, open and load (SOL) calibration standards, derive the full correction formula for measuring an unknown load impedance with the effects of the interconnect removed.

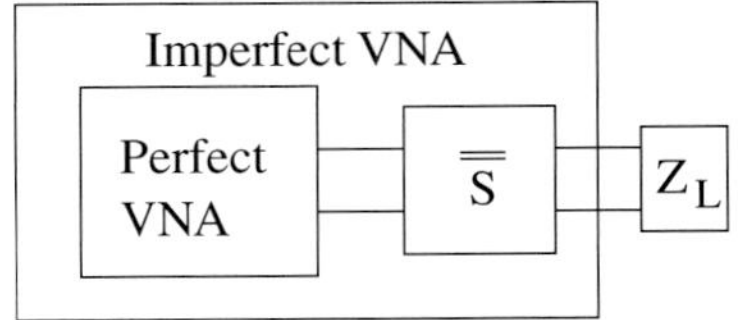

Figure 8.19. Model of a one-port vector network analyzer.

4. (a) Show how an $N \times N$ inductance matrix can be characterized with a TDR.

   (b) Show how an $N \times N$ capacitance matrix can be characterized with a TDR.

5. For a single line, derive an inductance extraction formula similar to (8.11) for the case when a known resistance is in series with the inductor. The resistor is assumed to be characterized separately with a DC resistance measurement.

6. Set up a SPICE simulation of a TDR based on Figure 8.6 with $Z_o = 50\Omega$ and a 1V 35ps ramp source and a 5pF capacitor load. Integrate the TDR response to find the load capacitance using (8.17) for varying lengths of cable, from

very short to very long. What are the effects of long cables on the integration and on the accuracy of the capacitance extraction?

7. A lossless 60Ω transmission line with 0.1ns of time of flight is shorted at one end. A 50Ω VNA drives the other end to extract the inductance of the closed loop.

   (a) What is the inductance of the loop?

   (b) What is the expected measured value of S_{11} at 100MHz?

   (c) Repeat for an open-circuit measurement for capacitance extraction.

8. Set up a SPICE simulation of a TDR based on Figure 8.6 with $Z_o = 50\Omega$ and a 1V 35ps ramp source, and demonstrate the extraction of the 2×2 inductance matrix defined by $L_1 = 10\text{nH}$, $L_2 = 9\text{nH}$, and $K = 0.35$. What are the effects of the long cables on the accuracy of the extraction?

9. Show that (8.5) is true.

Chapter 9

LUMPED MODELING

Lumped modeling is a principal starting point for modeling interconnects, including multiconductor transmission lines, packages, edge connectors, and sockets. For interconnects that are short compared to wavelength or rise time, the lumped approximation holds, and a single lumped stage is a sufficient model. Otherwise, wideband models are required (see chapter 10).

Modeling technique can also be used to circumvent difficulties in collecting data. In packages, the connections to the device are often very finely pitched, with irregular grounds. In addition, plastic encapsulation and lids can limit accessibility. By developing techniques using more than one sample, probing of the package interior can be avoided.

Finally, topology can be exploited to avoid frequency-dependent models. In general, an N-port linear and time-invariant network is completely described by a frequency-dependent $N \times N$ impedance matrix, or equivalently, by admittance or S-parameter matrices. The data can be used directly in circuit simulation to include the effects of the N-port network. However, time-domain circuit simulators handle frequency-dependent models with difficulty. In addition, extrapolation and/or interpolation are always required with frequency-dependent data. By fitting the data to an assumed topology, frequency dependence can often be eliminated or reduced to negligible levels to produce faster, more stable simulations.

9.1 Transmission Line Introduction

The input impedance of a loaded lossy transmission line is

$$Z_{\text{in}} = Z_o \frac{Z_L + Z_o \tanh(\gamma d)}{Z_o + Z_L \tanh(\gamma d)}, \tag{9.1}$$

repeated here from chapter 2, equation (2.66). There are two unknowns: Z_o and γ. If measurements are taken at both ends of the transmission line, then the reflection coefficient and the transmission coefficient serve as the two knowns to find the two unknowns. Therefore, one approach to characterize the transmission line is to perform measurements at each end.

A second approach relies on the application of known loads at one end with measurements taken at the other. Since there are two unknowns, measurements with two different known loads are required. While any two significantly different loads can be used, open and short circuits are the simplest to implement. For the short-circuit load, (9.1) simplifies to

$$Z_{\text{in}} = Z_o \tanh(\gamma d).$$

The lumped approximation is used to eliminate the tanh function by requiring $\gamma d \ll \pi/2$, then $\tanh(\gamma d) \approx \gamma d$. Combining this approximation with the lossy definitions for Z_o and γ from (2.26) and (2.21), respectively, yields

$$Z_{\text{in}} = R + j\omega L,$$

where $R = rd$ and $L = \ell d$ are the total resistance and inductance of the transmission line. The transmission line is lumped into a single resistance and inductance, termed *lumped elements*, and this approximation is good while the frequency is sufficiently low. As frequency becomes too high, the lumped approximation breaks down and wideband modeling (see chapter 10) is required. Calculations of bandwidth limitations are covered in section 9.6. For the open-circuit load, let $Z_L \to \infty$, then (9.1) simplifies to

$$Z_{\text{in}} = \frac{Z_o}{\tanh(\gamma d)}.$$

Applying the lumped approximation for a lossy line and inverting yields

$$Y_{\text{in}} = G + j\omega C,$$

where $G = gd$ and $C = cd$ are the total conductance and capacitance of the line. The transmission line is completely characterized once R, L, G, and C are found. The model is good while the lumped approximation holds. Generally, R, L, G, and C are frequency-dependent.

For example, consider a lossless 1cm-long 100Ω transmission line with an effective dielectric constant of 4. The phase velocity is $3 \times 10^8/\sqrt{4}$m/s, so the TOF$=0.01/(3 \times 10^8/\sqrt{4}) = 67$ps. From (2.11) and (2.12), the inductance per-unit-length is 670nH/m, while the capacitance per-unit-length is 67pF/m. Multiplying the per-unit-length values by 1cm, the lumped model of the transmission line is 6.7nH in series with 0.67pF to ground. For a lossless line, $|\gamma| = \omega\sqrt{\ell c}$, so the lumped approximation requires that the frequency be limited by $f \ll 1/(4d\sqrt{\ell c}) = 3.7$GHz.

In summary, under the lumped approximation an interconnect can be characterized either with two measurements on one sample, where access at both ends is required, or with single measurements on two samples, requiring access only at one end. With the two-sample technique, measurement of the input impedance of a shorted sample yields R and L, while measurement of the input admittance of an open sample yields G and C. Extension of the two-sample approach to multiple conductors is treated in the next section, followed by a one-sample extension.

9.2 Multiconductor Modeling with Two Samples

The two-sample process demonstrated in section 9.1 for a single transmission line can be generalized for multiple conductor interconnects. For bandwidth estimation, extension of the methodology in section 9.1 to N conductors requires the simultaneous solution of N differential equations, followed by the application of the lumped approximation to obtain a multiconductor lumped model. This approach is not feasible, so instead, the lumped model is assumed from the start. Unfortunately, an explicit derivation of the bandwidth limit of the lumped approximation does not follow when using an assumed topology; however, the general principle presented in section 9.1 for the single line can be assumed to hold for each individual line in the multiline case.

An intuitive topology for an N-conductor point-to-point interconnect is shown in Figure 9.1, where each conductor is modeled with series resistance and partial self-inductance and shunt capacitance to ground, and each pair of conductors is coupled with partial mutual inductance and capacitance. A common return path is provided.

The model must satisfy Kirchhoff's voltage and currents laws, which for this circuit are respectively given by

$$\overline{V}^L - \overline{V}^R = (\overline{\overline{R}} + j\omega\overline{\overline{L}})\overline{I}^L \tag{9.2}$$

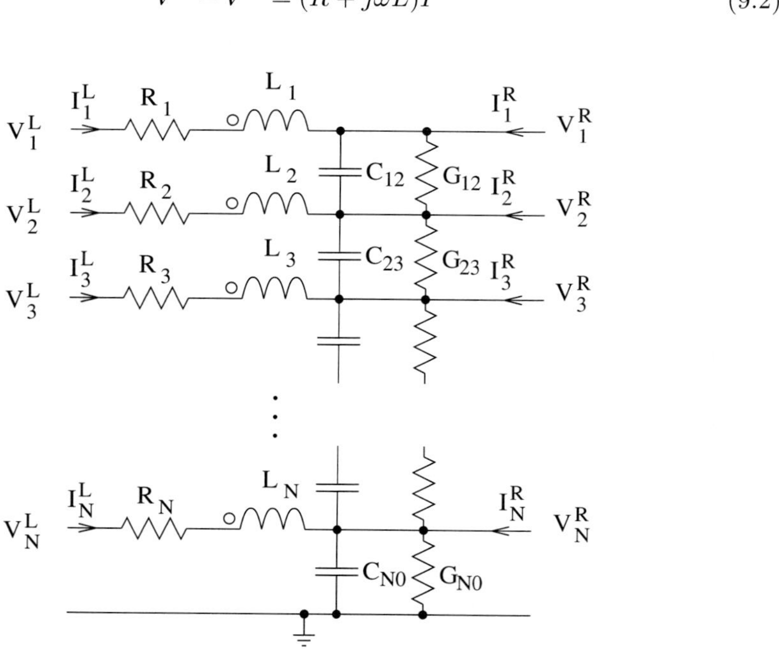

Figure 9.1. Lumped model for a multiconductor interconnect. All of the mutual inductors and most of the capacitors and conductances are omitted for clarity.

and

$$\overline{I}^L + \overline{I}^R = (\overline{\overline{G}} + j\omega\overline{\overline{C}})\overline{V}^R. \tag{9.3}$$

Two samples are used to find $(\overline{\overline{R}} + j\omega\overline{\overline{L}})$ and $(\overline{\overline{G}} + j\omega\overline{\overline{C}})$.

For the first sample, shorting the right side of the model to ground forces $\overline{V}^R = 0$, so (9.2) and (9.3) simplify to

$$\overline{V}^L = (\overline{\overline{R}} + j\omega\overline{\overline{L}})\overline{I}^L \tag{9.4}$$

and

$$\overline{I}^L = -\overline{I}^R.$$

Note that Kirchhoff's current law is automatically satisfied. Equation (9.4) shows that the shorted sample can be described by the N-port impedance matrix, $\overline{V}^L = \overline{\overline{Z}}\,\overline{I}^L$. Any measurement or simulation method that can produce the impedance matrix for a multiport circuit is adequate to find the resistance and inductance matrix. Given the impedance matrix, then

$$\overline{\overline{R}} = \mathrm{Re}(\overline{\overline{Z}}) \tag{9.5}$$

and

$$\overline{\overline{L}} = \mathrm{Im}(\overline{\overline{Z}})/\omega. \tag{9.6}$$

For the second sample, opening the left side of the model forces $\overline{I}^L = 0$, so (9.2) and (9.3) simplify to

$$\overline{V}^L = \overline{V}^R$$

and

$$\overline{I}^R = (\overline{\overline{G}} + j\omega\overline{\overline{C}})\overline{V}^R. \tag{9.7}$$

For the open sample, Kirchhoff's voltage law is automatically satisfied. Equation (9.7) shows that the open sample can be described by the N-port admittance matrix,

$\overline{I}^R = \overline{\overline{Y}}\,\overline{V}^R$, and any method that finds the admittance matrix gives the conductance and capacitance matrices as

$$\overline{\overline{G}} = \mathrm{Re}(\overline{\overline{Y}})$$

and

$$\overline{\overline{C}} = \mathrm{Im}(\overline{\overline{Y}})/\omega.$$

In summary, two samples can be used to produce a lumped model of a multiconductor interconnect. A sample with short circuits on one side can be characterized exclusively with measurements or simulations at the terminals on the other side to find the resistance and inductance matrices. Similar measurements on an open sample can be used to find the conductance and capacitance matrices. In practice, the open and short circuits are applied to the same side, while the measurements are taken from the other. The left/right distinction forced by the lumped model does not actually matter since the electrical effects are distributed over the interconnect.

A main advantage of the two-sample method is that it is often difficult to access one side of the interconnect. BGAs often have die connections on the top and solder ball connections on the bottom, so access is convenient only from one side or the other. QFPs and other encapsulated packages make access to the dieside of the interconnect difficult due to the encapsulation; access to these interconnects is most convenient from the outside of the package. By building two samples, access can be provided where it is convenient to minimize the need for special fixturing.

As discussed in chapter 8, impedance matrices are easier to measure than admittance matrices due to the use of open circuits on unprobed ports. The impedance matrices of the two samples can be measured and then the admittance matrix for the capacitance and conductance models can be found through matrix inversion.

9.3 Multiconductor Modeling with One Sample

When all of the ports are accessible, as is often the case in simulation, the complete lumped model can be obtained from a characterization of one sample at all the ports. Two formulations exist that lead to π- and T-network topologies.

9.3.1 π-Network Topology

The π-network topology of a lumped-model interconnect is shown in Figure 9.2. Kirchhoff's voltage law for this circuit requires that

$$\overline{V}^L - \overline{V}^R = (\overline{\overline{R}} + j\omega\overline{\overline{L}})\overline{I}, \tag{9.8}$$

where $\overline{I}$ is an intermediate variable to be eliminated shortly. Kirchhoff's current law requires that

$$\overline{I}^L = \overline{I} + (\overline{\overline{G}}^L + j\omega\overline{\overline{C}}^L)\overline{V}^L \tag{9.9}$$

and

$$\overline{I}^R = -\overline{I} + (\overline{\overline{G}}^R + j\omega\overline{\overline{C}}^R)\overline{V}^R. \tag{9.10}$$

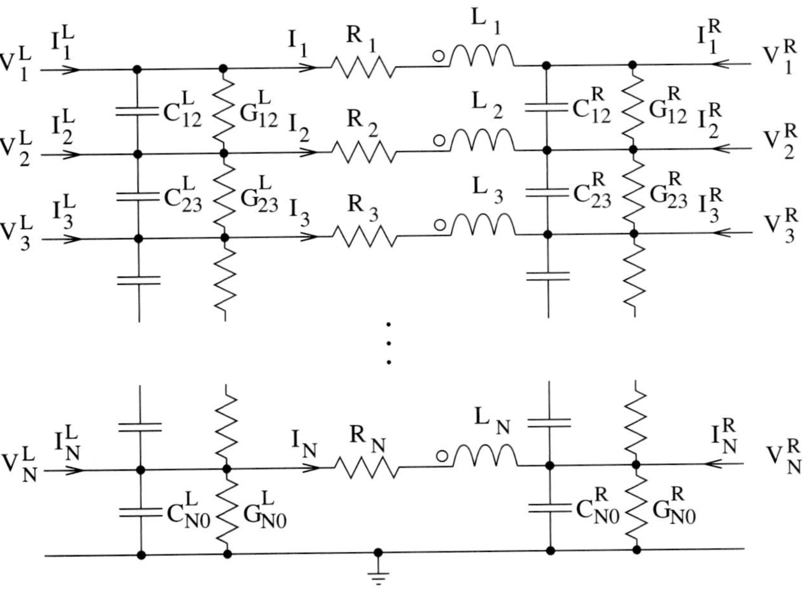

Figure 9.2. π-network topology of a lumped-model multiconductor interconnect. All of the mutual inductors and most of the capacitors and conductances are omitted for clarity.

Using (9.8) to eliminate $\overline{I}$ from (9.9) and (9.10) and assembling into matrix form yields

$$\begin{bmatrix} \overline{I}^L \\ \overline{I}^R \end{bmatrix} = \begin{bmatrix} \overline{\overline{Y}}^L + \overline{\overline{Z}}^{-1} & -\overline{\overline{Z}}^{-1} \\ -\overline{\overline{Z}}^{-1} & \overline{\overline{Y}}^R + \overline{\overline{Z}}^{-1} \end{bmatrix} \begin{bmatrix} \overline{V}^L \\ \overline{V}^R \end{bmatrix},$$

where $\overline{\overline{Y}}^L = \overline{\overline{G}}^L + j\omega\overline{\overline{C}}^L$, $\overline{\overline{Y}}^R = \overline{\overline{G}}^R + j\omega\overline{\overline{C}}^R$, and $\overline{\overline{Z}} = \overline{\overline{R}} + j\omega\overline{\overline{L}}$. Therefore, given the admittance matrix of the interconnect, the lumped model can be found by inspection.

9.3.2 T-Network Topology

The T-network topology of a lumped-model interconnect is shown in Figure 9.3. Kirchhoff's current law for this circuit requires that

$$\overline{I}^L + \overline{I}^R = (\overline{\overline{G}} + j\omega\overline{\overline{C}})\overline{V}, \tag{9.11}$$

where $\overline{V}$ is an intermediate variable to be eliminated shortly. Kirchhoff's voltage law requires that

$$\overline{V}^L - \overline{V} = (\overline{\overline{R}}^L + j\omega\overline{\overline{L}}^L)\overline{I}^L \tag{9.12}$$

and

$$\overline{V}^R - \overline{V} = (\overline{\overline{R}}^R + j\omega\overline{\overline{L}}^R)\overline{I}^R. \tag{9.13}$$

Using (9.11) to eliminate $\overline{V}$ from (9.12) and (9.13) and assembling into matrix form yields

$$\begin{bmatrix} \overline{V}^L \\ \overline{V}^R \end{bmatrix} = \begin{bmatrix} \overline{\overline{Z}}^L + \overline{\overline{Y}}^{-1} & \overline{\overline{Y}}^{-1} \\ \overline{\overline{Y}}^{-1} & \overline{\overline{Z}}^R + \overline{\overline{Y}}^{-1} \end{bmatrix} \begin{bmatrix} \overline{I}^L \\ \overline{I}^R \end{bmatrix}, \tag{9.14}$$

where $\overline{\overline{Z}}^L = \overline{\overline{R}}^L + j\omega\overline{\overline{L}}^L$, $\overline{\overline{Z}}^R = \overline{\overline{R}}^R + j\omega\overline{\overline{L}}^R$, and $\overline{\overline{Y}} = \overline{\overline{G}} + j\omega\overline{\overline{C}}$. Therefore, given the impedance matrix of the interconnect, the lumped model can be found by inspection.

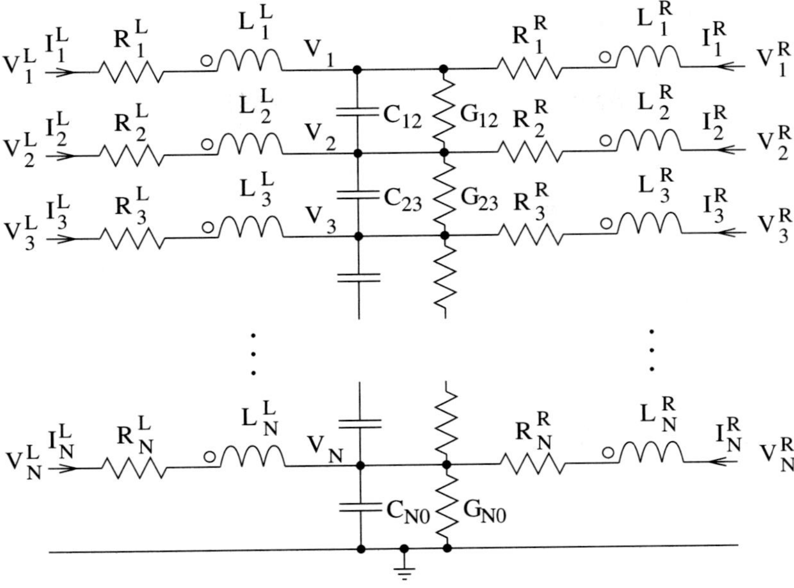

Figure 9.3. T-network topology of a lumped-model multiconductor interconnect. All of the mutual inductors and most of the capacitors and conductances are omitted for clarity.

9.3.3 Practical Issues

In principle, the complete π- or T-model is available once the $2N$-port Y- or Z-matrices are known for the N-line interconnect. If the interconnect consists of a precisely π or T topology, perhaps constructed directly from lumped components, then the complete model will be accurately derived.

In actual interconnects, the electrical behavior is distributed in nature, so the π- and T-networks are approximations. In practice, the complete π- or T-model is not resolvable with data at a single frequency. For the π-model, $\overline{\overline{C}}^L$ and $\overline{\overline{C}}^R$ are individually incorrect, but the sum $\overline{\overline{C}}^L + \overline{\overline{C}}^R$ accurately represents the total lumped capacitance of the interconnect. Similarly for the T-model, the sum $\overline{\overline{L}}^L + \overline{\overline{L}}^R$ is the total inductance for the interconnect. Similar relationships hold for the resistance and conductance matrices. Once these combinations are made, the π- and T-models form identical L models that match the one produced with two samples, as outlined in section 9.2.

9.4 Internal Nodes

Packages targeted for high-performance applications will usually have one or more planes to reduce inductance. The die will typically make several connections to the planes for reduced inductance, and for the same reason, several connections from the package to the system board or socket will also be made. The resulting topology has a node internal to the package, as shown in Figure 9.4 for a BGA-style package with three connections to the plane from the die and four to the system.

The topology for internal nodes fits the T-network. A generalized T-network is shown in Figure 9.5, where the model is broken into three sections to enable arbitrary connections. The three sections are reattached with two incidence matrices that define how the sections come together. The voltages are related according to

$$\overline{V}^a = \overline{\overline{Q}}^a \overline{V}^d$$

and

$$\overline{V}^b = \overline{\overline{Q}}^b \overline{V}^d,$$

where the incidence matrices $\overline{\overline{Q}}^a$ and $\overline{\overline{Q}}^b$ are constructed by inspection. For example, in Figure 9.6, three left leads join four right leads at two common nodes, and the incidence matrices are

$$\overline{\overline{Q}}^a = \begin{bmatrix} 1 & 0 \\ 1 & 0 \\ 0 & 1 \end{bmatrix} \tag{9.15}$$

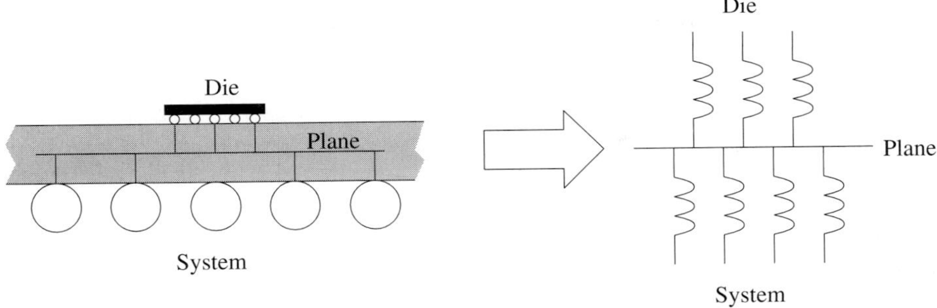

Figure 9.4. A BGA-style package inductance model with an internal node.

Figure 9.5. Generalized T-network for modeling of internal nodes. Note that most mutual terms are omitted for clarity.

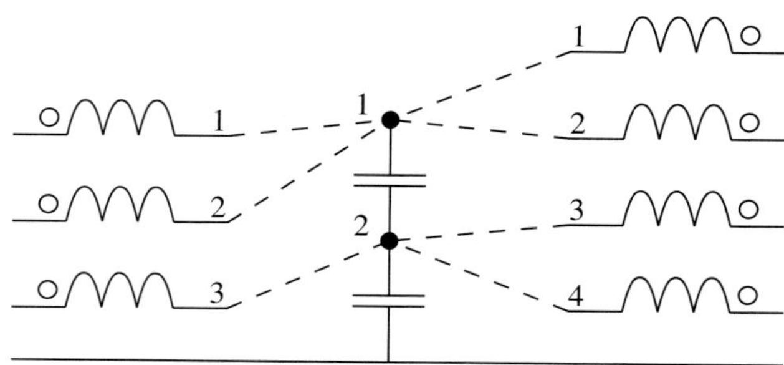

Figure 9.6. Example connections for construction of the incidence matrices $\overline{\overline{Q}}^a$ in (9.15) and $\overline{\overline{Q}}^b$ in (9.16).

and

$$\overline{\overline{Q}}^b = \begin{bmatrix} 1 & 0 \\ 1 & 0 \\ 0 & 1 \\ 0 & 1 \end{bmatrix}. \tag{9.16}$$

Assuming there are N "L" branches, M "R" branches, and P internal nodes, then $\overline{\overline{Q}}^a$ is an $N \times P$ matrix and $\overline{\overline{Q}}^b$ is $M \times P$. For the currents,

$$\overline{I}^c = \left(\overline{\overline{Q}}^a\right)^T \overline{I}^L$$

and

$$\overline{I}^d = \left(\overline{\overline{Q}}^b\right)^T \overline{I}^R.$$

The voltage drops across the "L" and "R" branches are

$$\overline{V}^L - \overline{V}^a = \overline{\overline{Z}}^L \overline{I}^L$$

and

$$\overline{V}^R - \overline{V}^b = \overline{\overline{Z}}^R \overline{I}^R,$$

respectively, where $\overline{\overline{Z}}^L = \overline{\overline{R}}^L + \jmath\omega\overline{\overline{L}}^L$ and $\overline{\overline{Z}}^R = \overline{\overline{R}}^R + \jmath\omega\overline{\overline{L}}^R$. Summing the currents at the internal nodes yields

$$\overline{I}^c + \overline{I}^d = \overline{\overline{Y}}\,\overline{V}^d,$$

where $\overline{\overline{Y}} = \overline{\overline{G}} + \jmath\omega\overline{\overline{C}}$.

Eliminating $\overline{V}^a$, $\overline{V}^b$, $\overline{V}^d$, $\overline{I}^c$, and $\overline{I}^d$ yields

$$\begin{bmatrix} \overline{V}^L \\ \overline{V}^R \end{bmatrix} = \overline{\overline{Z}} \begin{bmatrix} \overline{I}^L \\ \overline{I}^R \end{bmatrix},$$

where

$$\overline{\overline{Z}} = \begin{bmatrix} \overline{\overline{Z}}^L + \overline{\overline{Q}}^a\overline{\overline{Y}}^{-1}\left(\overline{\overline{Q}}^a\right)^T & \overline{\overline{Q}}^a\overline{\overline{Y}}^{-1}\left(\overline{\overline{Q}}^b\right)^T \\ \overline{\overline{Q}}^b\overline{\overline{Y}}^{-1}\left(\overline{\overline{Q}}^a\right)^T & \overline{\overline{Z}}^R + \overline{\overline{Q}}^b\overline{\overline{Y}}^{-1}\left(\overline{\overline{Q}}^b\right)^T \end{bmatrix}. \tag{9.17}$$

For the case of straight-through connections with no internal nodes, then $N = M = P$, $\overline{\overline{Q}}^a$ and $\overline{\overline{Q}}^b$ are $N \times N$ identity matrices, and (9.17) simplifies to (9.14).

The incidence matrices are not square, so some manipulation is required to solve for the lumped matrices given $\overline{\overline{Z}}$. Matrix inversions can be performed after forming the normal equations from nonsquare incidence matrices, then $\left(\overline{\overline{A}}^T\overline{\overline{A}}\right)^{-1}$ exists even when $\overline{\overline{A}}$ is not square. If $\overline{\overline{Z}}$ is partitioned as

$$\overline{\overline{Z}} = \begin{bmatrix} \overline{\overline{Z}}_{11} & \overline{\overline{Z}}_{12} \\ \overline{\overline{Z}}_{21} & \overline{\overline{Z}}_{22} \end{bmatrix},$$

then $\overline{\overline{G}}$ and $\overline{\overline{C}}$ can be found as

$$\overline{\overline{G}} = \mathrm{Re}\left[\left(\left(\left(\overline{\overline{Q}}^a\right)^T\overline{\overline{Q}}^a\right)^{-1}\left(\overline{\overline{Q}}^a\right)^T\overline{\overline{Z}}_{12}\overline{\overline{Q}}^b\left(\left(\overline{\overline{Q}}^b\right)^T\overline{\overline{Q}}^b\right)^{-1}\right)^{-1}\right],$$

$$\overline{\overline{C}} = \mathrm{Im}\left[\left(\left(\left(\overline{\overline{Q}}^a\right)^T\overline{\overline{Q}}^a\right)^{-1}\left(\overline{\overline{Q}}^a\right)^T\overline{\overline{Z}}_{12}\overline{\overline{Q}}^b\left(\left(\overline{\overline{Q}}^b\right)^T\overline{\overline{Q}}^b\right)^{-1}\right)^{-1}\right]\Big/\omega.$$

With these known, then

$$\overline{\overline{R}}^{L} = \mathrm{Re}\left[\overline{\overline{Z}}_{11} - \overline{\overline{Q}}^{a}\left(\overline{\overline{G}} + j\omega\overline{\overline{C}}\right)^{-1}\left(\overline{\overline{Q}}^{a}\right)^{T}\right],$$

$$\overline{\overline{L}}^{L} = \mathrm{Im}\left[\overline{\overline{Z}}_{11} - \overline{\overline{Q}}^{a}\left(\overline{\overline{G}} + j\omega\overline{\overline{C}}\right)^{-1}\left(\overline{\overline{Q}}^{a}\right)^{T}\right] \Big/ \omega,$$

$$\overline{\overline{R}}^{R} = \mathrm{Re}\left[\overline{\overline{Z}}_{22} - \overline{\overline{Q}}^{b}\left(\overline{\overline{G}} + j\omega\overline{\overline{C}}\right)^{-1}\left(\overline{\overline{Q}}^{b}\right)^{T}\right],$$

and

$$\overline{\overline{L}}^{R} = \mathrm{Im}\left[\overline{\overline{Z}}_{22} - \overline{\overline{Q}}^{b}\left(\overline{\overline{G}} + j\omega\overline{\overline{C}}\right)^{-1}\left(\overline{\overline{Q}}^{b}\right)^{T}\right] \Big/ \omega.$$

The source for $\overline{\overline{Z}}$ would likely be an $(M+N) \times (M+N)$ S-parameter matrix produced through either simulation or measurement. With $\overline{\overline{S}}$ known, then chapter 4, equation (4.8) can be used to compute $\overline{\overline{Z}}$.

9.5 Frequency Dependence

The lumped models discussed in this chapter can be constructed from data at a single frequency. When the model topology is a close fit to the actual interconnect, the model derived at one frequency accurately reproduces the terminal behavior at other frequencies over some usable bandwidth. However, if the match is poor, then the model is accurate only for a narrow band of frequencies centered on the single frequency from which the model was constructed. The model can be recomputed at multiple frequencies to extend the fit, but then the model's component values are frequency-dependent and can require additional modeling to run in a time-domain simulator such as SPICE.

If frequency-dependence is encountered in a model, a change in topology and a new model may enable the elimination of the frequency dependence. However, for any given topology there is the extraction problem to obtain the model component values from the data. This chapter presents a few lumped models with extraction methods, but general topologies will usually not have a straightforward extraction method. A very common approach is optimization, where the component values are varied until a suitable fit is achieved. While the approach is quite acceptable in most instances, the model is rarely unique and is often nonphysical. In other words, it represents a black box that simply reproduces the terminal characteristics. If the data does not completely cover the bandwidth of the signals to be simulated, including DC, then care must be taken to ensure that the model extrapolates well.

As a simple example, consider the reactance of a capacitor. If the frequency-dependent reactance is fitted at a single frequency with a capacitor model, then the entire set of data can be regenerated from the model. However, if the reactance is instead fitted at a single frequency with an inductor model, the model fits the data only at that frequency. In Figure 9.7, the reactance of a 5pF capacitor is plotted along with that of an inductor, which can exactly fit the capacitor at one frequency. A -100nH inductor fits at 225MHz, but at all other frequencies, the inductor does not fit. The nonphysical negative inductance value is a big hint that there is a problem with the model.

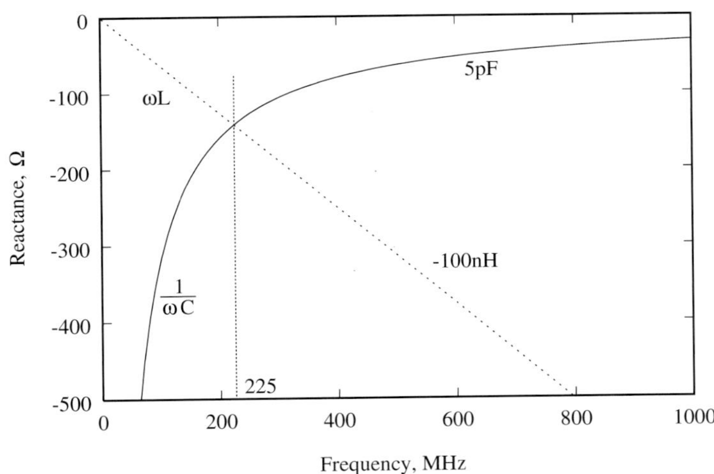

Figure 9.7. An inductor can accurately model a capacitor at only one frequency.

9.6 Iterative Impedance and Bandwidth

Several modeling options based on T, π, and L topologies have been discussed so far. A general discussion of the bandwidth implications of these for the multiconductor case is very difficult. Instead, the single line case is explored, and it is assumed that the results are applicable to the multiple line case. Of the many ways bandwidth can be studied, iterative impedance offers a particularly convenient framework.

When loaded by an impedance equal to its characteristic impedance (a matched load), the input impedance of a transmission line is equal to that impedance. For a lumped model to behave like a transmission line, it must also exhibit this property. The iterative impedance of a lumped interconnect model is the load impedance giving rise to an equal input impedance.

The input impedance of the single-section T-network lumped model in Figure 9.8a is

$$Z_{\text{in}} = j\omega L/2 + \frac{1}{j\omega C + \frac{1}{j\omega L/2 + Z_L}}. \tag{9.18}$$

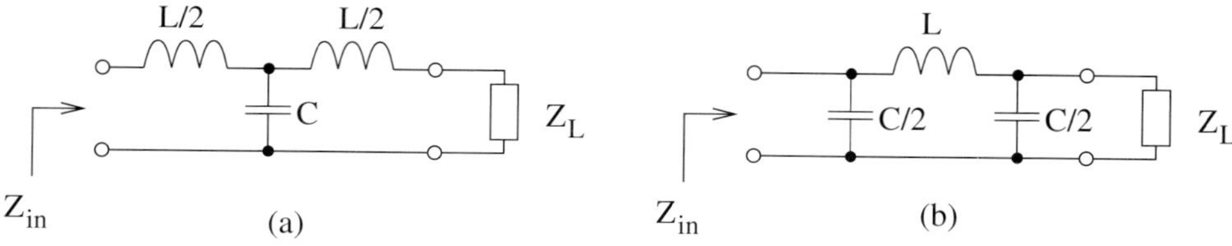

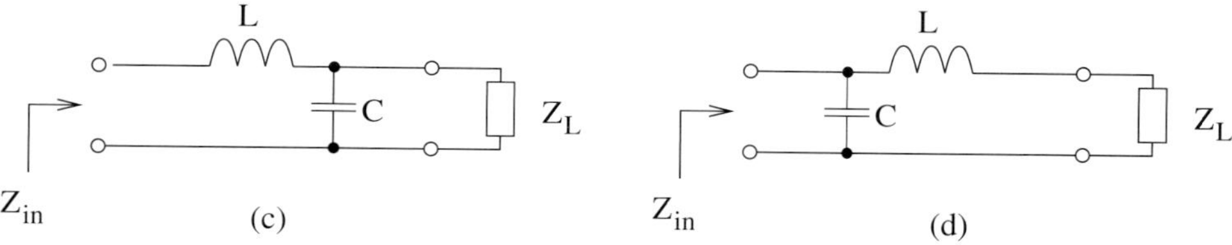

Figure 9.8. The four possible single-section lumped models for a single lossless interconnect.

The iterative impedance, Z_i, is defined when $Z_i = Z_{\text{in}} = Z_L$, then (9.18) simplifies to

$$Z_i = \sqrt{\frac{L}{C}}\sqrt{1 - \left(\frac{\omega}{\omega_c}\right)^2}, \tag{9.19}$$

where $\omega_c = 2/\sqrt{LC}$. If the T-network models a transmission line, (9.19) shows that at low frequencies, the iterative impedance equals the transmission line characteristic impedance, $\sqrt{L/C}$. However, as the frequency begins to approach ω_c, the iterative impedance starts dropping. For the iterative impedance to stay within 5% of the low-frequency value, the bandwidth of the model is

$$\omega < \frac{0.62}{\sqrt{LC}}. \tag{9.20}$$

For the π-network in Figure 9.8b, the input impedance is

$$Z_{\text{in}} = \frac{1}{j\omega C/2 + \frac{1}{j\omega L + \frac{1}{j\omega C/2 + \frac{1}{Z_L}}}}.$$

Setting $Z_i = Z_{\text{in}} = Z_L$ to find the iterative impedance yields

$$Z_i = \frac{\sqrt{\frac{L}{C}}}{\sqrt{1 - \left(\frac{\omega}{\omega_c}\right)^2}},$$

where again $\omega_c = 2/\sqrt{LC}$. As for the T-network, at low frequencies the iterative impedance equals the transmission line characteristic impedance $\sqrt{L/C}$. As frequency increases, the iterative impedance also increases, so for the iterative impedance to stay within 5% of the low-frequency value, the bandwidth is limited according to (9.20).

For the L-network in Figure 9.8c, the iterative impedance is

$$Z_i = \sqrt{\frac{L}{C}}\left(\sqrt{1 - \left(\frac{\omega}{\omega_c}\right)^2} + j\frac{\omega}{\omega_c}\right). \tag{9.21}$$

As before, at low frequencies the iterative impedance equals the transmission line characteristic impedance, $\sqrt{L/C}$, but as frequency increases, the iterative impedance becomes complex with a decreasing real part and increasing imaginary part.

For the magnitude of the error to be below some small number α, then

$$\left|\sqrt{1-\left(\frac{\omega}{\omega_c}\right)^2}-1+j\frac{w}{w_c}\right|<\alpha,$$

or

$$\omega<\frac{2\sqrt{1-\left(1-\frac{\alpha^2}{2}\right)^2}}{\sqrt{LC}}.$$

For $\alpha = 0.05$, then

$$\omega<\frac{0.1}{\sqrt{LC}}.$$

Therefore, because of the reactive component to the iterative impedance, the L-network has about a factor of six reduction in bandwidth compared to the T- and π-networks.

For the other L-network in Figure 9.8d, the iterative impedance is

$$Z_i=\sqrt{\frac{L}{C}}\left(\sqrt{1-\left(\frac{\omega}{\omega_c}\right)^2}-j\frac{\omega}{\omega_c}\right),$$

which has the same error expression and bandwidth criteria as the prior L-network.

The results using iterative impedance show that the π- and T-networks should be used whenever maximum bandwidth is required. The bandwidth penalty for using the L-networks is about a factor of six.

For application of these results to compute bandwidth limits for multiconductor interconnects, single L and C values are required. For L it is essential to use an inductance and not a partial inductance—the L must represent a closed path. The most sensible choice is the inductance of the loop formed by the longest conductor returning on ground, where all ground leads are utilized. Because it maximizes C, the conservative choice for C is to take the longest conductor and assume that all other leads are grounded; then the C to use is just the diagonal term of the capacitance matrix.

From (9.20) it can be seen that the bandwidth of a π- or T-lumped model can be doubled by halving both L and C. For a transmission line, this can easily be achieved by simply halving the length, so a section of length d can be modeled by two lumped π- or T-sections of length $d/2$. Such a model is called a *multilumped transmission line model*, and any bandwidth requirement can be satisfied for any line length if enough sections are added. See also chapter 10, section 10.1.2.

9.7 Model Reduction

Interconnect models often have a substantial number of leads, but the model is often much larger than the needs of a particular simulation. Smaller models enable faster circuit simulation, so it is advantageous to simplify the interconnect model to incorporate known information. Any manipulations of lumped models must rigorously account for all mutual terms.

9.7.1 Parallel Leads

Models often include many leads of the same function, such as ground, and these can be analytically paralleled together to reduce the size of the model. Leads are placed in parallel by forcing the voltages at their ports to be equal while satisfying Kirchhoff's current law. The conversion is derived in chapter 4, section 4.5.2, where the operation is cast in terms of the admittance matrix. If reducing a capacitance matrix, the operation in (4.20) can be applied directly. If reducing a resistance and/or inductance matrix, then the operation must be preceded and followed by matrix inversions to convert impedance to admittance and back.

9.7.2 Open, Short, and Matched Leads

Leads can be removed from a model by forcing them to be open-circuit, short-circuit, or terminated. The mathematical operations are given in chapter 4, section 4.5.1. To summarize, a lead is removed by deleting its row and column from the matrix. If the matrix is admittance, then the deletion produces a short circuit. Deletion from an impedance matrix produces an open circuit, and deletion from an S-parameter matrix terminates the line in the impedance used to normalize the matrix (typically

50Ω). Any of the three operations can be performed on any matrix by converting the matrix to the correct type, deleting rows and columns, then converting back.

Resistance/inductance matrices assume that one side of the interconnect is grounded (to form the closed loop necessary for characterization). Short-circuiting a lead results in both ends being grounded, so an isolated closed loop is formed that conducts eddy currents driven by the other leads through mutual inductances. Conversely, open-circuiting a lead leaves one end open and the other shorted to ground, so it will not support any current and has no inductive effect on the remaining leads. Recall that when constructing the model netlist, the grounded connections are ungrounded to construct an open-loop model suitable for interconnect work.

Capacitance matrices assume that one side of the interconnect is open-circuit (to float the leads for characterization). Open-circuiting a lead through row and column deletion results in both ends of the lead being open, so the conductor is floating, but with no static charge. Floating conductors collect charge and affect the voltages on surrounding leads through mutual capacitance. Short-circuiting a lead grounds the conductor and eliminates its ability to collect charge.

Matching the leads provides a termination to ground and can be useful for building in a driver impedance or termination resistance. The matching value can be selected by renormalizing the S-parameters using chapter 4, equation (4.15).

9.7.3 Excess Leads

Ideally, an interconnect model is constructed to precisely match the problem at hand, but sometimes the model is too large because a scaled down simulation is desired or a previously generated model is being used. For example, suppose a 16-line bus is to be simulated. Even if a 16-line model is specifically generated, a 2-line simulation may be desired for prototype work before scaling up to the larger simulation. Alternatively, a 32-line model may be available from previous work, but now just 16 lines are needed. What is a good procedure for reducing the size of the model to fit the smaller simulation?

The unused lines will affect the simulation, so the best strategy is to minimize their impact by terminating them in a way consistent with the simulation. For best accuracy, the lines can be terminated within the circuit simulation to capture the interaction between the termination parasitics and the interconnect. In many cases, ideal terminations can be incorporated directly within the interconnect model to eliminate unneeded leads for a smaller and faster simulation. Simulation runtime can be greatly improved with an analytically reduced model.

Different cases require different analytical model reductions. Matched terminations can be used to model both terminations and drivers holding low or high. Renormalization of the S-parameters will often be needed to achieve the proper impedance. Drivers holding in a high-impedance state and receivers can be modeled as open circuits, although the associated capacitances will be neglected.

9.7.4 Symmetry

Packages and interconnects often have considerable symmetry. For example, QFPs usually have three-way symmetry, where the entire package can be described from one-eighth of the layout. When characterizing packages in the laboratory, symmetry cannot be exploited to reduce the characterization effort. For example, the inductance matrix for a package that is symmetric about a centerline can be partitioned into three blocks as

$$\overline{\overline{L}} = \begin{bmatrix} \overline{\overline{L}}_A & \overline{\overline{L}}_B \\ \overline{\overline{L}}_B^T & \overline{\overline{L}}_A \end{bmatrix}.$$

If just half of the package is characterized, then $\overline{\overline{L}}_A$ is known but the coupling between the two sections, given by $\overline{\overline{L}}_B$, is not known. To complete the matrix, it would be necessary to extrapolate $\overline{\overline{L}}_B$ from the data in $\overline{\overline{L}}_A$.

In simulation, it is possible to introduce a perfect magnetic conductor (PMC) symmetry plane that numerically mirrors $\overline{\overline{L}}_A$ into the uncharacterized half. The key here is that the excitation is also mirrored, so that when one line is driven, its mirror is also driven. The final matrix consists of just $\overline{\overline{L}}_A$, which has built in the fact that its mirror is driven in tandem, and all coupling is correctly accommodated for this specialized excitation. If this mirrored excitation fits the modeling needs of the problem, then mirroring can dramatically reduce the characterization time. Not all simulators provide PMCs, so if one is not available, then symmetry cannot be exploited for model simplifications.

9.8 Approaches for Specific Interconnects

The theory developed in the chapter up to this point is sufficient to characterize a wide variety of interconnects. While an essentially infinite number of variations exist for characterizing any one interconnect, this section outlines simple approaches for two common interconnects. The techniques rely on the two-sample approach because the interiors are difficult to probe.

9.8.1 QFP

The QFP presents some difficulties in sample preparation. Ideally, two inserts are fabricated with bondable pad locations for the wirebonds. The first insert, used for characterizing inductance and resistance, simply shorts all pads together with a sheet of good conductor, and an aluminized die is ideal. The second insert, for capacitance, has all pads isolated; however, this sample is expensive due to the cost of the custom insert. The insert and wirebonds can be omitted, but then the model will not incorporate the contribution of the wirebonds to the capacitance. The two samples are fabricated with the inserts using the regular assembly process.

The samples are mounted over a ground plane at the approximate distance the part would normally experience. For capacitance, the distance is important for obtaining the correct capacitance to ground. For inductance, the distance is important for generating the correct strength of the eddy currents (which lower inductance) in the ground plane. A setup is suggested in Figure 9.9a. Each lead is probed with the probe tip on the lead and the probe ground on the ground plane. The two samples are set up and probed in exactly the same way except that the

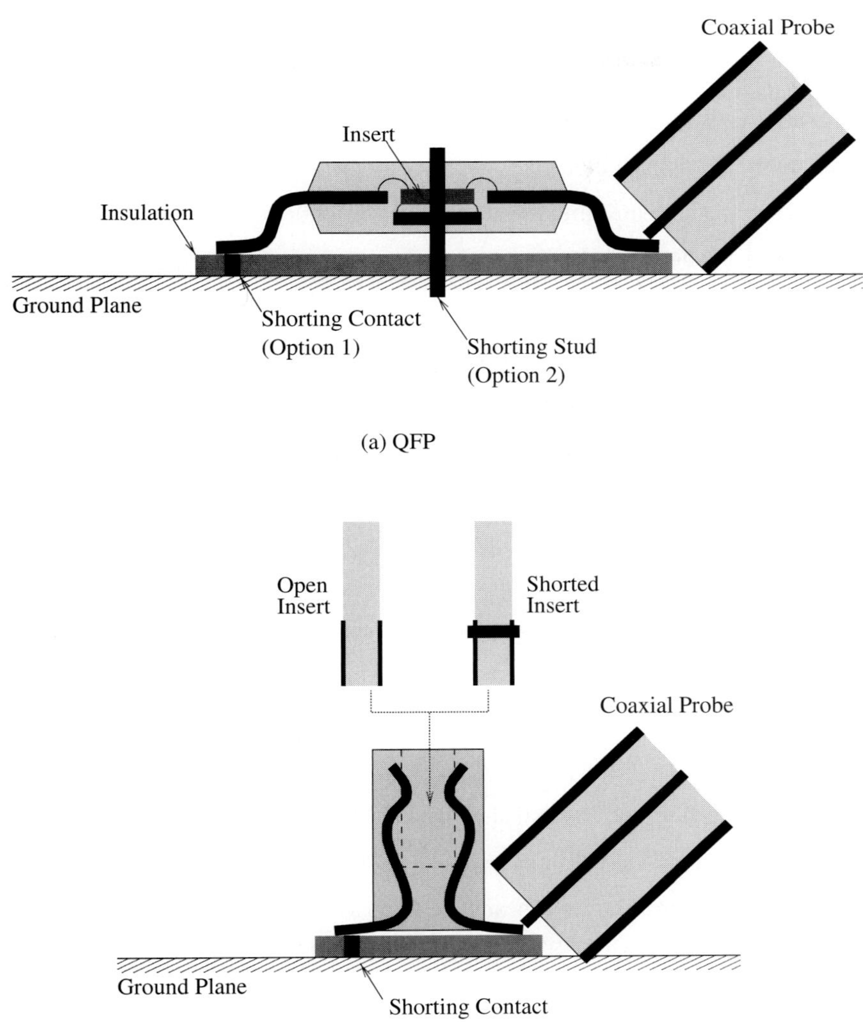

Figure 9.9. Suggested setups for characterizing QFP packages, edge connectors, and similar structures.

inductance sample requires a connection to the ground plane to complete the loop. The figure shows two approaches for making this connection for the inductance characterization.

In the first option, one or more of the leads of the package are shorted to ground. Since all of the leads are shorted together on the die insert, then a closed loop is formed by each of the remaining leads through this short to ground. The inductances of these loops and the mutual inductances between them are then characterized. The advantage of this option is that the sample needs no further preparation after sample construction. If the pin assignments of the package are known, it can be convenient to short all of the ground leads to the ground plane so that no renormalization is needed after the model is constructed, although the model is not then reusable for different ground pin assignments.

In the second option, a stud is inserted into the package to short the die insert to the ground plane. The advantage of this option is that the leads can be characterized to higher frequencies due to the lower loop inductances resulting from smaller loop areas. A second advantage is that every lead in the package can be characterized, so the model can be used for any pin assignment.

Using the setup in Figure 9.9, physical interference between the two coaxial probes can make characterization of two closely spaced leads difficult. The setup described below for BGAs can also be used to characterize QFPs and edge connectors, and that setup minimizes mechanical interference between the probes.

9.8.2 Edge Connectors

Edge connectors can be characterized in the same way as QFPs, except the two samples are formed by using different inserts. For capacitance measurements, an insert with all of the pads open-circuited is inserted into the connector with the connector isolated from the ground plane. It is important to use an insert so that the dielectric loading of the PCB is included. For inductance, the insert has all of the pads shorted together, and the connector has one or more pins shorted to the ground plane to complete the loop.

9.8.3 BGA

Sample preparation for wirebonded BGAs proceeds in the same manner as for QFPs. For flip-chip BGAs, sample preparation is particularly easy. Rather than use die inserts, the die site can be left open-circuited for the capacitance sample, while for the inductance sample, all connections at the die site can be shorted together with a sheet of good conductor soldered or epoxied (with conductive paste) into place.

Probing of BGA samples will typically be difficult without special hardware because of the irregular locations of the ground balls. One approach suggested in Figure 9.10 is to fabricate a printed circuit board with vias connecting landing pads on one side to probe pads on the other. The probe side has a gridded ground plane, isolated from each probe pad site, for contact by the probe ground. This setup enables probe contact to each BGA ball (through the via) while contacting the ground plane nearby. For the capacitance sample, all probe pads are left open-circuited with respect to the gridded ground plane. For the inductance sample the loop must be closed, and there are two options. In the first option, one signal lead

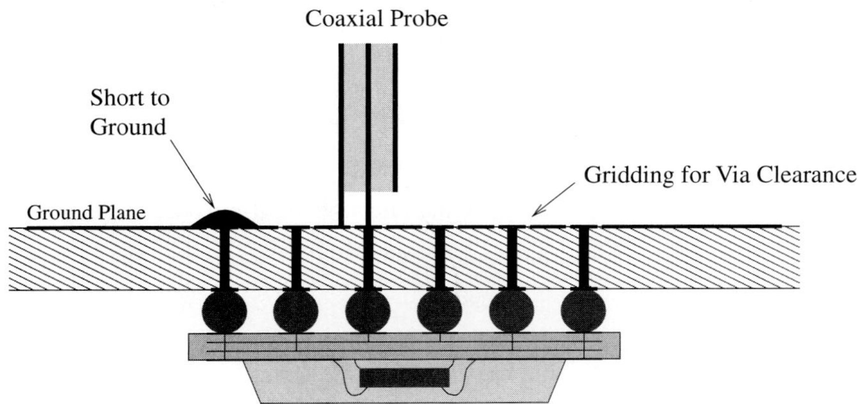

Figure 9.10. Measurement setup for characterizing a BGA. The short to ground is needed for inductance/resistance characterization only.

probe pad is shorted to the gridded ground, while in the second, all of the ground lead pads are shorted to the gridded ground. The advantages and disadvantages of each are the same as for the QFP setup.

The setup in Figure 9.10 is convenient for any package, socket, and connector. Its main drawback is the need for custom PCBs to mount the package. For BGAs, edge connectors, and sockets on a standard pitch, a single design suffices, but for QFPs a new card can be required for every characterization.

9.8.4 Internal Nodes

The BGA package in Figure 9.4 has an internal plane connecting several leads to the die (Path A) and several to the PCB (Path B). With the two-sample approach, it is not possible to resolve the individual partial inductances in Path A, since they are shorted together at both ends. Characterization will result in a model similar to Figure 9.13c, where the partial inductance of Path A is common to the individually characterized partial inductances of Path B and appears folded into the partial self- and mutual inductances of Path B.

9.9 General Topology

Prior sections show how an interconnect topology can be assumed and then fitted to a data set with a custom extraction methodology. For more complex networks, the interconnect topologies may be a poor fit, and a general-purpose topology can be required. Unfortunately, a general topology will usually require frequency-dependent components.

A circuit topology for implementing arbitrary admittance matrices is derivable using induction. For a 1×1 admittance matrix, a single admittance value is sufficient to implement the matrix. Assume that a network exists to implement an arbitrary $N \times N$ admittance matrix called $\overline{\overline{Y}}_N$. An additional port can be added to the circuit by connecting admittances between the new port and all N of the old ports plus an admittance to ground, then the $(N+1) \times (N+1)$ circuit in Figure. 9.11 is obtained. Solving Kirchhoff's current law at each node yields the relations

$$
\begin{aligned}
I_1 &= Y_1(V_1 - V_{N+1}) + \sum_{j=1}^{N} Y_{1j}V_j \\
I_2 &= Y_2(V_2 - V_{N+1}) + \sum_{j=1}^{N} Y_{2j}V_j \\
I_3 &= Y_3(V_3 - V_{N+1}) + \sum_{j=1}^{N} Y_{3j}V_j \\
&\vdots \\
I_N &= Y_N(V_N - V_{N+1}) + \sum_{j=1}^{N} Y_{Nj}V_j \\
I_{N+1} &= \sum_{j=1}^{N} Y_j(V_{N+1} - V_j) + Y_0V_{N+1}.
\end{aligned}
$$

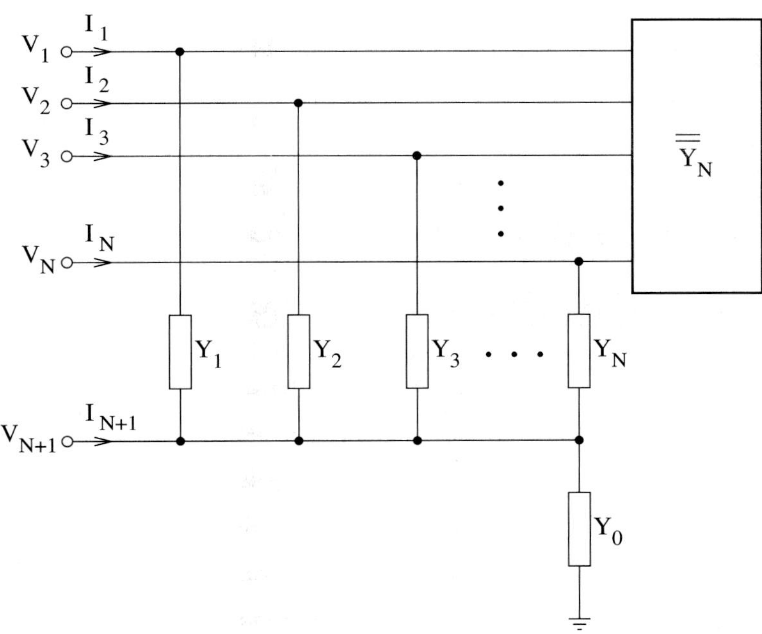

Figure 9.11. Augmented $N \times N$ admittance matrix to produce an $(N+1) \times (N+1)$ matrix.

In matrix form, this result can be rewritten as

$$\begin{bmatrix} I_1 \\ I_2 \\ \vdots \\ I_N \\ I_{N+1} \end{bmatrix} = \begin{bmatrix} Y_1 + Y_{11} & Y_{12} & \cdots & Y_{1N} & -Y_1 \\ Y_{21} & Y_2 + Y_{22} & \cdots & Y_{2N} & -Y_2 \\ \vdots & \vdots & \ddots & \vdots & \vdots \\ Y_{N1} & Y_{N2} & \cdots & Y_N + Y_{NN} & -Y_N \\ -Y_1 & -Y_2 & \cdots & -Y_N & Y_0 + Y_1 + \cdots + Y_N \end{bmatrix} \begin{bmatrix} V_1 \\ V_2 \\ \vdots \\ V_N \\ V_{N+1} \end{bmatrix}.$$

The new $(N+1) \times (N+1)$ admittance matrix is

$$\overline{\overline{Y}}_{N+1} = \begin{bmatrix} Y_1 + Y_{11} & Y_{12} & \cdots & Y_{1N} & -Y_1 \\ Y_{21} & Y_2 + Y_{22} & \cdots & Y_{2N} & -Y_2 \\ \vdots & \vdots & \ddots & \vdots & \vdots \\ Y_{N1} & Y_{N2} & \cdots & Y_N + Y_{NN} & -Y_N \\ -Y_1 & -Y_2 & \cdots & -Y_N & Y_0 + Y_1 + \cdots + Y_N \end{bmatrix}. \quad (9.22)$$

Each new entry in the $(N + 1)$th row and column includes a new and arbitrary admittance. Therefore, the new matrix can represent an arbitrary $(N+1) \times (N+1)$ admittance matrix as long as the $N \times N$ admittance matrix is also general. Since the process can start with the arbitrary 1×1 matrix, a general admittance matrix of any size can be built up by successively adding one port.

For any $N \times N$ admittance matrix, a circuit representation is constructed by tying each port to every other port, and to ground, with an admittance. The value of the admittance between port i and port j is just $-Y_{ij}$. The value of the admittance from port i to ground is just $\sum_{j=1}^{N} Y_{ij}$. Note the similarity to the implementation of the capacitance matrix.

As an example, consider the construction of a 2×2 network starting with the 1×1 in Figure 9.12a. Adding a second port yields the circuit in Figure 9.12b by connecting the second port to the first with the admittance Y_2 and to ground with Y_3. From (9.22), the new admittance matrix is the old with Y_2 added to Y_{11}, and new elements $Y_{12} = Y_{21} = -Y_2$ and $Y_{22} = Y_2 + Y_3$. The resulting admittance matrix

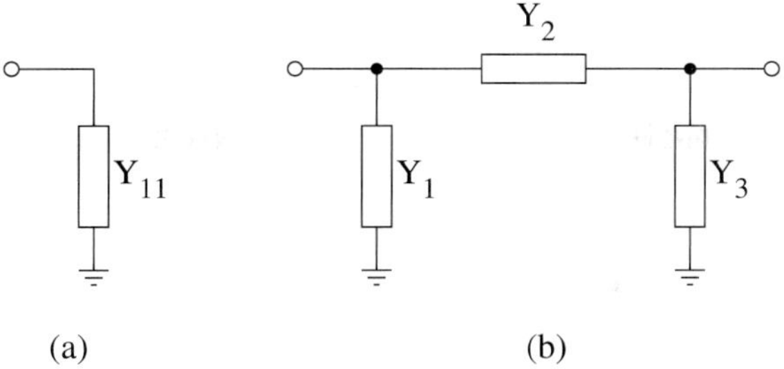

Figure 9.12. Construction of a 2×2 admittance network (a π-network) from a 1×1 network.

is

$$\overline{\overline{Y}} = \begin{bmatrix} Y_2 + Y_{11} & -Y_2 \\ -Y_2 & Y_2 + Y_3 \end{bmatrix}.$$

This result is derived differently in problem 13 of chapter 4.

9.10 Multidrop Nets

Multidrop nets connect together several points in a circuit. For example, in Figure 9.13a, four nodes in a circuit are connected together with a three-drop net, where one node is considered to be the input. Assuming that the lumped approximation holds for the net, an intuitive approach models each segment with a resistor and inductor to achieve the schematic in Figure 9.13b. However, the model's component values are not obvious from a multiport S-parameter (or equivalent) data set. An extraction strategy is needed, and optimization is often used. Alternatively, a general frequency-dependent model using the technique in section 9.9 can be constructed, but frequency-dependent models can be difficult to implement in time-domain simulators.

Or, an equivalent model can be extracted using the theory in section 9.2 to obtain a frequency-independent model (within limits) with a straightforward extraction methodology. By grounding node 1, the assumed input port, a three-port impedance

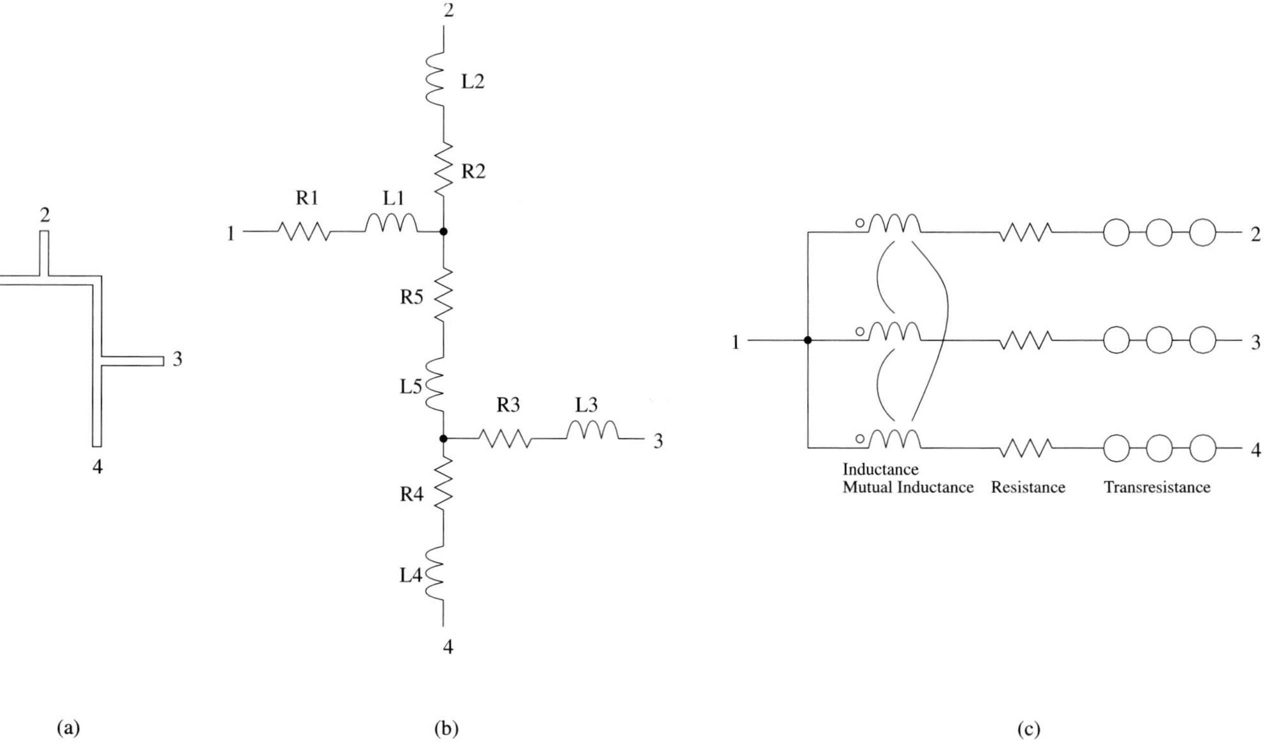

Figure 9.13. Multidrop net and modeling approaches: (a) a four-terminal, three-drop net, (b) intuitive model by sections, (c) model using a lumped extraction methodology.

matrix characterization yields a three-conductor resistive/inductive impedance model, similar to that in Figure 9.1, through the application of (9.5) and (9.6). By tying the ends of the interconnects together at node 1, the model in Figure 9.13c is obtained.

Planes often have multiple contact points, as shown in Figure 9.4, that can be treated as internal nodes. Planes can also be viewed as multidrop nets and modeled as such. Given the heavier mathematical load with internal node modeling (see section 9.4), multidrop net models for planes are simpler to produce.

9.11 Exercises

1. A five-lead interconnect with a common grounded return is shorted to ground at the far end. Port 1 of a 50Ω VNA is applied at the near end between lead 3 and ground, while port 2 is applied between lead 5 and ground. All other leads remain open. At 100MHz, the measured S-parameters are

$$\overline{\overline{S}} = \begin{bmatrix} -0.98697 + \jmath 0.124425 & 0.00682198 + \jmath 0.0492758 \\ 0.00682198 + \jmath 0.0492758 & -0.981598 + \jmath 0.148791 \end{bmatrix}.$$

   Find and sketch a lumped model for these two leads.

2. (a) Without using vector notation shortcuts, write the mesh analyses equations for the three-conductor interconnect in Figure 9.14.

   (b) Rearrange the equations and show how the results are equivalent to (9.2).

3. (a) Without using vector notation shortcuts, write the nodal analyses equations for the three-conductor interconnect in Figure 9.15.

   (b) Rearrange the equations and show how the results are equivalent to (9.3).

4. A lossless 150Ω transmission line with 0.1ns of delay is to be modeled with a π-network. Derive and sketch the circuit. Over what bandwidth can this model be used? What is the fastest rise time that the model supports?

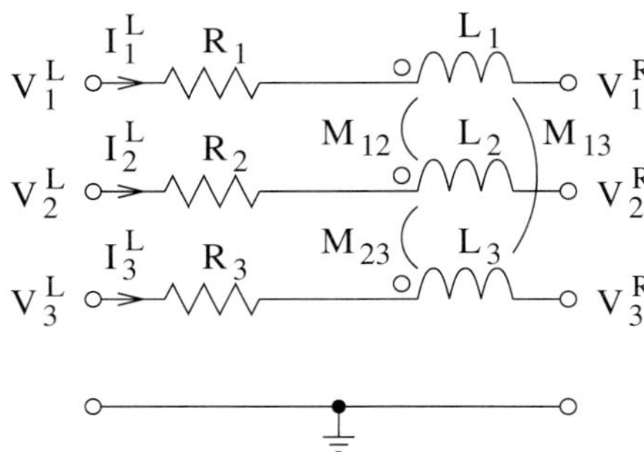

Figure 9.14. A three-conductor interconnect lumped model with resistance and inductance, for problem 2.

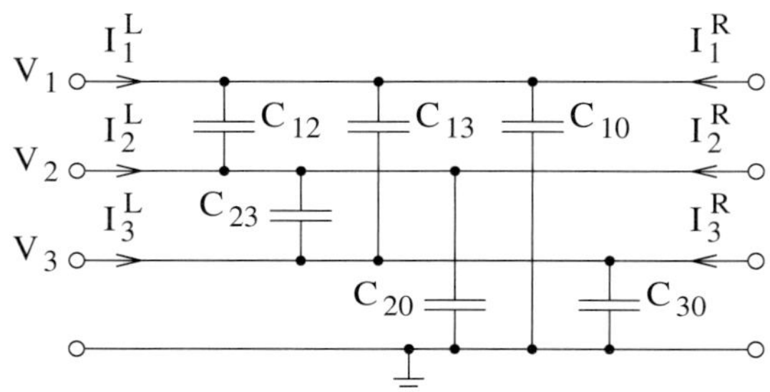

Figure 9.15. A three-conductor interconnect lumped model with capacitance, for problem 3.

5. For a lossless 60Ω transmission line with $\epsilon_{\text{eff}} = 4$, what is the longest section that can be modeled with a single T- or π-model for signals with edge rates down to 2ns?

6. For a lossless transmission line, the lumped approximation requires that $\tan(\beta d) \approx \beta d$, where $\beta = \omega\sqrt{\ell c}$. Define the error in the approximation as $\epsilon = (\beta d - \tan(\beta d))/\tan(\beta d)$, then what is the allowed bandwidth for $|\epsilon| < 5\%$? How does the result compare to (9.20)?

7. Apply the iterative impedance concept directly to the input impedance of a

transmission line given by (2.66) in chapter 2. What happens, and what does the result mean?

8. A branched network is shown in Figure 9.16.

   (a) Use the two-sample method to find a 2×2 inductance matrix and a capacitor value that describe the network.

   (b) Sketch a model for the circuit using the topology of Figure 9.1.

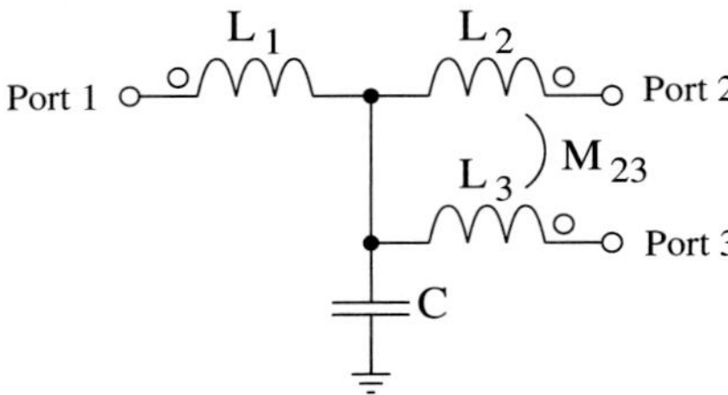

Figure 9.16. A branched network, for problem 8.

Chapter 10

WIDEBAND MODELING

Single lumped models with fixed component values are usable over a very broad bandwidth, from DC to an upper frequency limit where the lumped approximation breaks down. To capture high-frequency detail, including skin effects and propagation delays, more complex models are required. Enhanced models that exceed the frequency range of fixed-value single lumped models are termed *wideband models.* Transmission lines are describable in precise mathematical terms, so several specialized wideband techniques are available for them, including multilumped models and modal decompositions.

General networks described by multiport frequency-dependent data can be implemented using a variety of techniques. A frequency-dependent circuit can be extracted from the network's admittance matrix, but many simulators do not work well, if at all, with frequency-dependent components. Alternatively, a ladder network with frequency-independent components can be synthesized after fitting the data with a rational function approximation. Or, a circuit topology can be guessed and then fitted to the data either manually or semiautomatically.

Failure to include sufficient bandwidth in interconnect models can lead to serious defects in simulation results. Timing can be affected by inaccurate delay. Noise margins can be exceeded due to excessive overshoot, undershoot, and ringing. Lack of frequency-dependent losses can cause edge rates that are too fast, leading to overestimations of crosstalk and simultaneous switching noise. Lack of high-frequency losses can also lead to nonphysical, high-frequency oscillations in the simulation.

10.1 Transmission Line Lumped Modeling

Because of the uniformity along its length, transmission line lumped models can be subdivided into sections to increase bandwidth. While most simulators include a transmission line model, lumped models of transmission lines can speed up simulations and reduce convergence problems. Many interconnects form approximate transmission lines where lumped modeling can be very useful. Bandwidth issues for transmission lines provides good insight for many interconnects.

10.1.1 Limits of Lumped Modeling

Lumped models are usable only up to a maximum frequency where the lumped approximation fails. In chapter 9, section 9.6 it is shown that the maximum frequency for lumped π- and T-models for less than 5% error is (repeated here from (9.20))

$$\omega < \frac{0.62}{\sqrt{LC}}. \tag{10.1}$$

Assuming that $f_{max} = 1/t_r$, where t_r is the shortest edge rate in the digital signal, then (10.1) implies that

$$t_r > 10\sqrt{LC}. \tag{10.2}$$

Any simulation with the model must restrict all signals traversing the lumped model to edge rates greater than or equal to this value. If the π- or T-model represents an actual transmission line, then the delay (time of flight) of the transmission line is $t_d = \sqrt{LC}$ and (10.2) yields

$$t_r > 10t_d. \tag{10.3}$$

Alternatively, since $d = v_p t_d$, then

$$t_r > 10\frac{d}{v_p}, \tag{10.4}$$

where v_p is the phase velocity of the transmission line. The two inequalities in (10.1) and (10.3) provide limits on the model computed in either the frequency or time domains. Both state that the rise/fall time must be long compared to the time of flight of the transmission line for the lumped model to be valid.

10.1.2 Multilumped Models

Lumped models can be cascaded to create wide bandwidth interconnect models. If a line is two times too long for a lumped model, then simply cut it in half and model the two parts separately. Because each section is half the length of the original, the bandwidth is doubled. This process can be extended to provide the needed bandwidth for an interconnect of any length. However, large numbers of lumped sections can lead to slow runtimes.

As an example, consider a 15cm-long transmission line with an effective dielectric constant of 4 that will need to support signals with an edge rate as fast as 1ns. The phase velocity is $3 \times 10^8/\sqrt{4}$m/s, so (10.4) provides that $10^{-9} > 10d/1.5 \times 10^8$, or $d < 1.5$cm. Since the line needs to be 15cm long, then a cascade of 10 π- or T-models is required. The effects of lumping for this example are clearly visible in Figure 10.1, where different levels of lumping are compared to the full transmission line simulation using the SPICE transmission line model. The results using the

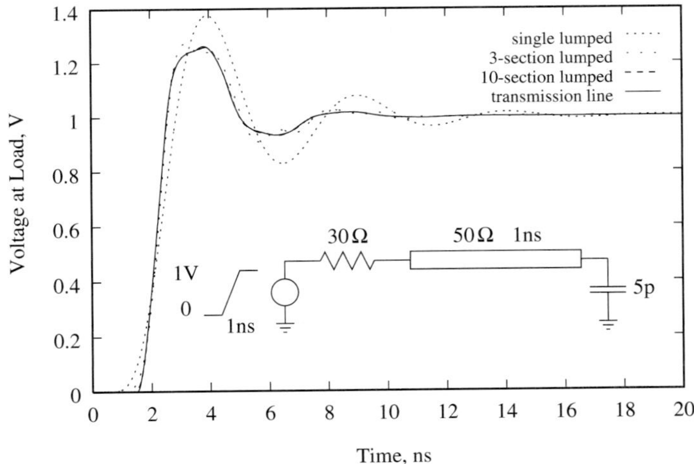

Figure 10.1. Simulation of a transmission line with a multilumped model can be very accurate.

multilumped model with 10 lumps is almost indistinguishable from that using the transmission line model.

Transmission line models can sometimes cause convergence problems and/or slow runtimes. In these cases, simulation performance can be improved, sometimes dramatically, by replacing the transmission line model with a multilumped model.

Time-Domain Reflectometry

Time-domain reflectometry (TDR) offers a convenient method to obtain multilumped models under certain circumstances. When losses are low and coupling is minimal, a TDR can provide information on the impedance and total delay of sections of an interconnect. Using impedance profile extraction (chapter 8, section 8.4.3) or layer peeling (chapter 8, section 8.4.4), an accurate wideband multilumped model of an interconnect can be constructed.

2D Simulation of 3D Structures

Long interconnects routed on an orthogonal grid (Manhattan routing) over unbroken ground planes can be accurately modeled using two-dimensional decomposition. As suggested in Figure 10.2, the structure is sliced to produce short interconnects that satisfy (10.2) or a similar limit. Each slice is then simulated with a 2D electromagnetic simulator to find the inductance, capacitance, and resistance matrices per-unit-length for all of the conductors. The results from the slices are cascaded to produce a model of the whole interconnect structure.

Many commercial simulators are based on this technique, and they are very accurate as long as the underlying assumptions are satisfied. Compared to full 3D simulation, the technique is very fast and can solve very large problems. Full-scale printed circuit boards are easily accommodated.

10.2 Coupled Transmission Lines

Coupled transmission lines often occur in bus structures and areas of high wiring congestion. Because of the strong crosstalk, accurate simulation requires a model

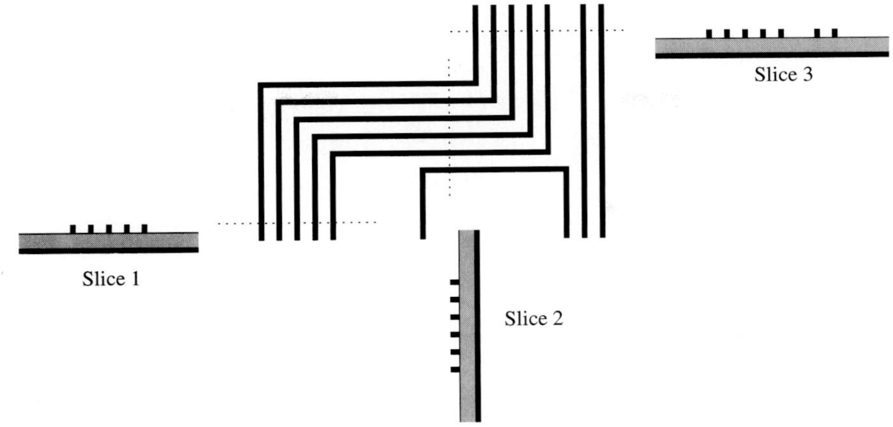

Figure 10.2. A section of orthogonal routing can be sectioned to enable characterization by repeated 2D analyses.

of the full coupled structure. Multilumped models can be used, but the mathematics of coupled transmission lines enable an elegant solution based on modal decomposition, with powerful applications in modeling and theory. The equations of coupled transmission lines are derived first, followed by a derivation of the modal decomposition and then some applications.

10.2.1 Telegrapher Equations

Consider the lumped L-network in Figure 10.3. Note that this figure is very similar to Figure 9.1 in chapter 9, except that the currents $\overline{i}^R$ are reversed to facilitate cascading multiple sections. Kirchhoff's voltage law requires

$$\overline{v}^R - \overline{v}^L = -\overline{\overline{R}}\,\overline{i}^L - \overline{\overline{L}}\frac{\partial \overline{i}^L}{\partial t},$$

while Kirchhoff's current law requires

$$\overline{i}^R - \overline{i}^L = -\overline{\overline{G}}\,\overline{v}^R - \overline{\overline{C}}\frac{\partial \overline{v}^R}{\partial t}.$$

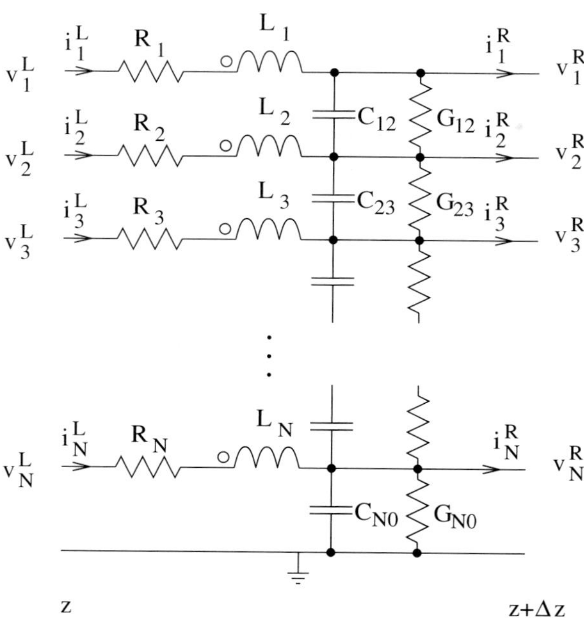

Figure 10.3. A differential section of a multiconductor transmission line suitable for cascading. All of the mutual inductors and most of the capacitors and conductances are omitted for clarity.

Divide through each by the length of the segment, Δz, to obtain

$$\frac{\overline{v}^R - \overline{v}^L}{\Delta z} = -\frac{\overline{\overline{R}}}{\Delta z}\,\overline{i}^L - \frac{\overline{\overline{L}}}{\Delta z}\frac{\partial \overline{i}^L}{\partial t},$$

and

$$\frac{\overline{i}^R - \overline{i}^L}{\Delta z} = -\frac{\overline{\overline{G}}}{\Delta z}\,\overline{v}^R - \frac{\overline{\overline{C}}}{\Delta z}\frac{\partial \overline{v}^R}{\partial t}.$$

Let $\Delta z \to 0$ to shrink the segment to an infinitesimal length to obtain

$$\frac{\partial \overline{v}}{\partial z} = -\overline{\overline{r}}\,\overline{i} - \overline{\overline{\ell}}\frac{\partial \overline{i}}{\partial t}, \tag{10.5}$$

and

$$\frac{\partial \overline{i}}{\partial z} = -\overline{\overline{g}}\,\overline{v} - \overline{\overline{c}}\frac{\partial \overline{v}}{\partial t}, \tag{10.6}$$

where $\bar{\bar{r}}$, $\bar{\bar{\ell}}$, $\bar{\bar{g}}$, and $\bar{\bar{c}}$ are the per-unit-length resistance, inductance, conductance, and capacitance matrices, respectively. Equations (10.5) and (10.6) are collectively called the *telegrapher equations.* By incorporating the telegrapher equations directly into the simulator, simulation times of long transmission lines can be significantly reduced compared to multilumped models.

The most common transmission line model in SPICE-like simulators is a single line with frequency-independent per-unit-length matrices. Therefore, skin effects and crosstalk cannot be modeled. In many circumstances, the dominant electrical effects are provided by the impedance and delay of the transmission line, so this implementation of the telegrapher equations is still quite useful.

More complex implementations of the telegrapher equations are the focus of ongoing research projects, and many viable formulations have been proposed. As time goes by, the transmission line capabilities built into circuit simulators will advance. One of the mature techniques for simulating multiconductor interconnects is modal decomposition.

10.2.2 Modal Decomposition

In chapter 5, section 5.8, modal decomposition of inductance matrices is used to replace the full and dense $N \times N$ inductance matrix with N linearly independent circuits. The terminal voltages and currents are broken into components corresponding to the modes of the inductance matrix (the eigenvectors). The circuit behavior of each mode is computed separately using the modal inductances (the eigenvalues), and the results of all the modes are recombined to obtain the complete circuit behavior. With a modal decomposition, the effect of the mutual inductances is completely simulated while not actually using a mutual inductance circuit element.

The same approach can be used to solve the telegrapher equations. The per-unit-length inductance and capacitance matrices are processed to produce a modal decomposition. Each mode represents a single isolated transmission line that is analyzed separately. The overall response of the coupled lines is then the recombined modal results. With a modal decomposition, the circuit response is based on simulations using single uncoupled transmission lines, and it is not necessary to implement a transmission line solver to cover the general N-line coupled case; the single-line solver can just be reused. In total, N conductors plus a common return path leads to N eigenvalues, N modes, and N decoupled transmission lines.

Consider the loss-free case with $\overline{\overline{r}} = \overline{\overline{g}} = 0$. Eliminating $\overline{i}$ between (10.5) and (10.6) yields the voltage wave equation

$$\frac{\partial^2 \overline{v}}{\partial z^2} = \overline{\overline{\ell}}\,\overline{\overline{c}}\frac{\partial^2 \overline{v}}{\partial t^2}. \tag{10.7}$$

The product $\overline{\overline{\ell}}\,\overline{\overline{c}}$ forms a new square matrix that is not symmetric even though $\overline{\overline{\ell}}$ and $\overline{\overline{c}}$ are symmetric. However, $\overline{\overline{\ell}}\,\overline{\overline{c}}$ can still be diagonalized as

$$\overline{\overline{\ell}}\,\overline{\overline{c}} = \overline{\overline{T}}_V \overline{\overline{\lambda}}\,\overline{\overline{T}}_V^{-1}, \tag{10.8}$$

where $\overline{\overline{\lambda}}$ contains the eigenvalues of $\overline{\overline{\ell}}\,\overline{\overline{c}}$, and $\overline{\overline{T}}_V$ is constructed as in chapter 5, equation (5.57), using the eigenvectors of $\overline{\overline{\ell}}\overline{\overline{c}}$. Note that (5.58) does not hold because $\overline{\overline{\ell}}\overline{\overline{c}}$ is not symmetric. Substituting (10.8) into (10.7) and multiplying through by $\overline{\overline{T}}_V^{-1}$ yields

$$\frac{\partial^2 \overline{\overline{T}}_V^{-1}\overline{v}}{\partial z^2} = \overline{\overline{\lambda}}\frac{\partial^2 \overline{\overline{T}}_V^{-1}\overline{v}}{\partial t^2}.$$

Defining the modal voltages as

$$\overline{v}_m = \overline{\overline{T}}_V^{-1}\overline{v}, \tag{10.9}$$

then

$$\frac{\partial^2 \overline{v}_m}{\partial z^2} = \overline{\overline{\lambda}}\frac{\partial^2 \overline{v}_m}{\partial t^2}, \tag{10.10}$$

which are decoupled since $\overline{\overline{\lambda}}$ is diagonal.

If $\overline{v}$ is eliminated between (10.5) and (10.6) instead of $\overline{i}$, then a current wave equation is found as

$$\frac{\partial^2 \overline{i}}{\partial z^2} = \overline{\overline{c}}\,\overline{\overline{\ell}}\frac{\partial^2 \overline{i}}{\partial t^2}.$$

Note that $\bar{\bar{c}}$ and $\bar{\bar{\ell}}$ are reversed compared to (10.7). The matrix products $\bar{\bar{c}}\bar{\bar{\ell}}$ and $\bar{\bar{\ell}}\bar{\bar{c}}$ have the same eigenvalues but not the same eigenvectors. This follows since a matrix and its transpose have the same determinant, i.e., $|\bar{\bar{A}}^T| = |\bar{\bar{A}}|$, and $(\bar{\bar{c}}\bar{\bar{\ell}})^T = \bar{\bar{\ell}}^T\bar{\bar{c}}^T = \bar{\bar{\ell}}\bar{\bar{c}}$ since $\bar{\bar{c}}$ and $\bar{\bar{\ell}}$ are symmetric.

The modal currents are then defined as

$$\bar{i}_m = \bar{\bar{T}}_I^{-1}\bar{i}, \tag{10.11}$$

where $\bar{\bar{T}}_I$ is constructed from the eigenvectors of $\bar{\bar{c}}\bar{\bar{\ell}}$. The wave equation for each modal current is then

$$\frac{\partial^2 \bar{i}_m}{\partial z^2} = \bar{\bar{\lambda}}\frac{\partial^2 \bar{i}_m}{\partial t^2}, \tag{10.12}$$

where $\bar{\bar{\lambda}}$ is shared with (10.10) and

$$\bar{\bar{c}}\bar{\bar{\ell}} = \bar{\bar{T}}_I\bar{\bar{\lambda}}\bar{\bar{T}}_I^{-1}. \tag{10.13}$$

Since $\bar{\bar{\lambda}}$ is diagonal, (10.10) and (10.12) are decoupled. Rather than solve one $N \times N$ multiconductor transmission line problem, the modal decomposition allows the simultaneous solution of N single transmission line problems.

Single transmission lines are characterized by their characteristic impedance, Z_o, and their total delay, $t_d = d/v_p$, where d is the line length and v_p is the phase velocity or speed of light on the line. So far, it has been shown that multiconductor transmission lines can be decomposed into decoupled single transmission lines that each propagate a single mode. It is now time to find the characteristic impedances and delays for the decoupled transmission lines.

Using (10.9) and (10.11) to replace the voltages and currents in the telegrapher equations given by (10.5) and (10.6) (with $\bar{\bar{r}} = \bar{\bar{g}} = 0$) yields

$$\frac{\partial \bar{v}_m}{\partial z} = -\bar{\bar{T}}_V^{-1}\bar{\bar{\ell}}\bar{\bar{T}}_I\frac{\partial \bar{i}_m}{\partial t} \tag{10.14}$$

and

$$\frac{\partial \bar{i}_m}{\partial z} = -\bar{\bar{T}}_I^{-1}\bar{\bar{c}}\bar{\bar{T}}_V\frac{\partial \bar{v}_m}{\partial t}. \tag{10.15}$$

Define new modal inductance and capacitance matrices as

$$\overline{\overline{\ell}}_m = \overline{\overline{T}}_V^{-1} \overline{\overline{\ell}}\, \overline{\overline{T}}_I \tag{10.16}$$

and

$$\overline{\overline{c}}_m = \overline{\overline{T}}_I^{-1} \overline{\overline{c}}\, \overline{\overline{T}}_V. \tag{10.17}$$

If $\overline{\overline{\ell}}_m$ and $\overline{\overline{c}}_m$ are diagonal, then the needed single transmission line parameters are available by inspection.

From (10.8),

$$\begin{aligned}
\overline{\overline{\lambda}} &= \overline{\overline{T}}_V^{-1} \overline{\overline{\ell}}\, \overline{\overline{c}}\, \overline{\overline{T}}_V \\
&= \overline{\overline{T}}_V^{-1} \overline{\overline{\ell}}\, \overline{\overline{T}}_I \overline{\overline{T}}_I^{-1} \overline{\overline{c}}\, \overline{\overline{T}}_V \\
&= \overline{\overline{\ell}}_m\, \overline{\overline{c}}_m.
\end{aligned}$$

A similar manipulation on (10.13) yields $\overline{\overline{\lambda}} = \overline{\overline{c}}_m\, \overline{\overline{\ell}}_m$.

Eliminating $\overline{\overline{\ell}}_m$ yields

$$\overline{\overline{c}}_m\, \overline{\overline{\lambda}} = \overline{\overline{\lambda}}\, \overline{\overline{c}}_m.$$

Since matrices commute only when both are diagonal, $\overline{\overline{c}}_m$ must be diagonal ($\overline{\overline{\lambda}}$ is already known to be diagonal). From a similar derivation, $\overline{\overline{\ell}}_m$ can also be shown to be diagonal. Substituting the diagonal modal inductance and capacitance matrices into (10.14) and (10.15) leads to

$$\frac{\partial v_{mj}}{\partial z} = -\ell_{mj} \frac{\partial i_{mj}}{\partial t}, \qquad j = 1, 2, \cdots, N,$$

and

$$\frac{\partial i_{mj}}{\partial z} = -c_{mj} \frac{\partial v_{mj}}{\partial t}, \qquad j = 1, 2, \cdots, N.$$

These two equations are exactly the same as in chapter 2, (2.5) and (2.6), except that the voltage, current, inductance, and capacitance apply to modal quantities. The characteristic impedance of the jth modal transmission line is $Z_o = \sqrt{\ell_{mj}/c_{mj}}$, and its phase velocity is $v_p = 1/\sqrt{\ell_{mj} c_{mj}}$. The length of each modal transmission line is just that of the multiconductor transmission line.

A sign ambiguity exists in $\overline{\overline{T}}_V$ and $\overline{\overline{T}}_I$. Eigenvalue problems solve the equation $\overline{\overline{A}}\overline{x} = \lambda\overline{x}$. For any solution λ_i with its matching eigenvector $\overline{x}_i$, the vector $-\overline{x}_i$ is also a solution. When the eigenvectors are assembled into $\overline{\overline{T}}_V$ and $\overline{\overline{T}}_I$, these sign ambiguities can cause decomposition into nonphysical modes represented by negative modal inductances and capacitances on the diagonals of (10.16) and (10.17). The sign ambiguities can be corrected by computing $\overline{\overline{\ell}}_m$ and forcing all terms to be positive. Keeping the original $\overline{\overline{T}}_V$ and $\overline{\overline{\ell}}$, a corrected version of $\overline{\overline{T}}_V$ can be computed from (10.16) as

$$\overline{\overline{T}}_V = \overline{\overline{\ell}}\,\overline{\overline{T}}_I \overline{\overline{\ell}}_m^{-1}.$$

The corrected $\overline{\overline{c}}_m$ can then be computed from (10.17) using the original $\overline{\overline{T}}_I$ and $\overline{\overline{c}}$.

Homogeneous Materials

If all of the electric field lines see only a single dielectric, then the phase velocities of all of the modes are equal to

$$v_p = \frac{c_o}{\sqrt{\epsilon_r}},$$

where ϵ_r is the dielectric constant of the material and $c_o = 3 \times 10^8$m/s (speed of light in free space). In this case, $\overline{\overline{\lambda}}$ is known because all diagonal terms equal $1/v_p^2$, then (10.8) simplifies as

$$\begin{aligned} \overline{\overline{\ell}}\,\overline{\overline{c}} &= \overline{\overline{T}}_V \overline{\overline{\lambda}}\,\overline{\overline{T}}_V^{-1} \\ &= \frac{1}{v_p^2}\overline{\overline{T}}_V \overline{\overline{T}}_V^{-1} \\ &= \frac{1}{v_p^2}. \end{aligned}$$

If $\overline{\overline{c}}$ is known, then

$$\overline{\overline{\ell}} = \frac{1}{v_p^2}\overline{\overline{c}}^{-1}. \tag{10.18}$$

Similarly, $\overline{\overline{c}}$ can be found from $\overline{\overline{\ell}}$. Therefore, the computational load can be halved by computing only one matrix then using matrix inversion to find the other. Often the trick is to compute inductance from capacitance, but since partial inductance formulas are readily available, analytical modeling can often go the other way. However, the underlying assumption of (10.18) is that there are no losses, so the inductance must not include internal inductance.

Manual Implementation

While a hard-coded implementation provides the best performance, multiconductor transmission lines can be manually implemented in any circuit simulator. The voltages and currents at each end of the multiconductor transmission line are transformed into the modal voltages and currents, which are then propagated on decoupled transmission lines. It is assumed that the per-unit-length inductance and capacitance matrices are known.

Two eigenvalue problems must be solved. The eigenvectors of $\overline{\overline{\ell}}\,\overline{\overline{c}}$ are computed, orthogonalized and normalized through application of the Gram-Schmidt orthonormalization process. The eigenvectors are then assembled to form $\overline{\overline{T}}_V$. The process is repeated for $\overline{\overline{c}}\,\overline{\overline{\ell}}$, and $\overline{\overline{T}}_I$ is formed. With $\overline{\overline{T}}_V$ and $\overline{\overline{T}}_I$ known, the parameters of the decoupled lines, $\overline{\overline{\ell}}_m$ and $\overline{\overline{c}}_m$, are computed. Additional calculations may be required to eliminate sign ambiguities.

For an N-conductor transmission line (not counting the return path), there are N decoupled modal transmission lines in a modal decomposition. The voltages and currents at each end where the multiconductor transmission line attaches to the circuit must be coupled to the ends of the modal transmission lines. The approach described here takes the currents to the multiconductor transmission line and transforms them to modal currents, which are then applied as sources to the modal transmission lines. The modal voltages on these modal current sources are transformed to voltages and applied back to the circuit as sources where the multiconductor transmission line attaches.

An N-conductor transmission line is sketched in Figure 10.4. A circuit to model a multiconductor transmission line is shown in Figure 10.5. The two ends of the multiconductor transmission line are treated identically. For terminals 1 to N, the terminal voltages are computed from (10.9) using the modal voltages v_{m1} through v_{mN} taken from the modal transmission lines. N sources are required for each terminal and are connected in series to allow the circuit simulator to perform the summation. The terminal currents i_1 through i_N are detected and used to construct the modal currents i_{m1} through i_{mN} using (10.11). N sources are applied across each modal transmission line, where again the simulator performs the summation. For terminals $(N+1)$ to $2N$, the same transformations are applied with appropriate modifications.

10.2.3 Modal Decomposition Examples

Modal decomposition offers a rich framework for examining signal integrity questions in detail. A numerical example is followed by two analytical examples. Additional analytical work appears in chapter 11, section 11.3 on the subject of multiconductor termination.

Five-Conductor Transmission Line

A five-conductor multiconductor transmission line is shown in Figure 10.6. The inductance matrix in nH/m computed from a magnetostatic finite-element simulation is

$$\overline{\overline{\ell}} = \begin{bmatrix} 371.0 & 83.12 & 28.27 & 12.49 & 6.575 \\ 83.12 & 367.8 & 82.14 & 27.98 & 12.48 \\ 28.27 & 82.14 & 367.7 & 82.16 & 28.26 \\ 12.49 & 27.98 & 82.16 & 368.0 & 83.13 \\ 6.575 & 12.48 & 28.26 & 83.13 & 371.2 \end{bmatrix},$$

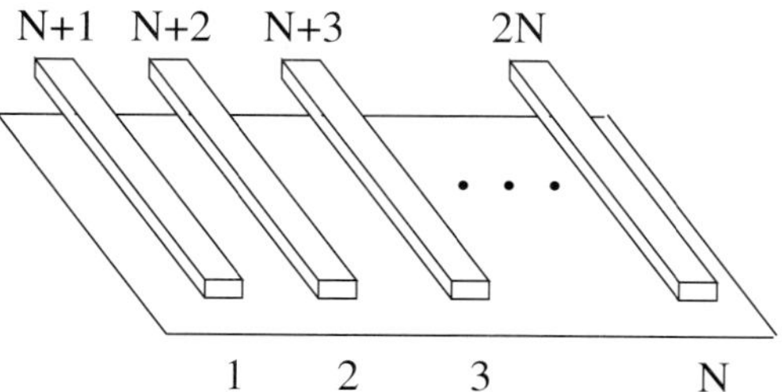

Figure 10.4. Representative N-conductor transmission line with two sets of N terminals.

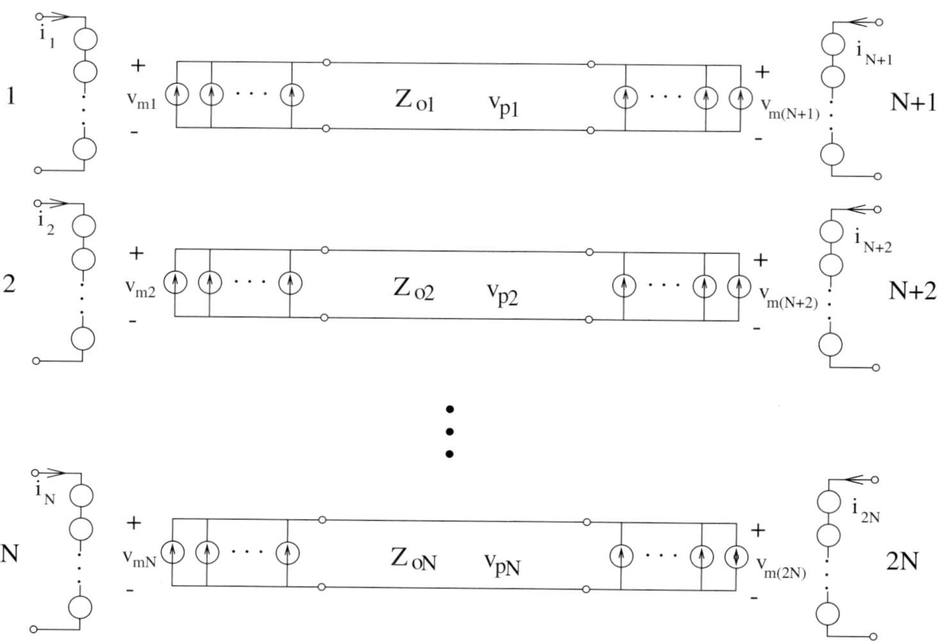

Figure 10.5. Modal decomposition using decoupled transmission lines to model an N-conductor transmission line.

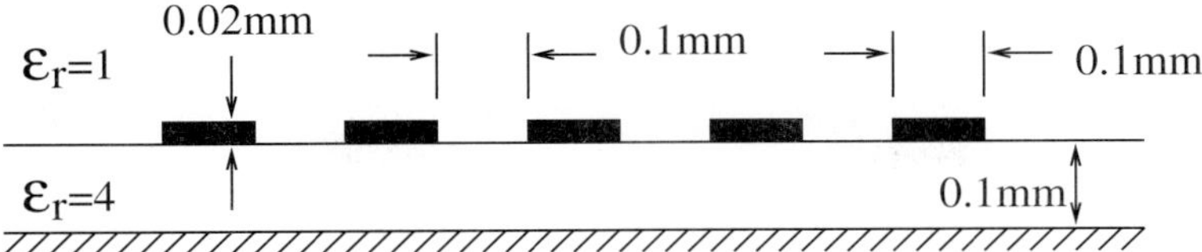

Figure 10.6. Cross section of a five-conductor multiconductor transmission line. All conductors have zero resistivity and all dielectrics are lossless.

while the capacitance matrix in pF/m from an electrostatic finite-element simulation is

$$\bar{\bar{c}} = \begin{bmatrix} 83.61 & -10.33 & -0.7328 & -0.2872 & -0.1612 \\ -10.33 & 85.24 & -10.21 & -0.6915 & -0.2867 \\ -0.7328 & -10.21 & 85.27 & -10.20 & -0.7307 \\ -0.2872 & -0.6915 & -10.20 & 85.23 & -10.34 \\ -0.1612 & -0.2867 & -0.7307 & -10.34 & 83.67 \end{bmatrix}.$$

The program in Appendix D takes this data and produces SPICE subcircuit models, also in the Appendix, that implement both a modal decomposition and a brute-force multilumped model. The multilumped model is lumped to accommodate edge rates down to 1ns. Both subcircuits are used in the test circuit in Figure 10.7 to compute near-end and far-end noise for an actively switching line and a quiet line. For the near-end waveforms, the computed results with 2ns edge rates on the voltage sources are shown in Figure 10.8, while the far-end waveforms are shown in Figure 10.9. In both cases the two models agree quite closely. By dropping the edge rates on the voltage sources down to 0.2ns, the lumped model no longer has sufficient bandwidth. The near-end waveforms for this case are shown in Figure 10.10, where ringing is observable on waveforms simulated using the lumped model. In this example, the lumped model does not introduce gross errors; however, the ringing does obscure some detail, leading to less dramatic depiction of pulse reflection and capacitive charging.

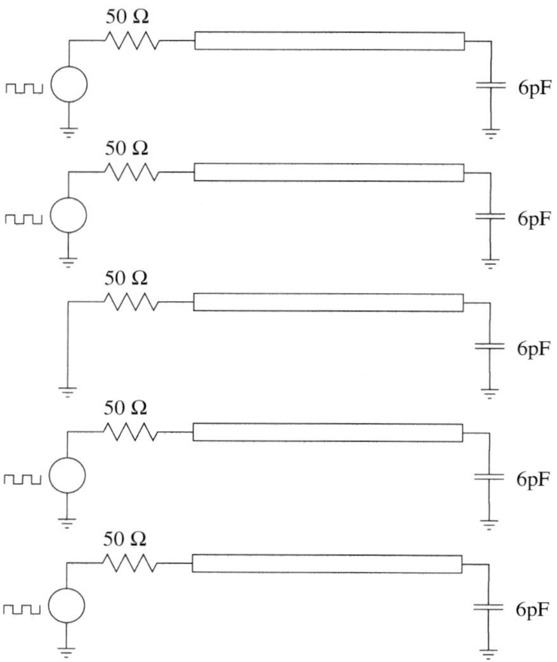

Figure 10.7. Schematic for simulations with the five-conductor transmission line models.

Even and Odd Modes

Consider a two-conductor transmission line composed of identical single lines that are close enough together to couple. Since the physical structure is symmetric, the inductance and capacitance matrices are symmetric, too. Writing

$$\overline{\overline{\ell}} = \begin{bmatrix} \ell & m \\ m & \ell \end{bmatrix}$$

and

$$\overline{\overline{c}} = \begin{bmatrix} c & -c_m \\ -c_m & c \end{bmatrix},$$

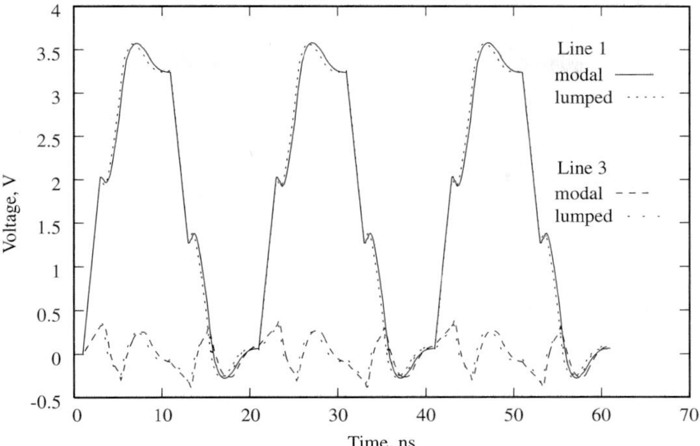

Figure 10.8. Waveforms at the inputs of lines 1 and 3 comparing SPICE simulations using modal decomposition and multilumped modeling.

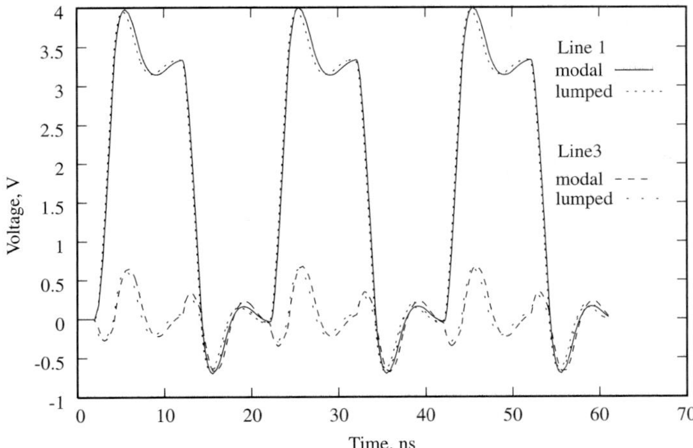

Figure 10.9. Waveforms at the outputs of lines 1 and 3 comparing SPICE simulations using modal decomposition and multilumped modeling.

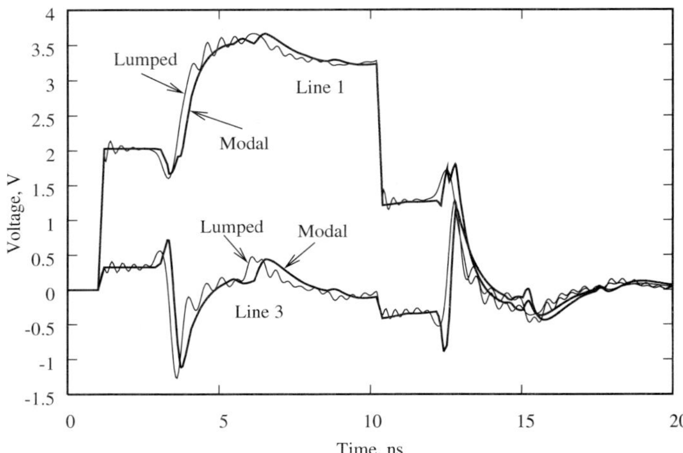

Figure 10.10. Waveforms at the inputs of lines 1 and 3 with insufficient bandwidth for the multilumped model. The ringing in the simulation using the lumped model is caused by a 0.2ns edge rate signal being driven into a model designed for edge rates down to 1ns.

then the eigenvalue problems for $\overline{\overline{\ell}}\,\overline{\overline{c}}$ and $\overline{\overline{c}}\,\overline{\overline{\ell}}$ yield

$$\overline{\overline{T}}_V = \overline{\overline{T}}_I = \begin{bmatrix} \frac{1}{\sqrt{2}} & -\frac{1}{\sqrt{2}} \\ \frac{1}{\sqrt{2}} & \frac{1}{\sqrt{2}} \end{bmatrix}. \tag{10.19}$$

Note that

$$\overline{\overline{T}}_V^{-1} = \overline{\overline{T}}_I^{-1} = \begin{bmatrix} \frac{1}{\sqrt{2}} & \frac{1}{\sqrt{2}} \\ -\frac{1}{\sqrt{2}} & \frac{1}{\sqrt{2}} \end{bmatrix}. \tag{10.20}$$

From (10.9), the modal voltages are

$$\begin{aligned} \overline{v}_m &= \begin{bmatrix} \frac{1}{\sqrt{2}} & \frac{1}{\sqrt{2}} \\ -\frac{1}{\sqrt{2}} & \frac{1}{\sqrt{2}} \end{bmatrix} \begin{bmatrix} v_1 \\ v_2 \end{bmatrix} \\ &= \begin{bmatrix} \frac{v_2+v_1}{\sqrt{2}} \\ \frac{v_2-v_1}{\sqrt{2}} \end{bmatrix}, \end{aligned} \tag{10.21}$$

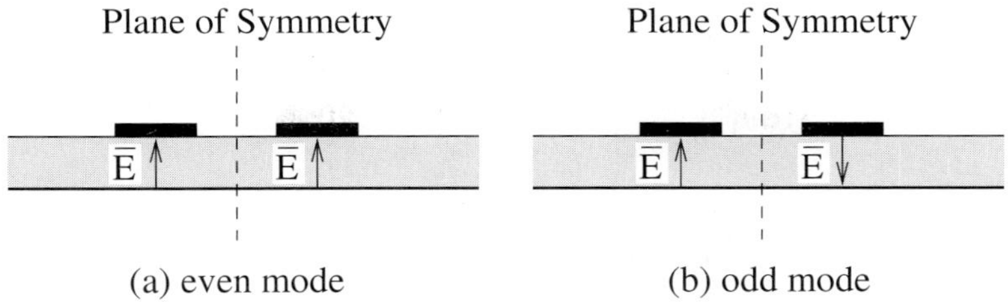

Figure 10.11. Example of mode excitation for symmetric coupled microstrip lines.

where v_1 and v_2 are the voltages driving lines 1 and 2, respectively. Two special cases are immediately apparent. If $v_1 = v_2$, then $v_{m2} = 0$, and only the first modal transmission line is excited. This case is depicted in Figure 10.11a, where the excitation is symmetric with respect to the plane of symmetry for these identical lines. Because the excitation is symmetric, the first mode is called the *even mode*. Similarly, if $v_1 = -v_2$, then $v_{m1} = 0$, and only the second modal transmission line is excited. As sketched in Figure 10.11b, the driving voltages in this case are antisymmetric, so the second mode is called the *odd mode*. If the coupled pair of lines are driven with any other voltages, then both the even and odd modes are excited.

The plane of symmetry can be exploited to compute even- and odd-mode phase velocities and impedances. For the even mode, the tangential magnetic field is zero on the plane of symmetry. A perfect magnetic conductor (PMC) boundary placed at the plane of symmetry can enforce a vanishing tangential magnetic field to enable the computation of the properties of the even mode. Similarly, a perfect electric conductor (PEC) can enforce a vanishing tangential electric field to facilitate computations for the odd mode. In addition to delineating the modes, the PMC and PEC boundaries cut the computational domain in half for substantial reduction in computation time and resources.

The modal inductance and capacitance matrices can be found using (10.16) and (10.17) to obtain

$$\bar{\bar{\ell}}_m = \begin{bmatrix} \ell + m & 0 \\ 0 & \ell - m \end{bmatrix} \tag{10.22}$$

and

$$\bar{\bar{c}}_m = \begin{bmatrix} c - c_m & 0 \\ 0 & c + c_m \end{bmatrix}, \tag{10.23}$$

which are diagonal, as expected. The even- and odd-mode characteristic impedances are

$$Z_{oe} = \sqrt{\frac{\ell + m}{c - c_m}}$$

$$Z_{oo} = \sqrt{\frac{\ell - m}{c + c_m}}$$

respectively, while the even- and odd-mode phase velocities are

$$v_e = \frac{1}{\sqrt{(\ell + m)(c - c_m)}} \tag{10.24}$$

$$v_o = \frac{1}{\sqrt{(\ell - m)(c + c_m)}}. \tag{10.25}$$

A schematic implementing the modal decomposition of the symmetric coupled transmission lines is shown in Figure 10.12 for a line of length d.

For weak coupling, when $\ell \gg m$ and $c \gg c_m$, then

$$Z_{oe} Z_{oo} = \sqrt{\frac{(\ell + m)(\ell - m)}{(c - c_m)(c + c_m)}} = \sqrt{\frac{\ell^2 - m^2}{c^2 - c_m^2}} \approx \frac{\ell}{c} = Z_o^2, \tag{10.26}$$

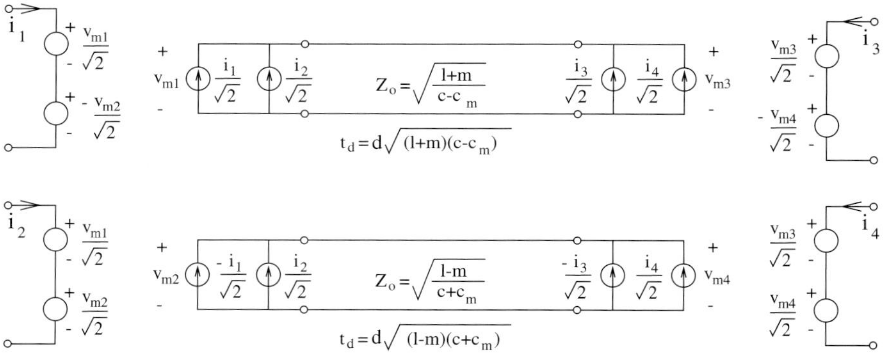

Figure 10.12. Schematic implementation of symmetric coupled transmission lines using modal decomposition into two isolated transmission lines.

where $Z_o = \sqrt{\ell/c}$ is the characteristic impedance of an isolated line. This relationship shows that increasing Z_{oe} always means decreasing Z_{oo}.

Crosstalk on a Symmetric Pair of Lines

Consider a pair of long, symmetric coupled transmission lines with the ports numbered as shown in Figure 10.4. For any excitation at ports 1 and 2, the modal voltages are given by (10.21).

Near-end Crosstalk

The modal currents are derived from impedance relationships as

$$\bar{i}_m = \begin{bmatrix} \frac{v_2+v_1}{\sqrt{2}Z_{oe}} \\ \frac{v_2-v_1}{\sqrt{2}Z_{oo}} \end{bmatrix}.$$

The terminal currents at ports 1 and 2 can be found from the modal currents as

$$\begin{aligned} \bar{i} &= \begin{bmatrix} \frac{1}{\sqrt{2}} & -\frac{1}{\sqrt{2}} \\ \frac{1}{\sqrt{2}} & \frac{1}{\sqrt{2}} \end{bmatrix} \begin{bmatrix} \frac{v_2+v_1}{\sqrt{2}Z_{oe}} \\ \frac{v_2-v_1}{\sqrt{2}Z_{oo}} \end{bmatrix} \\ &= \begin{bmatrix} \frac{v_2+v_1}{2Z_{oe}} - \frac{v_2-v_1}{2Z_{oo}} \\ \frac{v_2+v_1}{2Z_{oe}} + \frac{v_2-v_1}{2Z_{oo}} \end{bmatrix} \end{aligned} \tag{10.27}$$

Let port 2 be terminated in an impedance Z_2. The voltage at port 2 is then

$$v_2 = -i_2 Z_2 = -Z_2 \left(\frac{v_2 + v_1}{2Z_{oe}} + \frac{v_2 - v_1}{2Z_{oo}} \right),$$

where the negative sign accounts for the direction of current flow. Solving for v_2/v_1 to obtain a coefficient for the near-end crosstalk noise yields

$$K_{ne} = \frac{Z_{oe} - Z_{oo}}{Z_{oe} + Z_{oo} + 2\frac{Z_{oe}Z_{oo}}{Z_2}}. \tag{10.28}$$

It is readily apparent that the largest crosstalk occurs for port 2 open-circuited, then the familiar worst-case near-end crosstalk formula is

$$K_{ne,open} = \frac{Z_{oe} - Z_{oo}}{Z_{oe} + Z_{oo}}.$$

An endless variety of special cases can be considered. To consider just one, assume that $Z_2 = \sqrt{\ell/c}$ (the matched condition for the isolated line) and that the coupling is weak. The even-mode impedance is

$$\begin{aligned} Z_{oe} &= \sqrt{\frac{\ell + m}{c - c_m}} \\ &= \sqrt{\frac{\ell}{c}}\sqrt{\frac{1 + m/\ell}{1 - c_m/c}} \\ &\approx \sqrt{\frac{\ell}{c}}\frac{1 + \frac{1}{2}m/\ell}{1 - \frac{1}{2}c_m/c}, \end{aligned} \tag{10.29}$$

and similarly, the odd-mode impedance is

$$Z_{oo} \approx \sqrt{\frac{\ell}{c}}\frac{1 - \frac{1}{2}m/\ell}{1 + \frac{1}{2}c_m/c}. \tag{10.30}$$

Substituting these into (10.28) along with (10.26) and dropping second order small terms yields

$$K_{ne,weak/matched} = \frac{1}{4}\left(\frac{c_m}{c} + \frac{m}{\ell}\right), \tag{10.31}$$

which matches the heuristically derived result in (2.62).

Far-end Crosstalk

The modal voltages injected onto the coupled pair of transmission lines propagate to ports 3 and 4 with delays governed by the modal phase velocities. Assuming that all the ports are sufficiently matched so that reflected wave amplitudes are negligible, then the modal voltages at ports 3 and 4 can be written from (10.21) as

$$\begin{bmatrix} v_{m3} \\ v_{m4} \end{bmatrix} = \begin{bmatrix} \frac{v_2(t-d/v_e)+v_1(t-d/v_e)}{\sqrt{2}} \\ \frac{v_2(t-d/v_o)-v_1(t-d/v_o)}{\sqrt{2}} \end{bmatrix},$$

where d is the length of the coupled pair of lines. The voltages at ports 3 and 4 are

$$\begin{bmatrix} v_3 \\ v_4 \end{bmatrix} = \begin{bmatrix} \frac{1}{\sqrt{2}} & -\frac{1}{\sqrt{2}} \\ \frac{1}{\sqrt{2}} & \frac{1}{\sqrt{2}} \end{bmatrix} \begin{bmatrix} v_{m3} \\ v_{m4} \end{bmatrix}, \tag{10.32}$$

so

$$v_4 = \frac{v_2(t - d/v_e) + v_1(t - d/v_e)}{2} + \frac{v_2(t - d/v_o) - v_1(t - d/v_o)}{2}. \tag{10.33}$$

In a homogeneous medium, all of the transverse electromagnetic modes have the same phase velocity, so for a symmetric two-line system $v_e = v_o$. In this case, the voltages from line 1 cancel, leaving

$$v_4 = v_2(t - d/v_e),$$

which is just the time-delayed version of any signal driven at port 2. Specifically, there is no dependence on v_1, so there is no crosstalk. While a symmetric pair of transmission lines in a homogeneous medium has near-end crosstalk, there is no far-end crosstalk.

The far-end cancellation fails when $v_e \neq v_o$, and a pulse crosstalk waveform results. Assume that no signal drives port 2 so that just the crosstalk appears at port 4, then from (10.33)

$$v_4 = \frac{1}{2}\left(v_1(t - d/v_e) - v_1(t - d/v_o)\right). \tag{10.34}$$

With longer lines, less cancellation occurs as the two shifted versions of v_1 get farther out of alignment, and a larger pulse results. The generation of a crosstalk noise pulse is sketched in Figure 10.13 for a low-to-high transition. Here, the even mode is shown slower than the odd mode since, for microstrip-like lines, the even

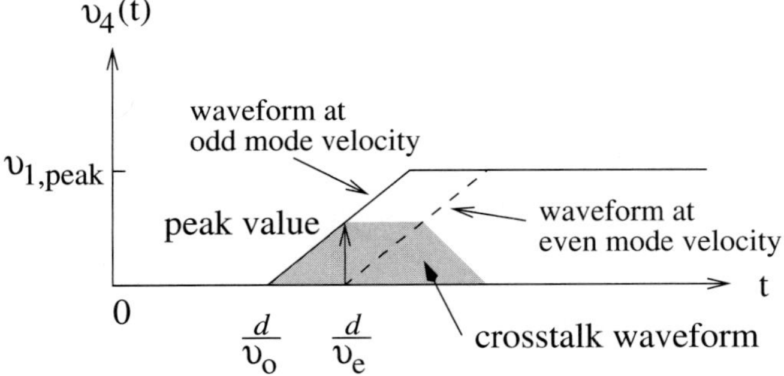

Figure 10.13. Sketch of far-end crosstalk resulting from unequal even- and odd-mode phase velocities after propagating a distance d.

mode has more electric flux in the dielectric, leading to a higher effective dielectric constant and lower phase velocity. For triangular edges and time shifts less than the rise time, then

$$\begin{aligned} v_{4,peak} &= \frac{1}{2}\left(v_1(t - d/v_e) - v_1(t - d/v_o)\right) \\ &= -\frac{1}{2}\Delta t \frac{dv_1}{dt} \\ &= \frac{1}{2}\left(\frac{d}{v_o} - \frac{d}{v_e}\right)\frac{dv_1}{dt}. \end{aligned} \tag{10.35}$$

For weak coupling,

$$v_e \approx \frac{1}{\sqrt{\ell c}} \frac{1}{1 + \frac{1}{2}m/\ell - \frac{1}{2}c_m/c}$$

and

$$v_o \approx \frac{1}{\sqrt{\ell c}} \frac{1}{1 - \frac{1}{2}m/\ell + \frac{1}{2}c_m/c}.$$

Substituting these into (10.35) yields the far-end crosstalk peak voltage

$$v_{4,peak} = \frac{1}{2}d\left(Z_o c_m - \frac{m}{Z_o}\right)\frac{dv_1}{dt}, \tag{10.36}$$

where $Z_o = \sqrt{\ell/c}$, the characteristic impedance of the isolated line. This is the same result derived heuristically in chapter 2, equation (2.61).

10.3 Skin Effect Models

The skin effect and current crowding are, in general, frequency-dependent phenomena. While the black box modeling technique in section 10.4 can be applied to model skin effects, in many cases simple models are sufficient. Two approaches are the PEEC method and ladder networks.

The PEEC method, discussed in chapter 7, section 7.3, can often be used to model skin effects and current crowding for short interconnects. To capture frequency dependence in a time-domain simulator, the PEEC model is not reduced using (7.6). Given frequency-dependent data, perhaps obtained through measurement, a PEEC model cannot be synthesized, since (7.6) is not reversible. Therefore, the PEEC method is good for analytical work given geometrical information, but it is not good for modeling data.

In chapter 7, section 7.4 it is shown that the frequency-dependent resistance and inductance of coaxial waveguide can be described by a ladder network consisting of resistors and inductors, as suggested in Figure 7.9b. For other types of transmission lines, the RL ladder network is still a good description and can be used to model frequency-dependence in interconnects. Except for the coaxial line, formulas based on geometrical information are not available to compute the component values of the RL ladder, so the component values must be obtained through optimization. Alternatively, a ladder can be synthesized using the technique in section 10.4. Deeper ladders provide better square root frequency dependence and wider frequency range.

10.4 Black Box Modeling

If nothing is known about a network, or the modeler chooses not to incorporate any knowledge, a black box model can be constructed that makes no claim to a physical basis but that accurately reproduces the network's terminal characteristics. For each element of a frequency-dependent admittance (or impedance, or S-parameter, etc.) matrix, a rational function is fitted over a specified bandwidth. Simply put, the rational function is a representation of the network in terms of its complex poles and zeros. The rational function can be used directly in a circuit simulator, if supported, or a network can be synthesized that implements the rational function in terms of standard circuit elements. With this very general method, any network can be modeled and incorporated into a simulation.

For example, consider the input impedance of a lossless transmission line shorted at the far end. The input impedance is given by

$$Z_{\text{in}} = \jmath Z_o \tan(kd), \tag{10.37}$$

where $Z_o = \sqrt{\ell/c}$, $k = \omega\sqrt{\ell c}$, and d are the line's characteristic impedance, propagation constant, and length. The input impedance is zero whenever $kd = n\pi$, and infinite whenever $kd = \frac{\pi}{2} + n\pi$. A lumped model consisting of a series inductance only provides zero input impedance at $\omega = 0$. As kd nears $\pi/2$, the input impedance of the model fails to capture the rapidly increasing input impedance because the impedance of the inductor increases only linearly with frequency. The bandwidth of the lumped model can be improved by adding a capacitance to ground; then the model works well until kd nears π. The bandwidth can be improved in increments of $\pi/2$ by alternately adding series inductance or shunt capacitance. In pole-zero terminology, the problem lies in the fact that (10.37) contains an infinite number of poles and zeros. Each added series inductor or shunt capacitor adds a zero and a pole. In general, the useful bandwidth of a model is then determined by how many poles and zeros are included.

Black box modeling can also be used to speed up simulations of existing circuits. A large linear circuit can be separately characterized and fitted with a rational function approximation. The new model is often called a *macro model* of the original circuit. Simulations with the macro model in place of the original circuit can achieve dramatic reductions in computation time with little to no loss of accuracy.

The method outlined in this section uses rational functions consisting of polynomials. Rational polynomial functions can be difficult to implement for several reasons. It can be difficult to determine the orders of the polynomials (i.e., the number of poles and zeros); the calculation of the polynomial coefficients can be numerically unstable and/or inaccurate if the orders of the polynomials become large or the bandwidth of the data becomes too great; and, the user must often write, debug, and validate software. Given these difficulties, a rational polynomial function may at times be the only option, and they can often work quite well.

This section presents a simple but complete method for synthesizing a network using rational polynomial function approximation and continued fractions. More advanced techniques are needed to overcome some of the method's limitations, but these are beyond the scope of this book. The method is a good introduction and still quite useful in its own right.

10.4.1 Single Port

Most of the detail required to implement a rational polynomial function approximation is illuminated by the one-port problem. Assume that the input admittance Y of a one-port network has been obtained at P frequencies. A general rational polynomial function for Y is

$$Y(s) = \frac{a_0 + a_1 s + a_2 s^2 + \cdots + a_M s^M}{b_0 + b_1 s + b_2 s^2 + \cdots + b_N s^N}. \tag{10.38}$$

If the data is acquired from a network that is lumped, linear, and time invariant, then the coefficients a_i and b_i are real.

Evaluating (10.38) at each of the P frequencies and cross-multiplying yields

$$\begin{aligned}
Y(s_1)b_0 + Y(s_1)s_1 b_1 + Y(s_1)s_1^2 b_2 + \cdots + Y(s_1)s_1^N b_N \\
= a_0 + s_1 a_1 + s_1^2 a_2 + \cdots + s_1^M a_M \\
Y(s_2)b_0 + Y(s_2)s_2 b_1 + Y(s_2)s_2^2 b_2 + \cdots + Y(s_2)s_2^N b_N \\
= a_0 + s_2 a_1 + s_2^2 a_2 + \cdots + s_2^M a_M \\
\vdots \\
Y(s_P)b_0 + Y(s_P)s_P b_1 + Y(s_P)s_P^2 b_2 + \cdots + Y(s_P)s_P^N b_N \\
= a_0 + s_P a_1 + s_P^2 a_2 + \cdots + s_P^M a_M.
\end{aligned} \tag{10.39}$$

To put (10.39) into a convenient matrix form, define the column vectors

$$\begin{aligned}
\overline{a} &= [\, a_0 \quad a_1 \quad \cdots \quad a_M \,]^T \\
\overline{b} &= [\, b_0 \quad b_1 \quad \cdots \quad b_N \,]^T,
\end{aligned}$$

and define the matrices

$$\overline{\overline{B}} = \begin{bmatrix}
Y(s_1) & Y(s_1)s_1 & Y(s_1)s_1^2 & \cdots & Y(s_1)s_1^N \\
Y(s_2) & Y(s_2)s_2 & Y(s_2)s_2^2 & \cdots & Y(s_2)s_2^N \\
Y(s_3) & Y(s_3)s_3 & Y(s_3)s_3^2 & \cdots & Y(s_3)s_3^N \\
\vdots & \vdots & \vdots & \ddots & \vdots \\
Y(s_P) & Y(s_P)s_P & Y(s_P)s_P^2 & \cdots & Y(s_P)s_P^N
\end{bmatrix},$$

and

$$\overline{\overline{A}} = \begin{bmatrix} 1 & s_1 & s_1^2 & \cdots & s_1^M \\ 1 & s_2 & s_2^2 & \cdots & s_2^M \\ 1 & s_3 & s_3^2 & \cdots & s_3^M \\ \vdots & \vdots & \vdots & \ddots & \vdots \\ 1 & s_P & s_P^2 & \cdots & s_P^M \end{bmatrix}.$$

Then, (10.39) can be recast as

$$\overline{\overline{B}}\,\overline{b} = \overline{\overline{A}}\,\overline{a}, \tag{10.40}$$

or equivalently,

$$\begin{bmatrix} \overline{\overline{A}} & -\overline{\overline{B}} \end{bmatrix} \begin{bmatrix} \overline{a} \\ \overline{b} \end{bmatrix} = 0. \tag{10.41}$$

So that $\overline{a}$ and $\overline{b}$ are computed as real numbers, (10.41) can be rearranged by breaking $\overline{\overline{A}}$ and $\overline{\overline{B}}$ into real and imaginary parts as

$$\overline{\overline{A}} = \overline{\overline{A}}_R + \jmath \overline{\overline{A}}_I$$

and

$$\overline{\overline{B}} = \overline{\overline{B}}_R + \jmath \overline{\overline{B}}_I.$$

Then (10.41) becomes

$$\begin{bmatrix} \overline{\overline{A}}_R & -\overline{\overline{B}}_R \\ \overline{\overline{A}}_I & -\overline{\overline{B}}_I \end{bmatrix} \begin{bmatrix} \overline{a} \\ \overline{b} \end{bmatrix} = 0, \tag{10.42}$$

which can be solved using real math.

There are several approaches to solving (10.42). In a direct approach, choose $b_0 = 1$ and plug into (10.42) to reduce it to the standard linear algebra problem $\overline{\overline{A}}\overline{x} = \overline{b}$. The result is

$$\begin{bmatrix} \overline{\overline{A}}_R & -\overline{\overline{B}}'_R \\ \overline{\overline{A}}_I & -\overline{\overline{B}}'_I \end{bmatrix} \begin{bmatrix} \overline{a} \\ \overline{b}' \end{bmatrix} = \begin{bmatrix} \overline{Y}_R \\ \overline{Y}_I \end{bmatrix}. \tag{10.43}$$

where

$$\overline{b}' = [\ b_1 \quad b_2 \quad \cdots \quad b_N\ ]^T,$$

$$\overline{Y} = [\ Y(s_1) \quad Y(s_2) \quad \cdots \quad Y(s_P)\ ]^T,$$

and

$$\overline{\overline{B}}' = \begin{bmatrix} Y(s_1)s_1 & Y(s_1)s_1^2 & \cdots & Y(s_1)s_1^N \\ Y(s_2)s_2 & Y(s_2)s_2^2 & \cdots & Y(s_2)s_2^N \\ Y(s_3)s_3 & Y(s_3)s_3^2 & \cdots & Y(s_3)s_3^N \\ \vdots & \vdots & \ddots & \vdots \\ Y(s_P)s_P & Y(s_P)s_P^2 & \cdots & Y(s_P)s_P^N \end{bmatrix}.$$

If $2P = M + N + 1$, then (10.43) can be solved by matrix inversion to find $\overline{a}$ and $\overline{b}'$. If $2P > M + N + 1$, then a least-squares solution is required, and many options are available including inversion of the normal equations, the QR algorithm, or the singular value decomposition.

The set of equations in (10.43) becomes ill-conditioned as the bandwidth increases and as the orders of the polynomials increase. The condition of the matrix can be dramatically improved by scaling the frequency by the maximum frequency. The problem should be solved with the best numerical algorithm available in high precision, and the QR algorithm can achieve good results. For any given application, the bandwidth and polynomial orders require close attention to ensure an accurate rational function approximation. Polynomial approximations are generally better behaved if they are overconstrained, so oversampling to achieve $2P > M + N + 1$ is desirable.

A fundamental question arises as to the values that should be selected for the polynomial orders, M and N. One strategy for automatic selection could start with $M = N = 1$, then compute the rational function approximation and the error between the approximation and the data. If the error is large, then increase M and/or N and repeat until the error drops below some specified cutoff value. For large values of M and/or N, numerical instability results in large error, so the process is not guaranteed to work. If a sufficient fit has not been achieved, the only recourse is to reduce the bandwidth and try again (or switch to a more sophisticated fitting method).

A different approach to solving (10.42) is to treat it as an eigenvalue problem. In the general case, $2P > M + N + 2$, so (10.42) can be squared up by forming the normal equations. Multiplying through by the transpose yields

$$\begin{bmatrix} \overline{\overline{A}}_R & -\overline{\overline{B}}_R \\ \overline{\overline{A}}_I & -\overline{\overline{B}}_I \end{bmatrix}^T \begin{bmatrix} \overline{\overline{A}}_R & -\overline{\overline{B}}_R \\ \overline{\overline{A}}_I & -\overline{\overline{B}}_I \end{bmatrix} \begin{bmatrix} \overline{a} \\ \overline{b} \end{bmatrix} = 0, \tag{10.44}$$

and the left-hand side is now a square $(M+N+2)\times(M+N+2)$ matrix. The standard eigenvalue problem is $\overline{\overline{A}}\overline{x} = \lambda\overline{x}$, so if λ is zero, then $\overline{\overline{A}}\overline{x} = 0$ and the form of (10.44) is recovered. The strategy is then to solve (10.44) as an eigenvalue problem, pick an eigenvalue equal to or sufficiently near zero, and then the eigenvector associated with the eigenvalue provides $\overline{a}$ and $\overline{b}$.

The eigenvalue method provides some help in choosing the polynomial orders. Choose M and N too small, and all of the eigenvalues are large. Choose them too large, and many near-zero eigenvalues will appear. A good solution is obtained when there is just one near-zero eigenvalue.

Even with careful selection of polynomial orders and careful computation of the polynomial coefficients, poles on the right-hand side of the complex plane can still appear. Interconnects are passive, hence stable, so no right-hand poles should be present in the model. Unfortunately, absence of right-hand poles cannot be guaranteed. However, right-hand poles can be easily computed and removed, so a general-purpose, robust method can be constructed. Note that absence of right-hand poles ensures stability but not passivity.

Right-Hand Pole Removal

The first step in right-hand pole removal is the computation of the poles of the rational polynomial function given by (10.38). The N zeros of the denominator can easily be found by solving for the eigenvalues of the $N \times N$ companion matrix

$$\begin{bmatrix} -\frac{b_{N-1}}{b_N} & -\frac{b_{N-2}}{b_N} & \cdots & -\frac{b_1}{b_N} & -\frac{b_0}{b_N} \\ 1 & 0 & \cdots & 0 & 0 \\ 0 & 1 & \cdots & 0 & 0 \\ \vdots & \vdots & \ddots & \vdots & \vdots \\ 0 & 0 & \cdots & 1 & 0 \end{bmatrix}.$$

Any zero with positive real part is discarded.

Once the right-hand poles are eliminated, the denominator polynomial must be reconstructed from the remaining zeros. Assume that $N^{'}$ of the original N zeros remain; then the new polynomial is

$$f(s) = \prod_{i=1}^{N^{'}} (s + p_i), \tag{10.45}$$

where p_i are the remaining zeros. Note that the p_i are either real or appear in complex conjugate pairs, so $f(s)$ is still described by a polynomial with real coefficients. The product in (10.45) can be iteratively constructed by noting that

$$(s+a)f(s) = (s+a)\sum_{i=0}^{M} b_i s^i \tag{10.46}$$

$$= ab_o + \sum_{i=1}^{M} (ab_i + b_{i-1})\, s^i + b_M s^{M+1}. \tag{10.47}$$

Repeated application with the starting polynomial set to $f(s) = s + p_1$ yields the new denominator polynomial.

The new denominator means that the numerator coefficients are no longer valid. Since the denominator is known, (10.38) can be rearranged as

$$Y^{'}(s) = \sum_{i=0}^{M} a_i s^i, \tag{10.48}$$

where

$$Y^{'}(s) = Y(s) \sum_{i=0}^{N} b_i s^i$$

is known. Evaluating (10.48) at P frequencies yields

$$\overline{Y}' = \overline{\overline{A}}\overline{a}, \tag{10.49}$$

where

$$\overline{Y}' = \begin{bmatrix} Y'(s_1) & Y'(s_2) & \cdots & Y'(s_P) \end{bmatrix}^T$$

and $\overline{\overline{A}}$ and $\overline{a}$ take the same definitions from (10.41). To solve using only real math, let $\overline{Y}' = \overline{Y}'_R + \jmath\overline{Y}'_I$, then (10.49) becomes

$$\begin{bmatrix} \overline{Y}'_R \\ \overline{Y}'_I \end{bmatrix} = \begin{bmatrix} \overline{\overline{A}}_R \\ \overline{\overline{A}}_I \end{bmatrix} \overline{a}. \tag{10.50}$$

Solution of this linear equation (and generally a least-squares problem) completes the rational function approximation with guaranteed absence of right-hand poles.

Once a rational function approximation has been constructed for a given bandwidth, it is important to utilize the function only over that bandwidth. Particular care must be observed when using ramp functions to provide excitation in SPICE-like simulations, since the corner at the transition from the triangular edge to the flat section produces significant frequency components to very wide bandwidths. It may be necessary to provide filtering to limit the bandwidth of ramp-like functions.

Continued Fractions

Rational polynomial functions can be rewritten as continued fractions. Since network analysis of ladder networks leads directly to continued fractions, then ladder networks can be synthesized from rational polynomial functions through continued fractions. To compute a continued fraction from a rational polynomial function, two formulas are needed. It is easily verified that

$$\frac{\sum_{i=0}^{N} a_i s^i}{\sum_{i=0}^{N} b_i s^i} = \frac{a_N}{b_N} + \frac{\sum_{i=0}^{N-1} (a_i - \frac{a_N}{b_N} b_i) s^i}{\sum_{i=0}^{N} b_i s^i} \tag{10.51}$$

and

$$\frac{\sum_{i=0}^{N} a_i s^i}{\sum_{i=0}^{N-1} b_i s^i} = \frac{a_N}{b_{N-1}} s + \frac{a_0 + \sum_{i=1}^{N-1} (a_i - \frac{a_N}{b_{N-1}} b_{i-1}) s^i}{\sum_{i=0}^{N-1} b_i s^i}, \tag{10.52}$$

which result from extracting the pole at infinity. These two equations lead to ladders in the First Cauer Form. A similar set of equations can be derived by extracting the pole at zero, leading to the Second Cauer Form. Because ladder networks are not unique, neither form is preferred.

Each of equations (10.51) and (10.52) reduces the order of the numerator polynomial by one. After extraction of a term, the remaining rational function can be inverted to enable the repeated application of the formulas. The process continues until the entire rational function has been reduced.

For example, assume that a rational function fits frequency-dependent admittance data and that $M = N$. Applying (10.51) extracts a constant admittance term that can be inverted to find a resistance for inclusion in a netlist. The remaining rational function represents an admittance in parallel to the resistor. Inverting the remaining rational function, it now describes impedance, and applying (10.52) extracts an inductance (sL). The remaining rational function represents an impedance in series with the inductor. Inverting the remaining rational function yields an admittance function, so we are back where we started except that the orders of the numerator and denominator have been reduced by one. This process is repeated to extract an RL ladder like the one in chapter 7, Figure 7.9b.

The formulas can be applied differently to obtain a different ladder network. Again assume that a rational function fits frequency-dependent admittance data, but let $N = M+1$. Applying (10.52) extracts a capacitance (sC), then immediately applying (10.51) (without inverting) extracts a parallel admittance. The remaining rational function represents admittance in parallel to the paralleled RC circuit. Inverting the remaining rational function yields an impedance, and applying (10.52) and (10.51) yields a series resistance and inductance. Inverting the remaining rational function restarts the process.

Example: Rational Function Approximation and Network Synthesis

Using the program in Appendix B, the frequency-dependent impedance is computed for copper coax with an inner radius of 100μm, an outer radius of 400μm, and an

Table 10.1. Rational function approximation for a coaxial line with an inner radius of 100μm, an outer radius of 400μm, and an outer conductor thickness of 99μm.

i	a_i	b_i
0	3.685456e-06	2.339143e-06
1	2.211692e-04	1.492312e-03
2	1.856831e-03	7.324409e-02
3	3.297555e-03	5.768429e-01
4	1.771712e-03	1.000000e+00
5	4.300401e-04	5.320013e-01
6	6.104166e-05	1.286278e-01
7	2.944794e-06	1.817411e-02
8	4.361208e-11	8.718464e-04

outer conductor thickness of 99μm. As discussed in chapter 7, section 7.4, a ladder network can be used to create a SPICE model that can reproduce the frequency-dependent terminal impedance while using only frequency-independent components.

Applying (10.43) to set up and solve for a rational polynomial function approximation to the simulated terminal admittance yields the expansion listed in Table 10.1 when the data is fitted over the frequency range from 0.106MHz to 1.06GHz. Note that frequency scaling of 1.06GHz is incorporated into Table 10.1.

The rational function is converted into a continued fraction through repeated application of (10.51) and (10.52), leading directly to the SPICE netlist.

```
.subckt ladder 1 out
Rp1 1 out 1.999094e+07
Ls1 1 2 2.782367e-07
Rp2 2 out 3.336287e+01
Ls2 2 3 4.055090e-09
Rp3 3 out 1.614620e+01
Ls3 3 4 7.901239e-09
Rp4 4 out 6.923177e+00
```

```
Ls4 4 5 -1.467985e-08
Rp5 5 out 2.205033e+01
Ls5 5 6 4.368958e-10
Rp6 6 out -1.810383e+01
Ls6 6 7 3.073400e-08
Rp7 7 out 5.538082e+00
Ls7 7 8 3.162878e-08
Rp8 8 out 2.493410e+00
Ls8 8 9 1.090311e-07
Rp9 9 out 1.302984e+00
.ends ladder
```

The input impedance of this network is computed in SPICE, converted to resistance and inductance, and plotted against the original computed data in Figure 10.14. The agreement between the synthesized network and the data is excellent. At low frequencies, resistance dominates reactance and is more closely fitted. The SPICE model nicely extrapolates the terminal behavior. Note that the ladder is stable (no right-hand poles) but active (negative lumped component values).

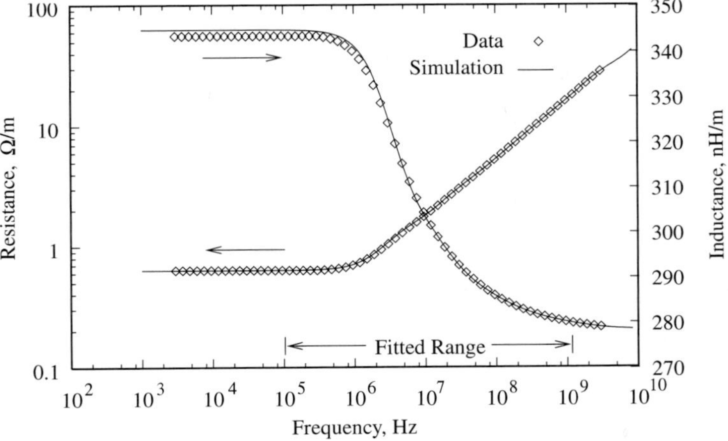

Figure 10.14. Resistance and inductance comparison between frequency-dependent data and a synthesized SPICE model for a coaxial transmission line.

10.4.2 Multiple Ports

An N-port network is described by a frequency-dependent $N \times N$ matrix. Fundamentally, a rational function approximation is required for each matrix element. Assuming reciprocity, then $N(N+1)/2$ rational function approximations are required. Each element can be fitted individually or simultaneously.

If each element is fitted individually, then the methodology for the one-port circuit can be applied to each matrix element. For each of the elements, the rational function approximations can be entirely different, or they can be constrained as desired. For example, the denominator polynomial can be computed once and then assumed to be the same for all remaining matrix elements by using the formulation in (10.50).

Alternatively, all of the matrix elements can be fitted simultaneously. The following sections demonstrate both techniques.

Full Matrix Fit

Assuming that each matrix element has the same poles, an admittance matrix for a Q-port circuit can be fitted as

$$\overline{\overline{Y}}(s) = \begin{bmatrix} \frac{\sum_{i=0}^{M} a_i^{11} s^i}{\sum_{i=0}^{N} b_i s^i} & \frac{\sum_{i=0}^{M} a_i^{12} s^i}{\sum_{i=0}^{N} b_i s^i} & \cdots & \frac{\sum_{i=0}^{M} a_i^{1Q} s^i}{\sum_{i=0}^{N} b_i s^i} \\ \frac{\sum_{i=0}^{M} a_i^{21} s^i}{\sum_{i=0}^{N} b_i s^i} & \frac{\sum_{i=0}^{M} a_i^{22} s^i}{\sum_{i=0}^{N} b_i s^i} & \cdots & \frac{\sum_{i=0}^{M} a_i^{2Q} s^i}{\sum_{i=0}^{N} b_i s^i} \\ \vdots & \vdots & \ddots & \vdots \\ \frac{\sum_{i=0}^{M} a_i^{Q1} s^i}{\sum_{i=0}^{N} b_i s^i} & \frac{\sum_{i=0}^{M} a_i^{Q2} s^i}{\sum_{i=0}^{N} b_i s^i} & \cdots & \frac{\sum_{i=0}^{M} a_i^{QQ} s^i}{\sum_{i=0}^{N} b_i s^i} \end{bmatrix}.$$

Applying the single-port methodology from (10.40), the ijth entry in the admittance matrix can be fitted as

$$\overline{\overline{B}}^{ij}\,\overline{b} = \overline{\overline{A}}\,\overline{a}^{ij},$$

where

$$\overline{a}^{ij} = [\ a_0^{ij} \quad a_1^{ij} \quad \cdots \quad a_M^{ij}\ ]^T,$$

$\overline{\overline{A}}$ is defined as before, and

$$\overline{\overline{B}}^{ij} = \begin{bmatrix} Y^{ij}(s_1) & Y^{ij}(s_1)s_1 & Y^{ij}(s_1)s_1^2 & \cdots & Y^{ij}(s_1)s_1^N \\ Y^{ij}(s_2) & Y^{ij}(s_2)s_2 & Y^{ij}(s_2)s_2^2 & \cdots & Y^{ij}(s_2)s_2^N \\ Y^{ij}(s_3) & Y^{ij}(s_3)s_3 & Y^{ij}(s_3)s_3^2 & \cdots & Y^{ij}(s_3)s_3^N \\ \vdots & \vdots & \vdots & \ddots & \vdots \\ Y^{ij}(s_P) & Y^{ij}(s_P)s_P & Y^{ij}(s_P)s_P^2 & \cdots & Y^{ij}(s_P)s_P^N \end{bmatrix}.$$

Assembling the equations for all ij into one expression for a Q-port admittance matrix yields

$$\begin{bmatrix} \overline{\overline{A}} & 0 & 0 & \cdots & 0 & -\overline{\overline{B}}^{11} \\ 0 & \overline{\overline{A}} & 0 & \cdots & 0 & -\overline{\overline{B}}^{12} \\ 0 & 0 & \overline{\overline{A}} & \cdots & 0 & -\overline{\overline{B}}^{13} \\ \vdots & \vdots & \vdots & \ddots & \vdots & \vdots \\ 0 & 0 & 0 & \cdots & \overline{\overline{A}} & -\overline{\overline{B}}^{QQ} \end{bmatrix} \begin{bmatrix} \overline{a}^{11} \\ \overline{a}^{12} \\ \overline{a}^{13} \\ \vdots \\ \overline{a}^{QQ} \\ \overline{b} \end{bmatrix} = 0. \tag{10.53}$$

Assuming a reciprocal admittance matrix, then there are $Q(Q+1)/2$ matrix entries for a Q-port matrix. The size of the matrix on the left-hand side of (10.53) is

$$PQ(Q+1)/2 \times [(M+1)Q(Q+1)/2 + N + 1],$$

while the left-hand column is of size

$$[(M+1)Q(Q+1)/2 + N + 1] \times 1$$

A square matrix requires that

$$P = 1 + M + \frac{2(N+1)}{Q(Q+1)},$$

which requires large N for large Q. In practice, an overdetermined system is required, but that is desirable anyway. Details remain regarding real math, right-hand pole removal, and so on.

Matrix Element Fit

A matrix can be fitted by individually modeling each matrix element. Since no additional derivation is required, the technique is demonstrated by modeling a lossy transmission line. Most circuit simulators include a model for the lossless transmission line, but few include a general model for the lossy case.

The exact equations for the 2×2 admittance matrix for a lossy transmission line are used to generate a data set that is then fitted with a rational polynomial function approximation and modeled with SPICE-compatible lumped elements.

As in the single-port example, the program in Appendix B is used to compute the frequency-dependent impedance for copper coax with an inner radius of 100μm, an outer radius of 400μm, and an outer conductor thickness of 99μm. These per-unit-length resistance and inductance values are used in chapter 4, equation (4.21) along with an assumed capacitance per-unit-length of 80pF/m to find the complex admittance matrix of a 3cm, 60Ω transmission line.

To model the two-port network in SPICE, the lumped π-network in chapter 4, Figure 4.8a is used with frequency-dependent admittances that are implemented as frequency-independent ladder networks. Since the transmission line is symmetric, $y_1 = y_3$. The results from problem 13 in chapter 4 show that the π-network can be constructed by inspection once the 2×2 admittance matrix is known.

Using the method in section 10.4.1 with right-hand pole removal, the data for y_1 is fitted over the frequency range from 3.3kHz to 3.0GHz, and the plots in Figure 10.15 show the excellent quality of the fit. The SPICE deck for y_1 is as follows:

```
.subckt y1 1 out
Rp1 1 out 1.448679e+03
Cp1 1 out 1.385524e-15
Rs1 1 t1 2.330182e+00
Ls1 t1 2 2.306115e-09
Rp2 2 out -1.696612e+04
Cp2 2 out 1.346553e-12
Rs2 2 t2 7.305309e-02
Ls2 t2 3 5.836348e-09
Rp3 3 out -3.113255e+04
```

```
Cp3 3 out 1.716403e-12
Rs3 3 t3 7.009935e-02
Ls3 t3 4 6.494891e-09
Rp4 4 out -2.145872e+04
Cp4 4 out 1.910516e-12
Rs4 4 t4 3.028198e-02
Ls4 t4 5 7.489118e-09
Rp5 5 out -1.041740e+04
Cp5 5 out 2.371987e-12
Rs5 5 t5 -8.263632e-02
Ls5 t5 6 1.041090e-08
Rp6 6 out -2.195519e+03
Cp6 6 out 4.457509e-12
.ends y1
```

Note the negative component values, meaning that the model is active, but since there are no right-hand poles, the network is stable.

Similarly for y_2, the data is fitted over a slightly smaller range of 3.3kHz to 1.05GHz to achieve better numerical stability, and the excellent fit is shown in Figure 10.16. For this example, an RL ladder outperforms an RLC ladder, and the resulting SPICE deck is:

```
.subckt y2 1 out
Rp1 1 out 2.252886e+01
Ls1 1 2 -1.164067e-10
Rp2 2 out -2.249623e+01
Ls2 2 3 -9.628880e-08
Rp3 3 out -6.111747e+01
Ls3 3 4 6.271309e-10
Rp4 4 out 6.071545e+01
Ls4 4 5 9.041897e-07
Rp5 5 out 2.978254e+02
Ls5 5 6 -2.304350e-09
Rp6 6 out -2.974121e+02
Ls6 6 7 -2.566824e-06
Rp7 7 out -2.410449e+03
Ls7 7 8 6.240773e-08
Rp8 8 out 2.262089e+03
```

```
Ls8 8 9 1.782445e-06
Rp9 9 out 1.245023e+00
Ls9 9 10 2.172213e-08
Rp10 10 out 2.247830e-01
.ends y2
```

Again, the network is stable but active.

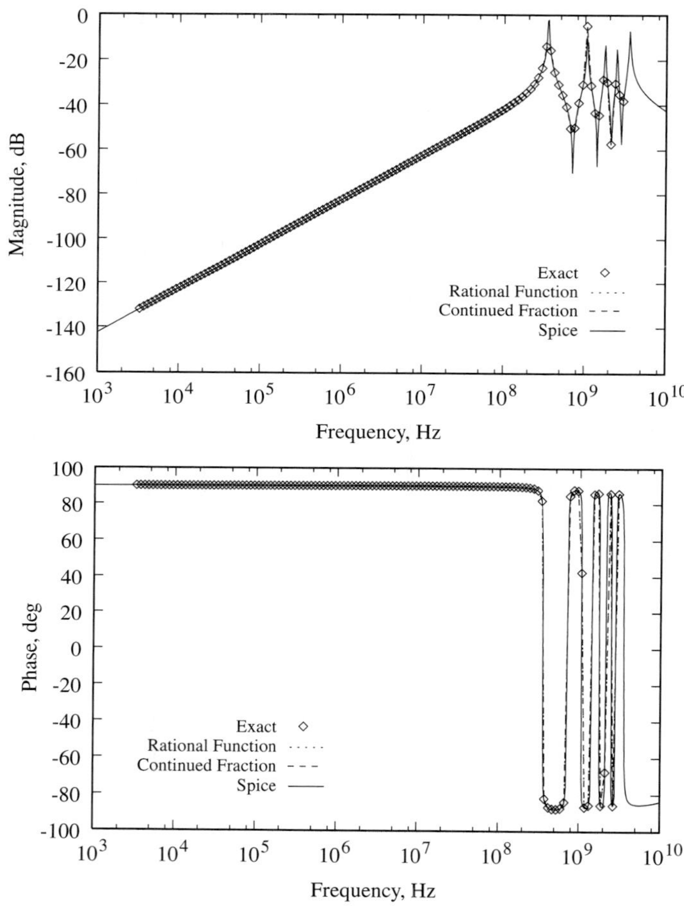

Figure 10.15. Fitted SPICE model of the one-port frequency-dependent admittance y_1.

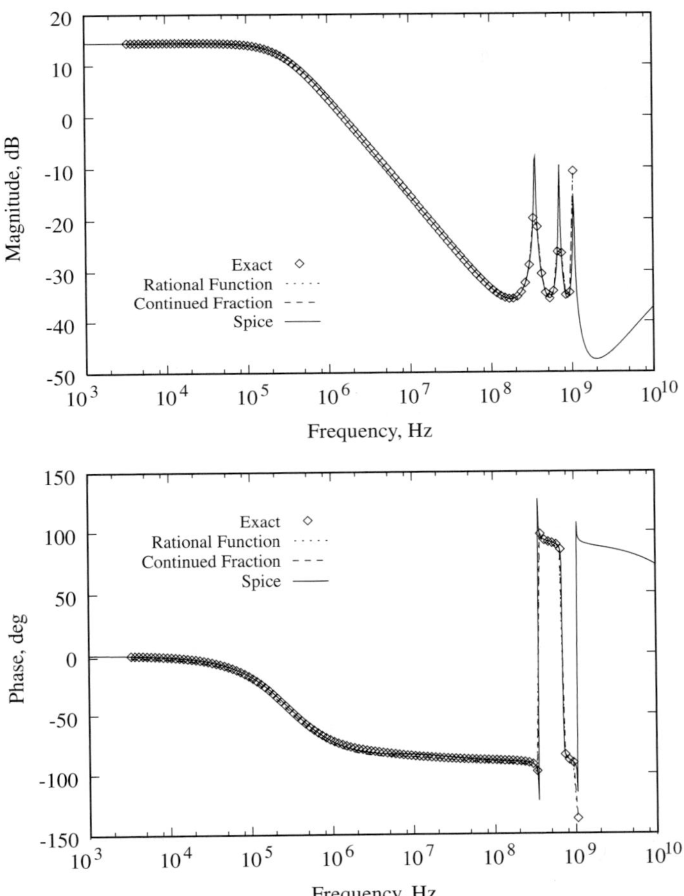

Figure 10.16. Fitted SPICE model of the one-port frequency-dependent admittance y_2.

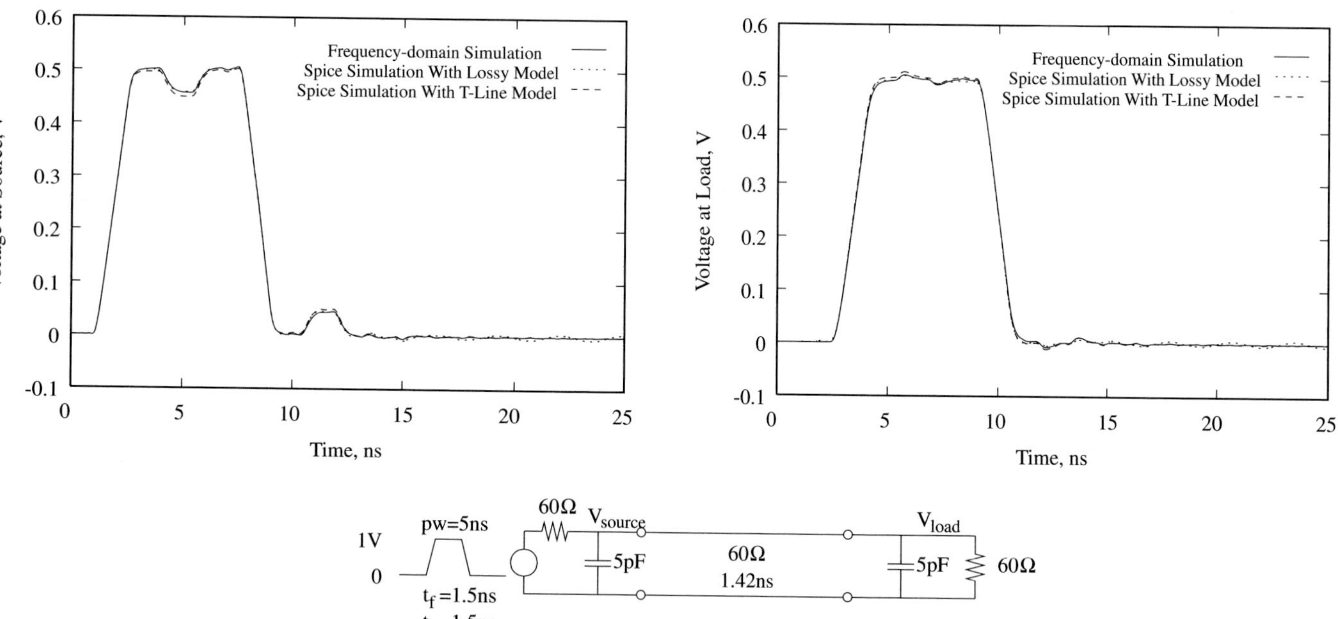

Figure 10.17. Comparison of simulation methods for a lossy transmission line.

Combining the SPICE decks for y_1 and y_2 into a π-network enables the simulation of the transmission line with frequency-dependent losses in a time-domain circuit simulator. For example, the voltages at the source and load of the above transmission line are shown in Figure 10.17 for the nearly matched case. Since this example contains no nonlinear elements, a frequency-domain simulation is also shown for comparison. The frequency-domain and lossy time-domain results are quite close, while the lossless simulation introduces some error. More dramatic differences can be easily demonstrated with a longer transmission line and/or higher losses.

10.5 Exercises

1. Derive a set of two continued fraction formulas that reduce rational functions by extracting the pole at zero. Application of these formulas leads to ladder networks of the Second Cauer Form.

2. Sketch a schematic that implements the following continued fraction, using only frequency-independent lumped elements:

$$Y(s) = 5s + \cfrac{1}{10 + \cfrac{1}{3s + \cfrac{1}{5 + 1/2s}}}.$$

3. Derive a formulation where a known DC value can be incorporated into a rational function approximation.

4. Derive a formulation where a known high-frequency asymptotic limit can be incorporated into a rational function approximation.

5. From chapter 2, equation (2.66) the input impedance of a loaded lossless transmission line is

$$Z_{\text{in}} = Z_o \frac{Z_L + \jmath Z_o \tan(\beta d)}{Z_o + \jmath Z_L \tan(\beta d)}.$$

(a) Expand the numerator and the denominator in infinite power series about βd and derive a rational function approximation for Z_{in}.

(b) How many poles and zeros does it take to represent a transmission line?

(c) Keep two terms in the numerator and denominator and derive a continued fraction approximation for Z_{in}.

(d) Sketch the circuit suggested by the continued fraction, assuming that the load is a frequency-independent resistor.

6. A lossless 75Ω transmission line with $\epsilon_{\text{eff}} = 9.8$ is 2.5cm long. Sketch a lumped model for the line that is suitable for signals with edge rates as short as 0.5ns.

7. Derive a modal decomposition for the multiconductor transmission line for the case of high resistance, where it is assumed that $\overline{\overline{\ell}} = \overline{\overline{g}} = 0$. On-chip interconnects are usually modeled as RC lines.

8. In a homogeneous medium, the phase velocities of all modes are identical. For the even and odd modes of a pair of symmetric transmission lines, show that

$$\frac{m}{\ell} = \frac{c_m}{c}$$

in a homogeneous medium. From the discussion in chapter 2, section 2.5.3, what does this mean regarding crosstalk on these two lines?

9. Show that (10.18) holds for a pair of symmetric transmission lines in a homogeneous medium.

10. (a) Find the modal characteristic impedances and phase velocities for three symmetric transmission lines described by

$$\overline{\overline{\ell}} = \begin{bmatrix} \ell & m & m \\ m & \ell & m \\ m & m & \ell \end{bmatrix}$$

and

$$\overline{\overline{c}} = \begin{bmatrix} c & -c_m & -c_m \\ -c_m & c & -c_m \\ -c_m & -c_m & c \end{bmatrix}.$$

(b) Sketch a circuit that implements these three coupled lines using three uncoupled transmission line models.

11. (a) Show that the near-end coupling coefficient to a small load, Z_2, for symmetric coupled transmission lines is given by

$$K_{ne} = \frac{1}{2} Z_2 \frac{Z_{oe} - Z_{oo}}{Z_{oe} Z_{oo}}.$$

(b) Show that for weak coupling, this expression simplifies to

$$K_{ne} \approx \frac{1}{2} Z_2 \sqrt{\frac{c}{\ell}} \left(\frac{m}{\ell} + \frac{c_m}{c} \right).$$

12. The input impedance of a lossless one-port network with a zero at infinity must be of the form[1]

$$Z(s) = K \frac{s(s^2 + \omega_1^2)(s^2 + \omega_3^2) \cdots (s^2 + \omega_{2N-1}^2)}{(s^2 + \omega_0^2)(s^2 + \omega_2^2) \cdots (s^2 + \omega_{2N}^2)},$$

where $0 \leq \omega_0 < \omega_1 < \omega_2 \cdots < \omega_{2N-2} < \omega_{2N-1} < \omega_{2N} < \infty$ and K is a positive constant. Derive a formulation similar to (10.40) to fit a rational function to P pairs of Z and s data.

1. Norman Balabanian and Theodore Beckart, *Linear Network Theory: Analysis, Properties, Design, and Synthesis.* Chesterland, Ohio: Matrix Publishers, Inc., 1981, p. 484.

Chapter 11

ENHANCING SIGNAL INTEGRITY

Up to this point, signal integrity is covered through presentation of basic phenomena and general analysis techniques. Many specific topics, such as termination schemes for minimizing reflections, are implicitly covered but require the reader to fill in the gaps. This chapter fleshes out a few major signal integrity topics both to spare the reader the need to extend the prior material and to serve as detailed examples.

11.1 Differential Signaling

The multiconductor interconnects discussed up to this point rely on one conductor per signal with a shared return path, such as a ground plane. This type of signaling is called *single-ended*. Voltages generated by one signal on the shared return path automatically show up in the others, so single-ended signaling tends to be noisy due to this common-mode noise. Note that SSN is common-mode noise.

It is possible to dedicate two conductors per signal with each line driven out of phase, leading to *differential* signaling on a differential pair. At the receive end, a differential amplifier recovers the signal as the difference in the voltages on the two lines. A differential pair still shares a common return path, but the common-mode noise introduced by the return path is rejected by the differential receiver. Therefore, differential signaling is very quiet compared to single-ended signaling and can operate to much higher signaling rates. Because differential signaling offers the highest signal integrity, it forms the foundation of many high-performance signaling standards, including LVDS and ECL.

The rejection of common-mode noise can be demonstrated using modal decomposition. If the signal waveform to be transmitted down a differential pair is called $v(t)$, then two new out-of-phase waveforms can be defined as

$$\begin{aligned} v_1(t) &= v_o - \frac{1}{2}v(t) \\ v_2(t) &= v_o + \frac{1}{2}v(t), \end{aligned}$$

where v_o is a constant voltage. The voltages v_1 and v_2 are launched at one end of a pair of symmetric coupled transmission lines as shown in Figure 11.1. According to chapter 10, equation (10.21), the voltages of the even and odd modes are

$$\begin{aligned} v_e(t) &= \sqrt{2}v_o \\ v_o(t) &= v(t)/\sqrt{2}. \end{aligned}$$

The even mode carries only a DC component, which generates no noise on reactive parasitics, while the odd mode carries a weighted version of the signal.

If the pair of transmission lines are tightly coupled (a proper design), then an external source of noise is impressed equally on both lines as common-mode noise. The signal then has high noise immunity to SSN, crosstalk, and EMI. The even mode

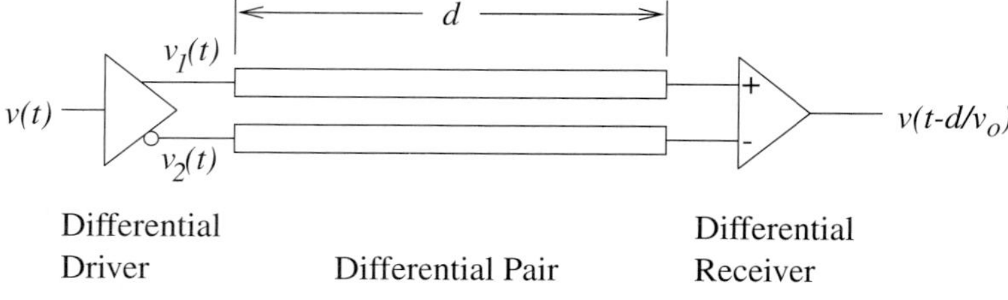

Figure 11.1. Differential signaling using a symmetric coupled pair of transmission lines.

collects the common-mode noise, while the odd mode remains relatively noise-free. Superimposing common-mode noise onto the far-end mode voltages yields

$$\begin{aligned} v_e(t+\Delta t) &= \sqrt{2}v_o + v_{\text{noise}}(t) \\ v_o(t+\Delta t) &= v(t-d/\nu_o)/\sqrt{2}, \end{aligned}$$

where ν_o is the odd-mode phase velocity. Based on (10.32), the terminal voltages at the far end are

$$\begin{aligned} v_1(t+\Delta t) &= \frac{v_e(t+\Delta t) - v_o(t+\Delta t)}{\sqrt{2}} \\ &= v_o + v_{\text{noise}}(t)/\sqrt{2} - v(t-d/\nu_o)/2 \\ v_2(t+\Delta t) &= \frac{v_e(t+\Delta t) + v_o(t+\Delta t)}{\sqrt{2}} \\ &= v_o + v_{\text{noise}}(t)/\sqrt{2} + v(t-d/\nu_o)/2. \end{aligned}$$

At the far end, a differential receiver is used to detect only the difference between the signals on the two lines, yielding the output

$$\begin{aligned} v_{\text{received}} &= v_2(t+\Delta t) - v_1(t+\Delta t) \\ &= v(t-d/\nu_o). \end{aligned}$$

Therefore, differential signaling enables the transmission of a signal at the odd-mode phase velocity without being corrupted by common-mode noise.

Compared to single-ended signaling, differential signaling requires more complex drivers and receivers plus double the interconnect leads. However, the immunity to SSN allows a reduction in the number of power and ground pins in packages, sockets, and connectors. In addition, the excellent noise immunity may allow the deletion of some bypass capacitors.

Breaks in the symmetry of the differential pair cause mode conversion between the even and odd modes. Common-mode noise is injected onto the differential signal whenever symmetry is disrupted. For maximum performance, all components must maintain excellent symmetry. Each line must look just like the other.

Finally, some common-mode noise does get through the receiver to appear as noise on the signal. Good differential receivers have a high common-mode rejection ratio to maximize the rejection of common-mode noise.

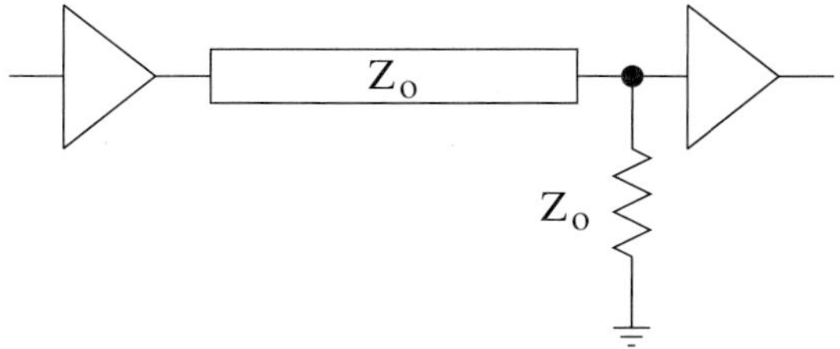

Figure 11.2. Basic far-end parallel termination.

11.2 Termination

When the load impedance does not match the characteristic impedance of a transmission line, a fraction of the signal is reflected from the load and propagates back down the line. While the reflected wave can be used to advantage in source-terminated second-incidence switching schemes, reflected waves usually serve no useful function and simply degrade signal quality.

A typical receiver has a small input capacitance (roughly 5pF) that leads to a reflection coefficient near +1. To minimize reflections, an impedance can be added in parallel to the receiver to create a reflection coefficient near zero. Basic far-end parallel termination at the receiver is shown in Figure 11.2. A resistance equal to the real part of the transmission line characteristic impedance is sufficient.

Basic termination is quite simple—add matched resistors in parallel to the loads. In practice, several factors, including parasitics and location, power dissipation, and multiconductor termination, can complicate termination design.

11.2.1 Parasitics and Location

Resistors have a small but significant series parasitic inductance. In addition, the PCB pads and vias required to mount a resistor introduce additional series inductance. Both sources of inductance should be minimized to keep the resistance as real as possible over the widest bandwidth. Small surface-mount resistors make excellent terminators, and a good PCB layout technique is suggested in Figure 11.3.

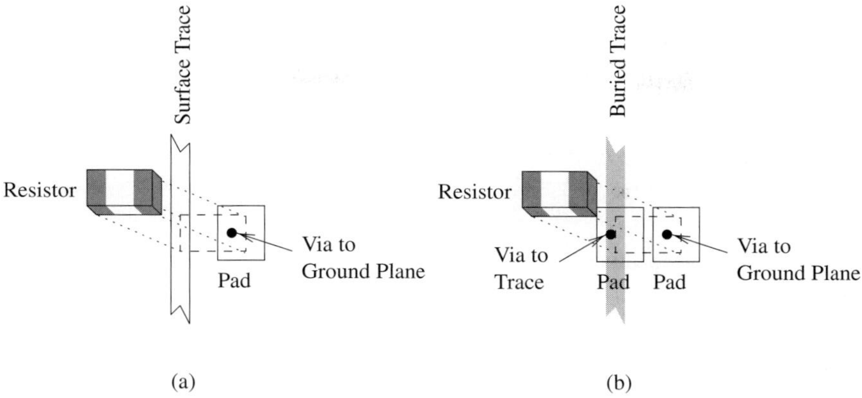

Figure 11.3. Low-inductance mounting schemes for: (a) surface routes, (b) buried routes.

Termination should be applied as close to the receiver as practical to avoid signal integrity loss from the unterminated stub between the termination and the receiver. The longer the stub, the greater the effect. For packaged devices, the ideal location is inside the package, but in practice, a location on the PCB near the package lead is best.

11.2.2 Static Power Dissipation and Current Carrying Requirements

Power is dissipated in parallel terminations even when there is no switching activity, and other than the costs associated with the acquisition and assembly of resistors, this is the main disadvantage of parallel termination. The topology of the termination can reduce power dissipation in some applications.

Four options for parallel termination of a transmission line of characteristic impedance Z_o are shown Figure 11.4, which includes the basic configuration in Figure 11.2. The transmission line is not shown, while for each case, a switch representation of a totem-pole driver is shown both holding high and holding low. The

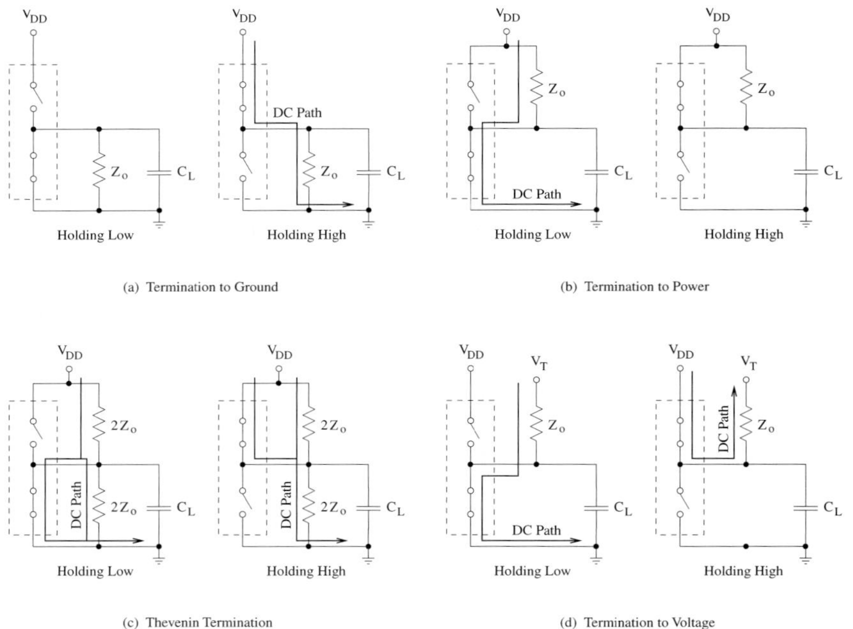

Figure 11.4. Four parallel termination schemes and DC current paths for totem-pole drivers.

driver impedance is assumed to be R when holding either high or low. The receiver is represented by C_L, which allows no DC current path. In all cases, sufficient bypass capacitance is assumed so that V_{DD} is a good AC ground.

In Figure 11.4a, the termination resistor is tied to ground, and when holding high the static power dissipation is $V_{DD}^2/(R+Z_o)$, while the DC power is zero when holding low. Therefore, there is an imbalance between the two states, and if a digital signal predominately holds low, power can be saved by termination to ground. The static power dissipation between the two states can be flipped by termination to V_{DD}, as shown in Figure 11.4b; then power can be saved if a signal predominately holds high.

One disadvantage of termination to only power or ground is the level of current that one of the totem-pole transistors must source. In Figure 11.4a, the upper transistor must be capable of continuously sourcing $V_{DD}/(R + Z_o)$. Similarly, the lower transistor in Figure 11.4b must be able to source the same amount. By using two termination resistors to create a Thevenin termination, as shown in Figure 11.4c, both holding-low and holding-high states dissipate $V_{DD}^2/(2Z_o + R||2Z_o)$, so neither state is preferred. Each transistor must be able to continuously source $V_{DD}/[2(R + Z_o)]$. Therefore, Thevenin termination requires the transistors to source half as much current as the single termination.

Balanced power dissipation is also achieved with termination to a voltage, as shown in Figure 11.4d, with $V_T = V_{DD}/2$. Power dissipation in both states is $V_{DD}^2/[4(R + Z_o)]$ with current-sourcing requirements of $V_{DD}/[2(R + Z_o)]$. The advantages of this termination include small parts count, lower power requirements, and a lower current-sourcing requirement. The major disadvantage is the need for a good low-impedance power supply voltage at $1/2V_{DD}$.

For Thevenin termination and termination to voltage, care must be taken not to leave the input to a receiver at $V_{DD}/2$ to avoid large crowbar currents and/or metastability. For both of these termination types, a failed driver or one set in a high-impedance output state allows the receiver input to drift to the voltage set by the termination network. Although it increases power dissipation, a safe design sets the termination voltage away from $V_{DD}/2$.

The power and current requirements of the various parallel termination schemes are summarized in Table 11.1. To facilitate comparison, normalized power and current are also provided, showing in particular that a termination to $V_{DD}/2$ achieves a quarter of the power (worst-case) and half the current-sourcing requirements of a single termination. Note that the current requirements are those for the driver and not for the power supply. Driver specifications must be checked to make sure that the current requirements of parallel termination can be met.

Static power dissipation can be eliminated by placing a capacitor in series with the termination resistor, as shown in Figure 11.5, for an unbalanced termination to ground. This scheme is often called *AC termination*. The capacitor blocks DC

Table 11.1. Summary of worst-case static power dissipation and current-sourcing requirements for various parallel termination schemes.

Type	Power	Normalized	Current	Normalized
To Power or Ground	$\frac{V_{DD}^2}{R+Z_o}$	1	$\frac{V_{DD}}{R+Z_o}$	1
Thevenin	$\frac{V_{DD}^2(R+2Z_o)}{4Z_o(R+Z_o)}$	$\frac{R+2Z_o}{4Z_o}$	$\frac{V_{DD}}{2(R+Z_o)}$	$\frac{1}{2}$
To Voltage $V_T = \frac{V_{DD}}{2}$	$\frac{V_{DD}^2}{4(R+Z_o)}$	$\frac{1}{4}$	$\frac{V_{DD}}{2(R+Z_o)}$	$\frac{1}{2}$

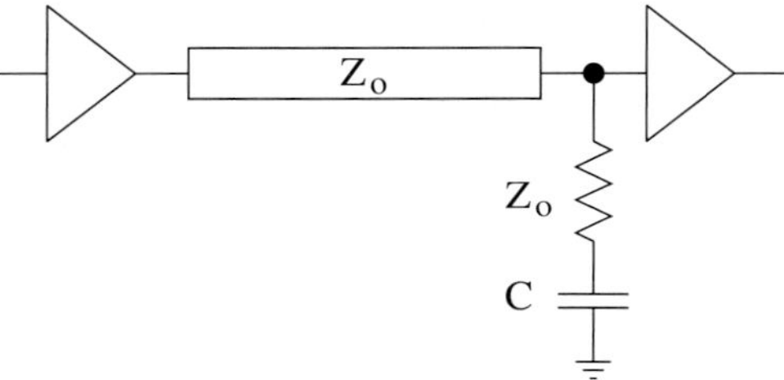

Figure 11.5. AC termination with series capacitance to eliminate static power dissipation.

current, so there is no static power dissipation. However, the capacitor decreases the effectiveness of the termination by reflecting low-frequency signal components and by providing additional capacitive loading on the transmission line. Optimization is required to balance termination effectiveness with capacitive loading, and in many cases no suitable value for the capacitance will be found.

11.2.3 Voltage Swing

Parallel termination creates a voltage divider between the driver source impedance and the termination resistance that limits the static voltage swings of totem-pole drivers. For the termination schemes in Figure 11.4, the static voltages holding low

Table 11.2. Summary of static voltage swings holding low and high with capacitive loading and parallel termination.

Type	Holding Low	Holding High
Single to Ground	0	$V_{DD}(1 - \frac{R}{R+Z_o})$
Single to V_{DD}	$V_{DD}(1 - \frac{Z_o}{R+Z_o})$	V_{DD}
Thevenin	$\frac{1}{2}V_{DD}(1 - \frac{Z_o}{R+Z_o})$	$\frac{1}{2}V_{DD}(1 + \frac{Z_o}{R+Z_o})$
To Voltage $V_T = \frac{V_{DD}}{2}$	$\frac{1}{2}V_{DD}(1 - \frac{Z_o}{R+Z_o})$	$\frac{1}{2}V_{DD}(1 + \frac{Z_o}{R+Z_o})$

and high are summarized in Table 11.2. To obtain the largest noise margins, the widest voltage swing is required. For each case, the voltage swing is maximized when $R \ll Z_o$. A small driver impedance causes a large reflection at the source and, from Table 11.1, high power dissipation while requiring large current-sourcing capability.

11.2.4 Diode Termination

Parallel termination can be implemented with diodes, as shown in Figure 11.6, to reduce static power dissipation. For the voltage range from $V_{SS} - V_T$ to $V_{DD} + V_T$,

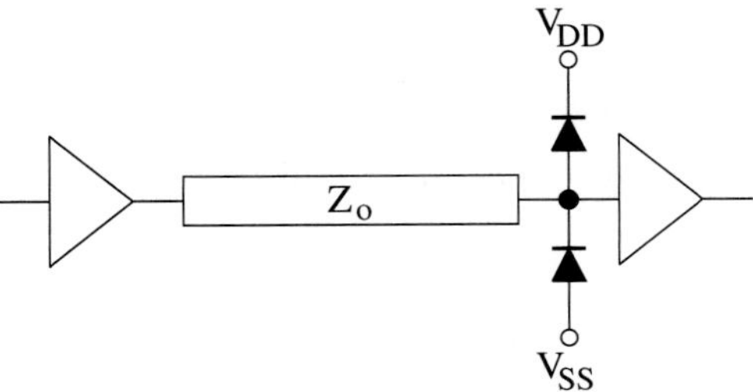

Figure 11.6. Diode termination.

where V_T is the threshold voltage of the diodes, the diodes are reverse-biased and only pass a very small leakage current. The transmission line is then unterminated and dissipates no static power. As the voltage drops below $V_{SS} - V_T$, the lower diode turns on and behaves like a resistor, with resistance determined by the slope of the IV curve. The transmission line is terminated in this resistance while the voltage stays below $V_{SS} - V_T$, so diodes with appropriate slope are needed. The behavior is similar while the voltage exceeds $V_{DD} + V_T$.

Diodes with very steep slope (low resistance) simply clip the waveforms at $V_{DD} + V_T$ and $V_{SS} - V_T$ and reflect high-frequency energy back towards the driver. Noise reduction is best achieved by dissipating unwanted energy. Hard-clipping diodes limit undershoot and overshoot but dissipate little energy, so a resistive element is needed elsewhere.

Standard ESD protection circuits include diode clamping in a configuration identical to diode termination. Therefore, diode termination is almost always present on packaged devices; however, the diode characteristics are selected to optimize ESD protection rather than signal integrity. Any diode termination added to enhance signal integrity requires diodes with a V_T lower than the ESD circuits.

11.2.5 Source Termination

In contrast to parallel termination at the load, source termination (see also chapter 2, section 2.2) allows the signal to reflect off of the load and travel back to the source, where it is absorbed by series termination. The series termination can be implemented as either the output impedance of the driver or as a resistor in series with the driver, in which case the driver must have an output impedance lower than the characteristic impedance of the transmission line. The resistor can be implemented within the driver itself, where it is often referred to as *ballast*, or it can be implemented externally on the PCB.

The voltage launched onto the transmission line is determined by the voltage divider between the driver impedance plus termination and the characteristic impedance of the transmission line. Generally, the voltage is

$$V_{\text{launch}} = V_{OH}\frac{Z_o}{Z_o + R_{\text{driver}} + R_{\text{term}}},$$

where V_{OH} is the high voltage output of the driver. To fully absorb the reflected wave, the driver plus termination must match the characteristic impedance of the transmission line, then $R_{\text{driver}} + R_{\text{term}} = Z_o$, and the voltage launched onto the transmission line is $1/2V_{OH}$.

Unless resistance has been added, the input impedance of receivers looks like a small capacitive load with a reflection coefficient near +1. The incident wave is almost fully reflected, and the incident and reflected waves combine to double the voltage of the incident wave at the receiver. For a source-matched driver, the full voltage swing is restored at the receiver. If parallel termination is also applied at the receiver, then the reflection coefficient is 0 and the voltage at the receiver swings to just $1/2V_{OH}$.

The advantage of source termination is that there is no static power dissipation. Since the matching resistance is in series with the load capacitance, a driver holding high or low supplies no current once the load capacitance is charged.

The power advantage of source termination diminishes for long transmission lines. Until the reflected wave returns to the driver, the load on the driver is strongly resistive. The driver sees a terminated circuit, and current consumption and power dissipation are equivalent to a far-end parallel terminated circuit. Once the reflected wave returns, the load appears capacitive and the current flow and power consumption stop. Therefore, power consumption peaks when the round-trip time equals half the clock period, assuming a 50% duty cycle. Beyond this point, power dissipation cycles up and down, depending on the timing of the return reflections.

An example of the length-dependence of power dissipation for source termination is provided in Figure 11.7, where a 3.3V 50Ω driver launches a 50MHz clock signal with a 50% duty cycle onto a 50Ω transmission line that is loaded at the far end with 5pF of capacitance. The average power dissipation for zero transmission line length is just $CV^2f = 2.72$mW for the capacitor plus 2mW for the driver (computed with

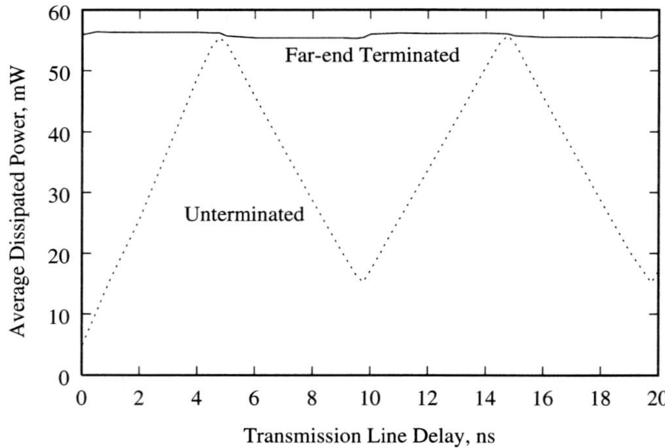

Figure 11.7. Example power dissipation by a source-terminated driver driving a long transmission line with the far end both terminated and unterminated. The circuit topology is from Figure 11.2.

no load), for a total of 4.72mW. As the line length increases, the average power dissipation increases until the line delay is almost 5ns, where the power equals the $V^2/(50+50)/2 = 54.5$mW static power dissipation of a far-end terminated net.

11.3 Multiconductor Termination

Strongly coupled multiconductor transmission lines should ideally be terminated in a resistor network that individually terminates each mode. Termination of each line as if it were in isolation can lead to significant reflections. For lossless coupled transmission lines, modal decomposition can be used to find the optimal termination network.

From (10.9) and (10.11), the voltages and currents on a multiconductor transmission line are $\overline{v} = \overline{\overline{T}}_V \overline{v}_m$ and $\overline{i} = \overline{\overline{T}}_I \overline{i}_m$, respectively. Since the modal transmission lines are decoupled, according to chapter 2, equations (2.16) and (2.17), the modal voltages and currents can be written in terms of forward- and backward-traveling waves, so

$$\overline{v}_m = \overline{v}_m^+ + \overline{v}_m^-$$

and

$$\overline{i}_m = \overline{\overline{k}}^{-1}(\overline{v}_m^+ - \overline{v}_m^-),$$

where

$$\overline{\overline{k}} = \begin{bmatrix} Z_{o1} & 0 & \cdots & 0 \\ 0 & Z_{o2} & \cdots & 0 \\ \vdots & \vdots & \ddots & \vdots \\ 0 & 0 & \cdots & Z_{oN} \end{bmatrix},$$

and Z_{oi} is the characteristic impedance of the ith mode.

Assume that the multiconductor transmission line is terminated with a network described by the impedance matrix $\overline{\overline{Z}}$. The voltages and currents are related by $\overline{v} = \overline{\overline{Z}}\,\overline{i}$. Substituting the total voltages and currents with the traveling wave representations of the modal voltages and currents yields

$$\overline{\overline{T}}_V(\overline{v}_m^+ + \overline{v}_m^-) = \overline{\overline{Z}}\,\overline{\overline{T}}_I\overline{\overline{k}}^{-1}(\overline{v}_m^+ - \overline{v}_m^-);$$

then by rearranging

$$\overline{v}_m^- = (\overline{\overline{Z}}\,\overline{\overline{T}}_I\overline{\overline{k}}^{-1} + \overline{\overline{T}}_V)^{-1}(\overline{\overline{Z}}\,\overline{\overline{T}}_I\overline{\overline{k}}^{-1} - \overline{\overline{T}}_V)\overline{v}_m^+.$$

The modal reflection coefficient is identified as

$$\overline{\overline{\Gamma}}_m = (\overline{\overline{Z}}\,\overline{\overline{T}}_I\overline{\overline{k}}^{-1} + \overline{\overline{T}}_V)^{-1}(\overline{\overline{Z}}\,\overline{\overline{T}}_I\overline{\overline{k}}^{-1} - \overline{\overline{T}}_V), \tag{11.1}$$

which is very useful for determining reflections of modes and conversions between them. In terms of a given load admittance, (11.1) can be rearranged to obtain

$$\overline{\overline{\Gamma}}_m = (\overline{\overline{T}}_I\overline{\overline{k}}^{-1} + \overline{\overline{Y}}\,\overline{\overline{T}}_V)^{-1}(\overline{\overline{T}}_I\overline{\overline{k}}^{-1} - \overline{\overline{Y}}\,\overline{\overline{T}}_V). \tag{11.2}$$

For perfect termination, all modes are completely absorbed at the load, so $\overline{\overline{\Gamma}}_m = 0$. From (11.1) this occurs when

$$\overline{\overline{Z}}\,\overline{\overline{T}}_I\overline{\overline{k}}^{-1} - \overline{\overline{T}}_V = 0$$

or

$$\overline{\overline{Z}} = \overline{\overline{T}}_V\overline{\overline{k}}\,\overline{\overline{T}}_I^{-1}. \tag{11.3}$$

A network implementing $\overline{\overline{Z}}$ is the optimal termination. The admittance of the matching network can be found by inverting (11.3) to obtain

$$\overline{\overline{Y}} = \overline{\overline{T}}_I\overline{\overline{k}}^{-1}\overline{\overline{T}}_V^{-1}. \tag{11.4}$$

11.3.1 Single Transmission Line

For a single transmission line, $\overline{\overline{T}}_V = \overline{\overline{T}}_I = 1$, and $\overline{\overline{k}} = Z_o$, so the termination impedance from (11.3) is simply Z_o, the characteristic impedance of the transmission line.

11.3.2 Differential Pair

For a symmetric transmission line pair terminated into a balanced π-network, the needed inputs are available from chapter 10, equations (10.19) and (10.20), and problem 13 in chapter 4 (with $y_3 = y_1$). Evaluating (11.4), the optimal termination network can be found, then equated to the admittance matrix of the π-network to fill out the circuit.

To view what happens to the even and odd modes, the modal reflection coefficient from (11.2) can be evaluated with

$$\overline{\overline{k}} = \begin{bmatrix} Z_{oe} & 0 \\ 0 & Z_{oo} \end{bmatrix},$$

to obtain

$$\overline{\overline{\Gamma}}_m = \begin{bmatrix} \Gamma_e & 0 \\ 0 & \Gamma_o \end{bmatrix} = \begin{bmatrix} \frac{1-y_1Z_{oe}}{1+y_1Z_{oe}} & 0 \\ 0 & \frac{1-y_1Z_{oo}-2y_2Z_{oo}}{1+y_1Z_{oo}+2y_2Z_{oo}} \end{bmatrix}. \tag{11.5}$$

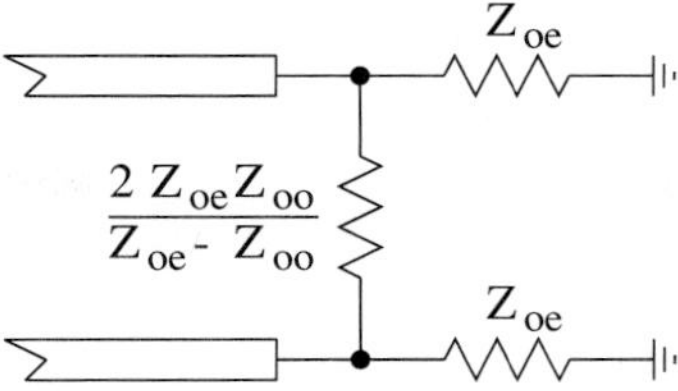

Figure 11.8. Optimal termination for a pair of symmetric transmission lines.

Note how there is no coupling between the modes since the off-diagonal terms are zero. The optimal termination network with $\overline{\overline{\Gamma}}_m = 0$ requires that $y_1 = 1/Z_{oe}$ and $y_2 = (Z_{oe} - Z_{oo})/(2Z_{oe}Z_{oo})$. The resulting schematic is shown in Figure 11.8.

Differential pairs are not typically terminated with the network in Figure 11.8, so it is interesting to investigate the reflections generated from imperfect termination. Three common termination schemes are shown in Figure 11.9, along with the hookup for the differential receiver. These terminations are discussed in the following sections.

Bridged Termination

As shown in Figure 11.9a, a common termination scheme for differential pairs is a single resistor bridging the lines, so $y_1 = 0$ and $y_2 = 1/(2Z_{oo})$. From (11.5) the modal reflection coefficients are

$$\overline{\overline{\Gamma}}_m = \begin{bmatrix} 1 & 0 \\ 0 & 0 \end{bmatrix},$$

and it can be seen that bridged termination leads to $\Gamma_e = +1$ and $\Gamma_o = 0$. The odd mode is completely absorbed, while the even mode is completely reflected. Since the signal is carried in the odd mode and the differential receiver rejects the even mode, extra noise on the even mode caused by bridged termination is inconsequential as long as the even-mode noise does not couple into the odd mode through mode conversion.

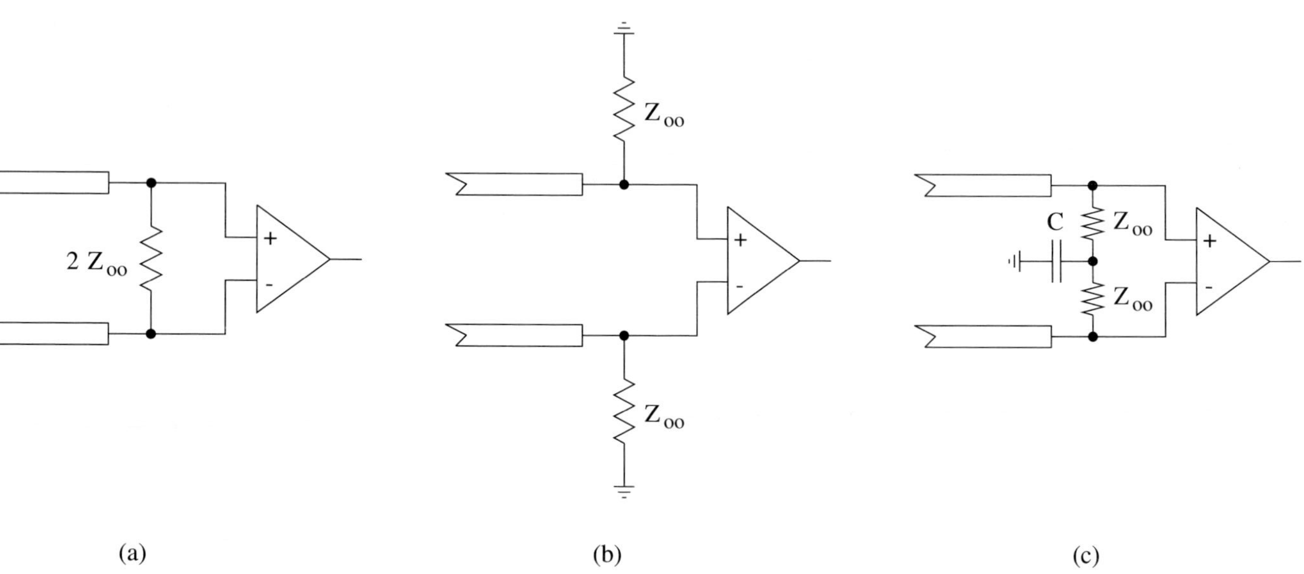

Figure 11.9. Imperfect termination for a differential transmission line pair: (a) bridged, (b) single-ended, (c) AC.

Single-Ended Termination

As shown in Figure 11.9b, the two transmission lines can be individually terminated. With $y_1 = 1/Z_{oo}$ and $y_2 = 0$, the modal reflection coefficients from (11.5) are

$$\overline{\overline{\Gamma}}_m = \begin{bmatrix} \frac{Z_{oo}-Z_{oe}}{Z_{oo}+Z_{oe}} & 0 \\ 0 & 0 \end{bmatrix},$$

so the odd mode is matched while the even-mode reflection is greatly reduced. Compared to the bridged termination, the extra damping of the even mode costs an additional resistor.

Differential drivers often provide single-ended termination. If bridged termination is used at the receiver, then the odd mode is absorbed at both ends for very good signal integrity, while the driver damps the even mode to keep it from becoming a noise and EMI problem.

AC Termination

The AC termination in Figure 11.9c is a T-network with the impedance matrix

$$\overline{\overline{Z}} = \begin{bmatrix} Z_{oo} + \frac{1}{\jmath\omega C} & \frac{1}{\jmath\omega C} \\ \frac{1}{\jmath\omega C} & Z_{oo} + \frac{1}{\jmath\omega C} \end{bmatrix},$$

using the results from problem 14 in chapter 4. Applying this to (11.1) with

$$\overline{\overline{k}} = \begin{bmatrix} Z_{oe} & 0 \\ 0 & Z_{oo} \end{bmatrix}$$

yields the modal reflection coefficient as

$$\overline{\overline{\Gamma}}_m = \begin{bmatrix} \frac{Z_{oo}-Z_{oe}+2\frac{1}{\jmath\omega C}}{Z_{oo}+Z_{oe}+2\frac{1}{\jmath\omega C}} & 0 \\ 0 & 0 \end{bmatrix}.$$

The odd mode is always matched, while the even mode has a frequency-dependent reflection coefficient. At low frequencies, $\Gamma_e = +1$, while at high frequencies, $\Gamma_e = (Z_{oo} - Z_{oe})/(Z_{oo} + Z_{oe})$ and a large fraction is absorbed. Compared to the bridged termination, even-mode attenuation is added without increasing static power dissipation. However, AC termination requires three components.

11.4 Power Distribution

Signal integrity in digital systems is directly affected by the quality of the power distribution system. Noise on the power supply manifests itself as rail collapse and SSN, with direct impact on noise margins and timing. There are several aspects of power distribution design, including PCB stackup; bypass capacitor selection and placement; voltage partitioning; package, socket, and connector selection; pin assignments, including selection of the signal-to-power and signal-to-ground ratios; and pin placement. The basic goal in power distribution is to minimize inductance and resistance while maximizing capacitance over a wide frequency range. Factors that complicate the design include cost (because excellent power distribution is easy to obtain with an unlimited budget), size limitations, nonideal component behavior (especially for capacitors), limits on pin counts, limits on the signal-to-power and signal-to-ground ratios, and limits on the layer count of the PCB. This section covers a few basics of design methodology, component behavior, and modeling techniques.

11.4.1 Target Impedance

The power distribution system must work as a low-impedance source over a bandwidth from DC to several harmonics of the clock frequency. Low impedance is needed to minimize the generation of noise, but it is also needed to minimize the radiation of electromagnetic energy and the EMI that would result. One method of specifying a power distribution system is in terms of a target impedance,[1] which is the maximum allowed impedance for the system to meet a specified noise level.

Consider a component operating at a voltage V_{DD} and dissipating an average of P watts. The average current is $I_{\text{avg}} = P/V_{DD}$. Assuming that the allowed ripple on the power supply is χV_{DD}, where typical values for χ are 5% to 10%, then the

1. This section is based on the work presented in the paper by Larry D. Smith, "Packaging and power distribution design considerations for a Sun Microsystems desktop workstation," *Proc. IEEE 6th Topical Meeting on Electrical Performance of Electronic Packaging*, Oct. 27–29, 1997, pp. 19–22.

target impedance is

$$\begin{aligned} Z_{\mathrm{T}} &= \frac{V_{\text{ripple}}}{I_{\text{avg}}} \\ &= \frac{\chi V_{DD}^2}{P}. \end{aligned}$$

The power distribution system is designed so that the impedance looking into the system at the site of the component is less than the target impedance over a specified bandwidth. The definition of target impedance is sketched in Figure 11.10.

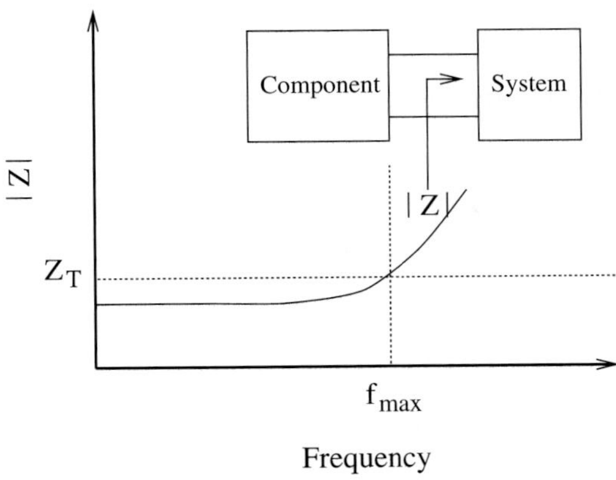

Figure 11.10. Target impedance sets the maximum impedance magnitude of the power distribution system as seen from the position of a component.

11.4.2 Design Overview

The target impedance of a power distribution system can be lowered using several techniques. The main areas where significant impact can be made are packaging, the PCB stackup, and bypass capacitors. The subject of regulated power supplies is beyond the scope of this book.

Packaging

Packaging components include integrated circuit packages, sockets, and edge connectors. Target impedance is lowered by choosing a packaging component that offers a shorter connection, for lower partial self-inductance, and/or by assigning more pins to power and ground connections. The number of power and ground pins should be equal unless the noise margins at logic low and high are significantly asymmetrical. The power and ground pins should be evenly interleaved across the component.

The most important design criteria for packaging is the signal-to-ground ratio. Lower ratios supply better impedance control and lower power-distribution inductance to support faster signaling. For a given signaling speed, different styles of packaging require different signal-to-ground ratios. For example, a BGA with planes can get by with a higher ratio than can a simple QFP.

The appropriate signal-to-ground ratio for a given application can be determined through detailed modeling and simulation of SSN. Good models of drivers, receivers, and packaging components are required, so the required level of effort is quite high. In most applications, the ratio is selected through experience with prior versions of the product or with similar products.

One package style is not inherently better than another as long as enough pins are available to lower the signal-to-ground ratio. BGAs are often seen in high-speed applications simply because they have more pins than QFPs.

PCB

The PCB stackup strongly affects the impedance of the power distribution system. Stackup design is dictated by PCB manufacturing technology, which is itself driven by cost since the PCB is often one of the most expensive components in a system. The basic elements of a multilayer PCB are two-sided PCBs called *cores*, dielectric sheets called *prepreg*, and copper sheets. A two-layer PCB consists of a single core by itself. A four-layer board can be made by laminating copper sheets onto prepreg, then laminating those onto each side of a core. A six-layer board can be made

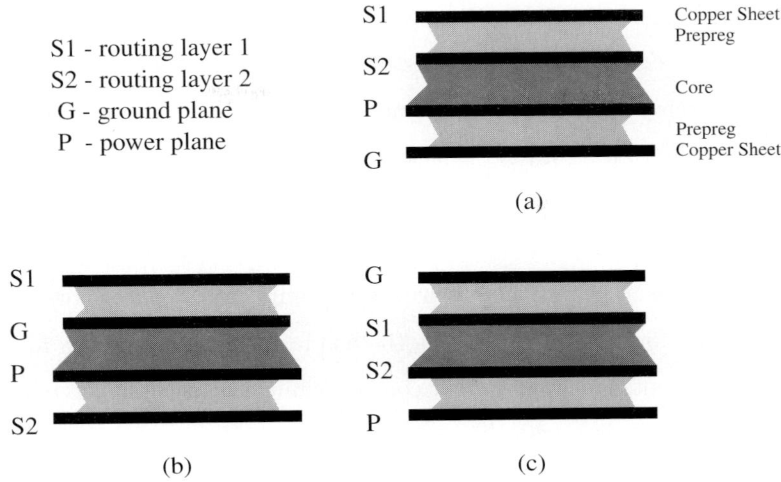

Figure 11.11. Possible stackups for a four-layer PCB.

by laminating two cores together with prepreg and then finishing like a four-layer board. Generally, an N-layer board has N wiring layers spaced by $N-1$ dielectrics, and N must be an even number to prevent warpage.

High-performance systems require power and ground planes plus routing layers; therefore, the simplest stackup for these systems is the four-layer PCB, which leaves two routing layers. The power and ground planes form a parallel-plate transmission line with characteristic impedance

$$Z_o = \sqrt{\frac{\mu_o}{\epsilon_r \epsilon_o}} \frac{H}{W} \tag{11.6}$$

(see problem 2 in chapter 2). Product size limits W, the size of the board, while cost constraints limit the choice of the dielectric constant ϵ_r (usually to just FR-4). The main variable under the designer's control is H, the separation between the planes. The lowest impedance is obtained when the power and ground planes are back-to-back.

A four-layer board has three possible stackups, shown in Figure 11.11. The stackup in Figure 11.11a has two problems. First, for a given line width, the characteristic impedance of a line routed on S2 is significantly lower than that for routing on S1, due to the closer proximity to the planes. Second, the PCB is susceptible to warping under temperature variations since the planes have a much higher loading of copper than the routing layers; it lacks symmetry. The stackup in Figure 11.11b is very common since it provides the lowest power distribution impedance with uniform characteristic impedance on the routing layers in a symmetric form. The alternative stackup in Figure 11.11c triples the impedance of the power distribution but provides for radiation shielding (if the planes are not punched full of holes to mount components) for better EMI control, while also providing symmetry and uniform characteristic impedance for routing.

If a circuit cannot be routed in four layers, then additional layers must be added in pairs. The assignment of layers must take into account routability, power distribution impedance, routing characteristic impedance, EMI, warpage, and cost.

Copper Weights

A major option on PCB construction is the thickness of the copper cladding. Thicker cladding has lower resistance and represents a good choice for power planes. Thinner cladding supports narrow lines with better width control, so thin cladding is good for signal layers to obtain higher routing density. Cladding thickness is specified by copper weight, where a specified weight of copper spread over a 1in $\times$ 1in area provides a certain thickness. The standard copper weights and their thicknesses are 0.5oz (0.72mil), 1oz (1.4mil) and 2oz (2.8mil), where 1mil=0.001in. The conductivity of copper is approximately 5.7×10^7S/m.

Bypass Capacitance

High-performance designs require bypass capacitance for four reasons: supplying current bursts for fast switching circuits, providing an AC connection between power and ground planes for return currents, controlling EMI, and lowering the impedance of the power distribution system. For all of these uses, capacitors connect the power and the ground planes. The difference between these uses is the size and quality of the capacitors and their locations.

To support the peak current needs of rapidly switching circuits, bypass capacitors accumulate charge over the clock cycle, then rapidly discharge during fast transitions. Power supplies cannot fulfill this role because voltage regulators respond too slowly and because the regulator is typically too far away. The distance issue is strictly related to the speed of light—the demand for the current must propagate to the source of the charge and then the current must propagate back. Capacitors fulfilling this role should be located very close to the component needing the current bursts and should be sized to supply sufficient capacity without picking up too much self-inductance, which limits the ability of the capacitor to quickly supply the current.

Currents return on the power distribution in such a way as to lower the total impedance. Low-frequency components spread over every available path to minimize resistance. High-frequency components crowd under signals on planes to minimize inductance and bridge from ground to power planes and back, as necessary, through the capacitive reactance between them. By adding bypass capacitors between the planes, current can more easily transition between the planes since the reactance is reduced, so generally, inductance is reduced and signal quality improves. Capacitors for this role should be located near the power pins of every high-speed component and should be sized with relatively small values to obtain low inductance.

EMI problems are reduced with a regular grid of capacitors that limits the size of current loops. The signal and return path form a loop that will radiate, so the larger the loop, the greater the radiation at a given frequency. By placing bypass capacitors on a fairly regular grid across a PCB, the size of any return path is limited by the grid size. Such a placement strategy is not strictly needed but helps avoid problems from unavoidable flaws in the design. Capacitors in this role should be evenly distributed and should be sized with relatively small values to obtain low inductance.

After bypass capacitor placement for the uses just discussed, the power distribution impedance may still exceed the target impedance. Additional bypass capacitors can be added to lower the impedance, and the size and location of the capacitors depends on the system design. Detailed modeling and simulation can help complete the design (see section 11.4.4), but due to the complexity of such an effort, it is not common. Product history and experience are often the only source of help in this area.

11.4.3 Capacitor Modeling

To avoid overly optimistic results, bypass capacitors should not be modeled as a pure capacitance. Real capacitors have a frequency response similar to the one in Figure 11.12, based on a series RLC circuit. Capacitors have series inductance and resistance due to contacts and the spatial dimensions of the capacitor plates. The inductance of mounting pads and vias are also significant and should be included in the capacitor model.

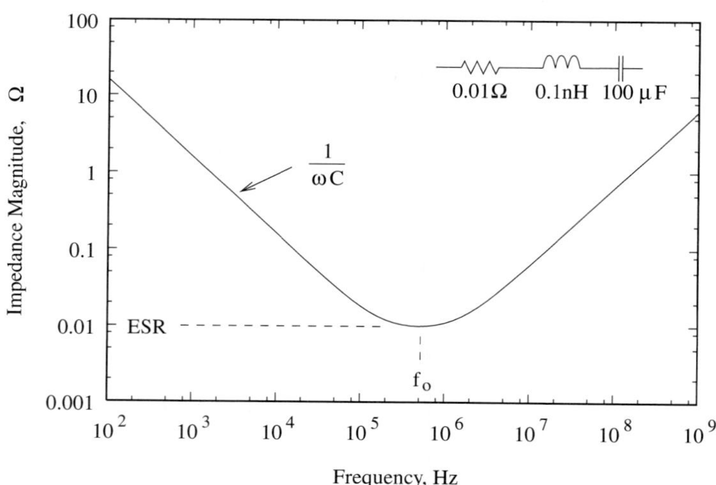

Figure 11.12. Example frequency response of a bypass capacitor modeled as a series RLC circuit.

The series RLC circuit resonates at the frequency

$$f_o = \frac{1}{2\pi\sqrt{LC}}.$$

Like all resonators, the impedance at the resonant frequency is purely resistive. Capacitors can be specified by providing the capacitance, the series resistance at resonance, called the *effective series resistance* (ESR), and f_o. Given these, the *effective series inductance* (ESL) can be found.

Smaller capacitors have lower inductance along with lower capacitance, so the resonant frequency is higher. Therefore, small capacitors are more useful for EMI and return path management. For supplying burst currents and for lowering Z_T, capacitors with larger capacity are needed; however, the lower self-resonant frequency can limit their usefulness. It may be necessary to use several smaller capacitors in parallel to increase capacitance while taking advantage of the higher f_o. (Note that N parallel capacitors have total capacitance of $N \times C$ and total inductance of L/N, so the resonant frequency (in radians) remains unchanged at $1/\sqrt{LC}$).

The parallel-plate capacitor formed by the power and ground planes in a multilayer PCB has a very low ESR and ESL while providing moderate amounts of capacitance. This capacitor is very effective in all of the bypass capacitance roles.

11.4.4 PCB Modeling

A simulation of the impedance of a power distribution system should include

1. The geometrical structure of the PCB
2. Component placement
3. Passive component models including resonance behavior

An effective and straightforward approach to achieve items 1 and 2 is to subdivide the PCB into small sections of transmission lines, as shown in Figure 11.13. Each subsection is a parallel-plate waveguide with the characteristic impedance given by (11.6) and phase velocity $c_o/\sqrt{\epsilon_r}$, where c_o is the speed of light (3×10^8m/s) and ϵ_r

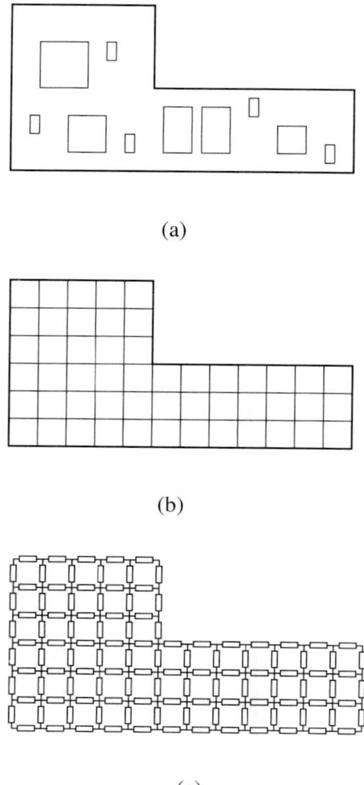

Figure 11.13. PCB power distribution system modeling through subdivision into transmission lines: (a) original PCB with components, (b) PCB marked for subdivision, (c) transmission line model.

is the dielectric constant of the PCB material. Actual transmission line models can be used in the simulation (which would be very slow), or the sections can be made small enough so that the lumped approximation holds and a network of resistors, inductors, and capacitors can be used.

Once the PCB has been gridded, component models can be added to account for packages, sockets, and bypass capacitors. The impedance looking into the model at the site of the components is computed and plotted, as in Figure 11.10, from DC to the upper frequency of interest to check for satisfaction of the target impedance. If the plot does not satisfy the Z_T specification, then component locations and values can be modified.

One of the reasons this technique is so effective is that the inductive coupling between transmission line subdivisions is negligible, and the capacitive coupling is zero, since all of the sections are shorted together. An extension of the technique to include signal-line routing would need to reintroduce coupling, which dramatically increases the complexity of the method.

An example calculation using this methodology is shown in Figure 11.14. The PCB is a four-layer board 1.6mm thick, so the power and ground planes are separated by 0.533mm. Assuming a dielectric constant of 3.4, then the wavelength at 500MHz (representing 5× a 100MHz board-level clock) is 0.325m. Limiting the transmission line lengths to 3.25cm allows them to be approximated with lumped components. To obtain sufficient granularity to locate components, the transmission line segments are sized to $5 \times 5\text{mm}^2$. A 10×20 grid of these models a $5 \times 10\text{cm}^2$ PCB. The characteristic impedance of each 5mm transmission line section is 21.8Ω, leading to lumped component values of L=0.67nH and C=1.41pF, using chapter 2, equations (2.14) and (2.15). An AC frequency sweep in SPICE enables the calculation of the impedance looking into the board at component sites, and the figure provides sample results. This simulation shows a reduction in peak impedance from almost 1000Ω to about 1Ω over the bandwidth of DC to 500MHz. A full-scale problem would include a full-size shaped PCB with all capacitor and component models in place and a good power supply model.

11.4.5 Core Noise Modeling

Off-chip drivers can create substantial noise on the core logic's power supply. This interaction cannot be directly simulated because cores have far too many transistors to simulate in SPICE. The core must be modeled to enable simulation.

A large digital component like a microprocessor is essentially a sea of switches (the transistors) in parallel with tiny capacitors (mostly gate and nwell capacitance

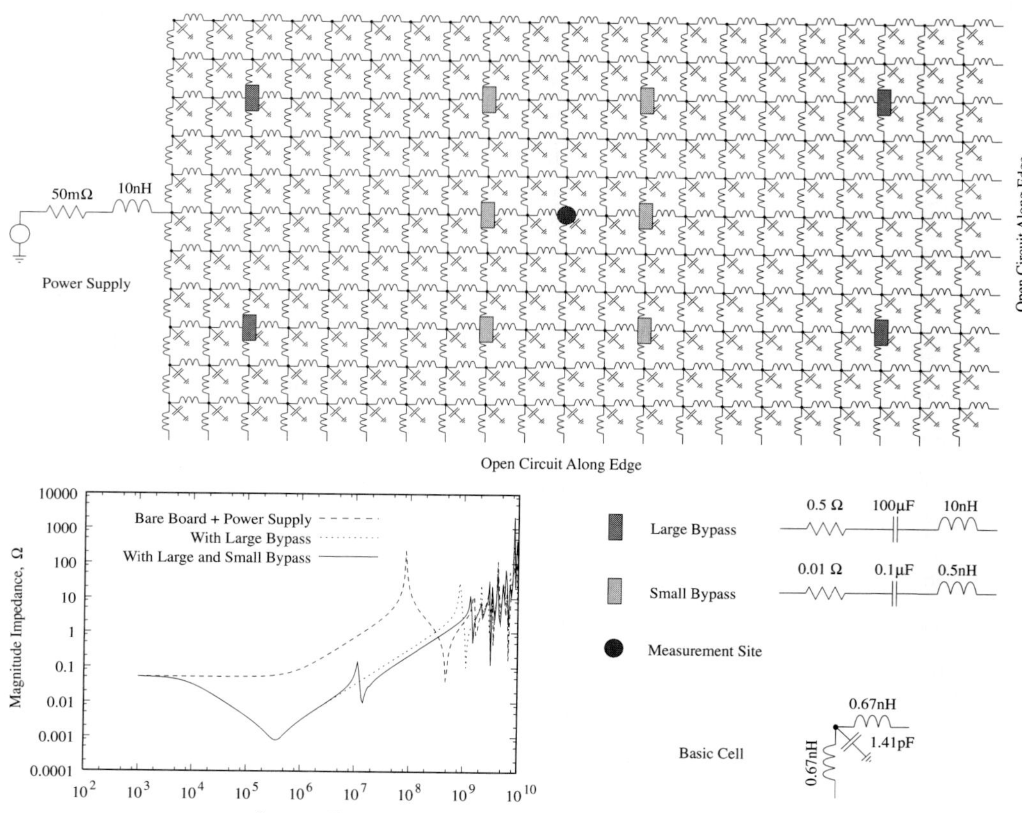

Figure 11.14. Example power distribution system modeling using an LC lumped approximation for a $5 \times 10\text{cm}^2$ PCB.

in CMOS). The switches operate over the course of a clock cycle, either charging or discharging the capacitors. The current supplied to or from the capacitors is supplied from two sources: on-chip capacitance not involved in switching (not all transistors switch at every clock cycle) and the power supply. Therefore, a reasonable model of the microprocessor suitable for system-level signal integrity simulations is a capacitor in parallel with a current source. This model is a Norton equivalent for the core.

A Norton equivalent model requires the capacitance of the core and the current profile of the source. The capacitance can be estimated from layout and process information, or it can be estimated from average power dissipation. As shown in chapter 1, equation (1.1), the average power dissipated in capacitive loads is

$$P = CV^2 f.$$

The power can be measured by monitoring the supply current and voltage. Knowing the power, supply voltage, and clock frequency, then the capacitance is

$$C = \frac{P}{V^2 f}.$$

For example, a core dissipating 5W at 500MHz with a 1.5V supply has approximately 4.4nF of core capacitance if all circuits are active all the time. Usually, much of the core is inactive on any given clock cycle. If on average 10% of the core transistors are active at a time, then the actual core capacitance is 10× higher than that predicted by average power dissipation. Continuing the example, for 10% switching activity, the core capacitance is actually 44nF.

The current profile for the Norton equivalent is much more difficult to estimate than capacitance, yet it is key to good simulations. A zero'th-order model can be constructed by assuming that the current ramps from zero to a peak, then back down during the course of a clock period, as suggested in Figure 11.15. Integrating the instantaneous power over a clock cycle yields the average power delivered by

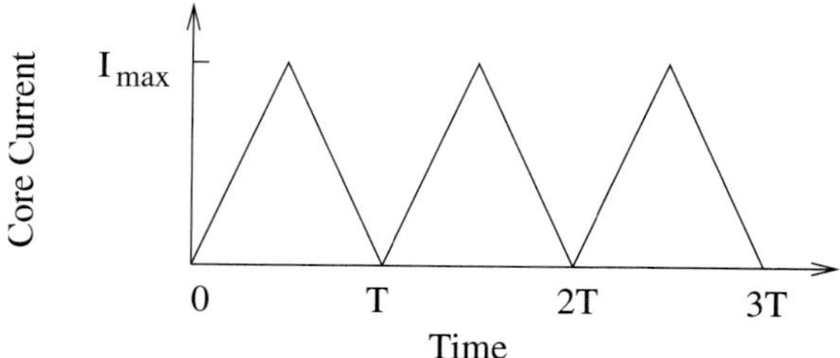

Figure 11.15. Simple model of current draw for a large core.

the power supply at voltage V as

$$\begin{aligned} P_{\text{avg}} &= \frac{1}{T}\int_0^T Vi(t)dt \\ &= \frac{1}{2}VI_{\text{max}}. \end{aligned}$$

The average power dissipation is used to fill out the model. For the example in the prior paragraph, the average current is 3.33A, so the peak current is 6.7A. Since the clock period is 2ns, the current slew rate is 6.7A/ns. The core's intrinsic capacitance supplies the very high edge-rate current while the power supply recharges the core capacitance at a much lower slew rate.

A Norton equivalent model of the core can be incorporated into a system level simulation by tying the model across the on-chip power and ground rails. Modifying the schematic in chapter 3, Figure 3.2 to add a Norton equivalent yields the schematic in Figure 11.16.

11.5 Advanced Packaging

There exists a huge variety of package styles, and when it comes to electrical performance, there is a natural tendency to try to rank the various styles from best to worst. However, it can be stated unequivocally that no package style has inherently higher performance than another. Rather, for a given application and its contraints (size, pin count, die size, reliability, and cost), one package may be a better electrical choice than another.

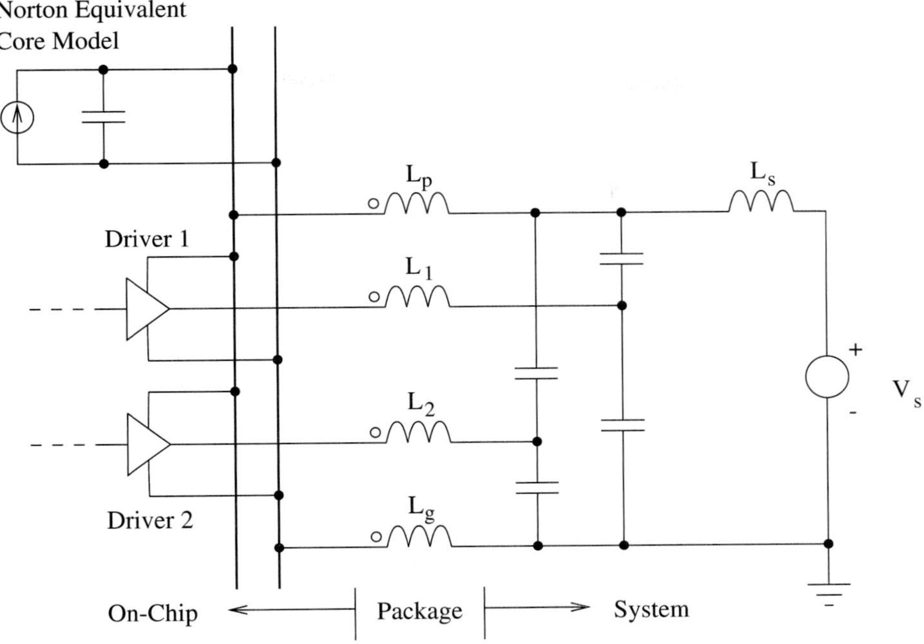

Figure 11.16. Simple system-level schematic with a Norton equivalent model of the core logic.

The fundamental issue is the number of power and ground pins that can be assigned. Package inductance can always be lowered to achieve better electrical performance by assigning more pins to power and ground. Pads can be added to the die until there is no more room, at which point the die must grow in size to accommodate more pads. At this point the die is said to be *pad limited*. Larger die are dramatically more expensive, so little to no die growth is normally allowed. Packages can accommodate more pins by growing in size with relatively little impact on cost, but larger packages can tolerate fewer thermal cycles, so reliability issues set an upper limit on package size, and hence on pin count.

Once the pin count is limited, higher performance can be achieved only through advanced packaging techniques to lower the inductance. These include packages with ground and/or power planes, area array packages (depending on design), and flip-chip die to eliminate wire bonds. Advanced packaging adds considerable cost without adding any inherent value to the silicon product—customers pay for functionality, and the best packaging can do is to not limit performance.

There are a few basic rules for the electrical design of packages. All connections should be as short as possible, and the order of importance in layout is ground, power, clock, asynchronous lines, fast digital lines, and finally, slow digital lines. Generally, the number and quality of power connections should equal that of the ground. Planes should not have any slots. The number of power and ground pins is determined by the ratio of the number of fast signals to the number of ground pins. The signal-to-ground ratio is typically set by experience with the product line or similar products. As clock speeds increase, the required ratio decreases. Detailed simulations can be used to explore the impact of the signal-to-ground ratio on new products and on products expecting either a large jump in speed or a redesign into lower-cost packaging or interconnects.

11.6 Exercises

1. For a differential driver swinging rail-to-rail with impedance R, derive the static power dissipations for the optimal, bridged, and single-ended termination schemes in section 11.3.2. Which dissipates more power?

2. Derive and sketch an optimal multiconductor termination scheme for the three-conductor transmission line described in chapter 10, problem 10.

3. LVDS uses current-mode signaling with differential detection of the voltage drop across the 100Ω resistor in the following schematic.

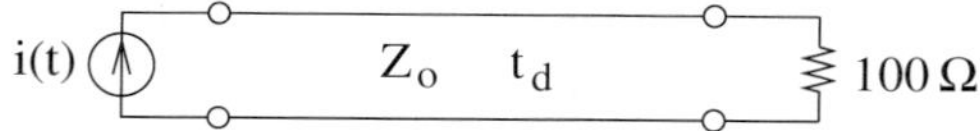

The current source provides ± 2.5mA for logic one and zero, respectively. Use even- and odd-mode analysis with the modal currents to find the detected voltage under the influence of common-mode noise.

4. (a) Show that the reflection coefficients for the even and odd modes of a symmetric differential pair are

$$\Gamma_e = 1$$
$$\Gamma_o = \frac{Z_o - Z_{oo}}{Z_o + Z_{oo}}$$

when using bridged termination with an impedance of $2Z_o$, where $Z_o = \sqrt{\ell/c}$.

   (b) How does this result compare to using bridged termination with $2Z_{oo}$?

   (c) For bridged termination with $2Z_o$, find Γ_o for $M = 0.1L$ and $C_m = 0.1C$ for the differential pair. Is this level of reflection acceptable?

5. A waveform is launched on the even mode of a symmetric differential pair towards an unbalanced π-network termination with $y_1 \neq y_3$ and $y_2 = 0$ (see chapter 4, Figure 4.8a). What is the voltage detected by a perfect differential receiver?

6. Derive a T-network that optimally terminates a symmetric differential pair.

7. A fully-cladded PCB forms a cavity that can resonate at high frequencies. If one entire edge is driven, the PCB acts like a low-impedance transmission line with input impedance

$$Z_{\text{in}} = -\jmath Z_o \cot(\beta d)$$

for the lossless case.

   (a) Compute the fundamental resonant frequency plus 2 harmonics of a square PCB that is 10cm on a side with dielectric thickness of 0.5mm. Assume that the dielectric constant is 4 and that the PCB is driven along one full edge.

   (b) Use the technique in section 11.4.4 to build a model of the PCB, assuming the PCB cladding is 0.5oz copper (0.18μm thick) with $\sigma = 5.7 \times 10^7$S/m.

Compute the input impedance while driving one full edge, and plot the result vs. frequency over a sufficient range to cover the fundamental plus 2 harmonics.

(c) Compare the computed results for the resonant frequencies to the analytical results and explain any differences.

(d) What are the ESR and ESL for the capacitor model?

(e) If the PCB is driven at only one point along the side, what happens?

Appendix A

SOLUTIONS TO SELECTED PROBLEMS

Chapter 2

2. The capacitance per-unit-length, neglecting fringing fields, is $c = \epsilon_r \epsilon_o W/H$. The phase velocity is $v_p = 1/\sqrt{\ell c} = 1/\sqrt{\epsilon_r \epsilon_o \mu_o}$. Solving for ℓ yields the required result. Solve for Z_o by finding $\sqrt{\ell/c}$. Note how ℓ does not depend on the dielectric constant.

3. (a) From (2.22) and (2.25), with the load at $z = 0$, then

$$V(0) = \mathrm{A} + \mathrm{B}$$

and

$$I(0) = \frac{1}{Z_o}(\mathrm{A} - \mathrm{B}),$$

so

$$Z_L = \frac{V(0)}{I(0)} = Z_o \frac{\mathrm{A}/\mathrm{B} + 1}{\mathrm{A}/\mathrm{B} - 1},$$

leading to

$$\frac{A}{B} = \frac{Z_L - Z_o}{Z_L + Z_o}.$$

The input impedance is

$$Z_{\text{in}} = \frac{V(-d)}{I(-d)} = Z_o \frac{\text{A/B}e^{\gamma d} + e^{-\gamma d}}{\text{A/B}e^{\gamma d} - e^{-\gamma d}}.$$

Substitute for A/B, multiply through by $Z_L - Z_o$, and collect terms to get

$$Z_{\text{in}} = Z_o \frac{Z_L(e^{\gamma d} + e^{-\gamma d}) + Z_o(e^{\gamma d} - e^{-\gamma d})}{Z_o(e^{\gamma d} + e^{-\gamma d}) + Z_L(e^{\gamma d} - e^{-\gamma d})}.$$

Then, note that $\tanh(x) = (e^x - e^{-x})/(e^x + e^{-x})$ to get the final result.

(b) Substitute $\gamma = \alpha + \jmath\beta$ with $\alpha = 0$ and $\tanh(\jmath z) = \jmath\tan(z)$ to get the result.

4. (a) At low frequencies, the capacitor looks like an open circuit, so low-frequency components of the incident wave are reflected with a reflection coefficient of $\Gamma = +1$. At high frequencies, the capacitor looks like a short circuit, so these components pass through unchanged. Therefore, the reflected signal is a low-passed version of the incident wave, while the transmitted signal is a pulse.

(b) Results from a SPICE simulation:

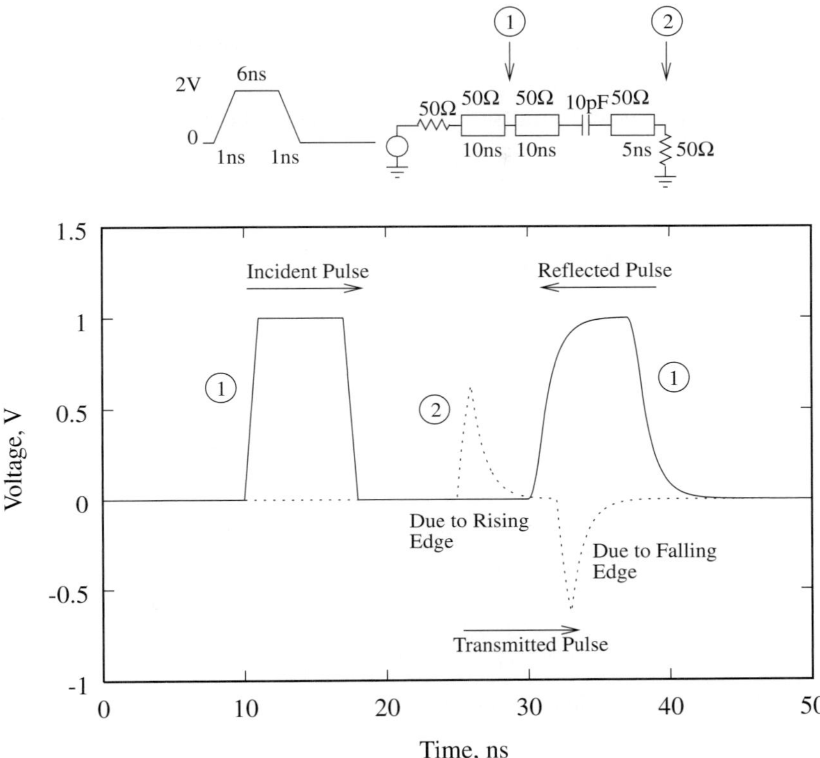

5. The current is continuous through the series capacitance, so $v_1^+ - v_1^- = v_2^+$. The voltage across the capacitor is $v_1^+ + v_1^- - v_2^+$, so applying the voltage-current relationship for the capacitor yields

$$v_2^+ = Z_o C \frac{d}{dt}(v_1^+ + v_1^- - v_2^+).$$

Eliminating v_1^- yields

$$v_2^+ = \tau \frac{d}{dt}(v_1^+ - v_2^-),$$

where $\tau = 2CZ_o$ is the time constant for the discontinuity. Taking the Laplace transform and solving for V_2^+ gives

$$V_2^+ = \frac{1}{s + 1/\tau},$$

so inverse transforming finds the transmitted pulse as

$$v_2^+ = e^{-t/\tau}u(t),$$

which is a positive pulse for a rising edge. The reflected waveform is

$$v_1^- = v_1^+ - v_2^+ = (1 - e^{-t/\tau})u(t).$$

The total voltages are

$$v_1 = v_1^+ + v_1^- = (2 - e^{-t/\tau})u(t)$$

and

$$v_2 = v_2^+ = e^{-t/\tau}u(t).$$

For weak reflections, $v_1^+ \gg v_2^+$, so

$$v_2^+ = \tau\frac{d}{dt}(v_1^+ - v_2^+) \approx \tau\frac{dv_1^+}{dt}.$$

Note that problem 4 does not involve a weak discontinuity, so the estimate of this formula for a transmitted voltage of 1V is significantly off from the simulated value of about 0.6V. For a discontinuity of 1pF, both SPICE and the approximate formula predict 0.1V transmitted pulses.

6. (a) At low frequencies, the inductor looks like a short circuit, so low-frequency components of the incident wave are reflected with a reflection coefficient of $\Gamma = -1$. At high frequencies, the inductor looks like an open circuit, so these components pass through unchanged. Therefore, the reflected signal is an inverted, low-passed version of the incident wave, while the transmitted signal is a pulse.

(b) Results from a SPICE simulation:

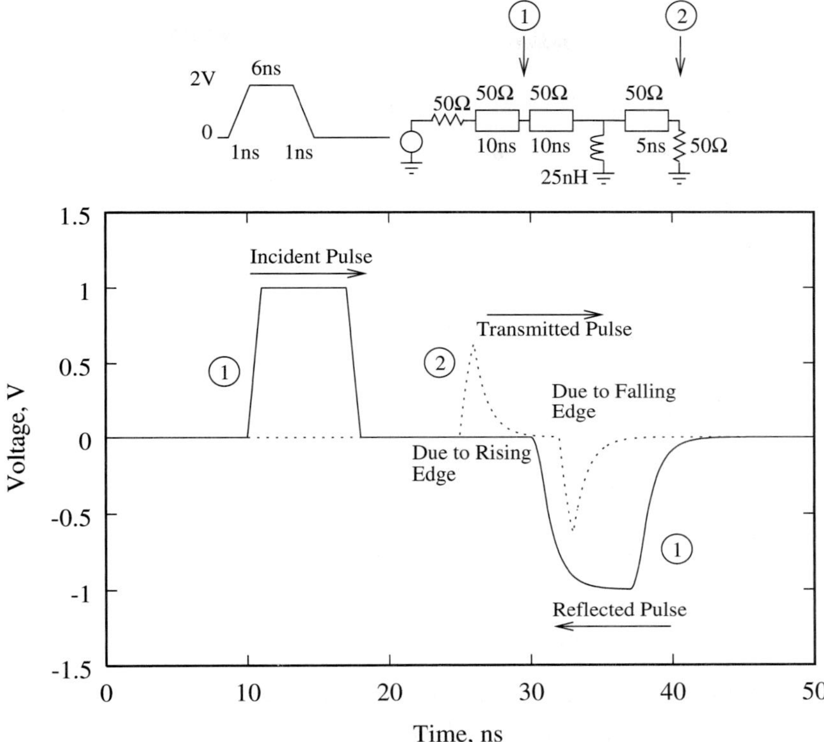

The large amplitude of the pulse transmitted through the shunt inductor shows that a good short circuit to ground can require a very low inductance connection.

7. Voltage continuity means that $v_1^+ + v_1^- = v_2^+$. The current into the inductor is $v_1^+/Z_o - v_1^-/Z_o - v_2^+/Z_o$, so applying the voltage-current relationship for the inductor yields

$$v_2^+ = \frac{L}{Z_o}\frac{d}{dt}(v_1^+ - v_1^- - v_2^+).$$

Eliminating v_1^- gives

$$v_2^+ = \tau\frac{d}{dt}(v_1^+ - v_2^+),$$

where $\tau = 2L/Z_o$ is the time constant of the discontinuity. The form for v_2^+ is the same as for problem 5, so the results are

$$v_2^+ = e^{-t/\tau}u(t),$$

$$v_1^- = v_2^+ - v_1^+ = -(1 - e^{-t/\tau})u(t),$$

$$v_1 = v_1^+ + v_1^- = e^{-t/\tau}u(t),$$

and

$$v_2 = v_2^+ = e^{-t/\tau}u(t).$$

For weak reflections, $v_1^+ \gg v_2^+$, so

$$v_2^+ = \tau\frac{d}{dt}(v_1^+ - v_2^+) \approx \tau\frac{dv_1^+}{dt}.$$

Note that problem 6 does not involve a weak discontinuity, so the estimate of this formula for a transmitted voltage of 1V is significantly off from the simulated value of about 0.6V. For a discontinuity of 2.5nH, both SPICE and the approximate formula predict 0.1V transmitted pulses.

8. Plotting the voltage at the receiver computed with SPICE for a number of driver strengths yields

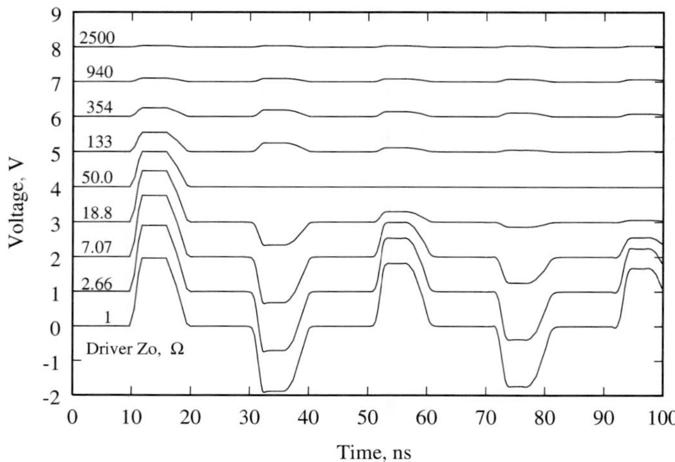

where a running 1V offset is applied to separate the curves. For low-impedance drivers, the driver is able to launch a strong pulse, but it also strongly reflects, with a reflection coefficient near -1, the full strength signal returning from the end of the unterminated line. As the driver strength drops, the initial pulse amplitude decreases, as does the magnitude of the reflection. At a driver strength of 50Ω, the amplitude is halved, but there are no reflections, so signal quality is maximized. This is the source termination case. As the impedance increases, the initial amplitude continues to drop and reflections pick up again, this time with a reflection coefficient of +1. Eventually, the driver is unable to launch a pulse with significant amplitude.

10. (a) The voltage at the source divides on the source resistance and the transmission line impedance. The launched voltage is $5\frac{70}{70+R}$. This voltage arrives at the small load and doubles. For first incidence switching, this voltage must equal 2, so it is required that $2 = 2 \cdot 5\frac{70}{70+R}$. Solving for R yields the source impedance as 280Ω.

    (b) For second incidence switching, the wave reflected from the load travels back to the source, where it is reflected with the reflection coefficient $\Gamma = (R-70)/(R+70)$, so the wave amplitude heading back towards the load is $\Gamma \cdot 5 \cdot 70/(70+R)$. At the load, this wave is fully reflected. The total voltage at the load is the sum of all four waves, and this must equal 2V. So, $2 = 5\frac{70}{70+R} + 5\frac{70}{70+R} + (R-70)/(R+70) \cdot 5 \cdot 70/(70+R) + (R-70)/(R+70) \cdot 5 \cdot 70/(70+R)$. Solving for R yields 8.89Ω and 551Ω. The largest impedance is then 551Ω.

11. A 70Ω transmission line 5ns long with a 5pF load is driven by a voltage source that rises from 0 to 5V in 1ns and holds there. The voltage source is connected to ground through a 5nH inductor. The source resistance is variable. Plotting several quantities of interest yields the plots:

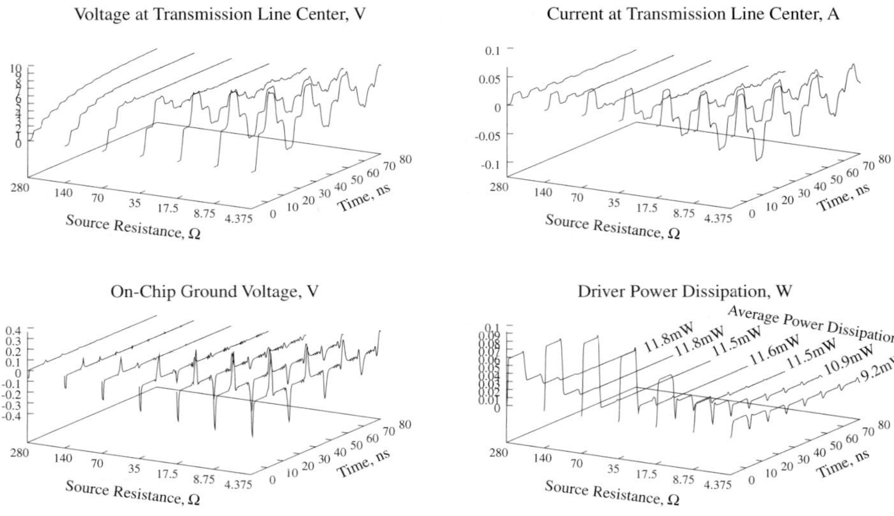

From problem 10, the maximum source resistance for first incidence switching is 280Ω. From the plots, dropping the resistance below this value has the following effects: voltage rises more rapidly and then begins to ring, but timing is not affected because all reach 2V at the same time; capacitive crosstalk is worse because of the higher voltage swings in a given amount of time; inductive crosstalk is worse because of the higher current swings in a given amount of time; SSN is worse because the voltage drop across the shared ground is larger; and the total and average power dissipation is constant (note that the waveforms have not completely settled for the lowest source impedances), but the instantaneous power required peaks at the source-matched condition.

12. The voltage launched onto the transmission line is $3 \cdot 75/(75+150) = 1$V. After 10ns, this wave arrives at the load, where it is fully reflected with a reflection coefficient of $+1$. After 10ns, the reflected wave arrives back at the source and is reflected again with a reflection coefficient of $(150-75)/(150+75) = 0.3333$, so the wave traveling back towards the load is 0.3333V. After 10ns, this wave arrives at the load, where it is fully reflected. At 35ns, the reflected wave is half way back towards the load. A bounce diagram and waveform sketch is:

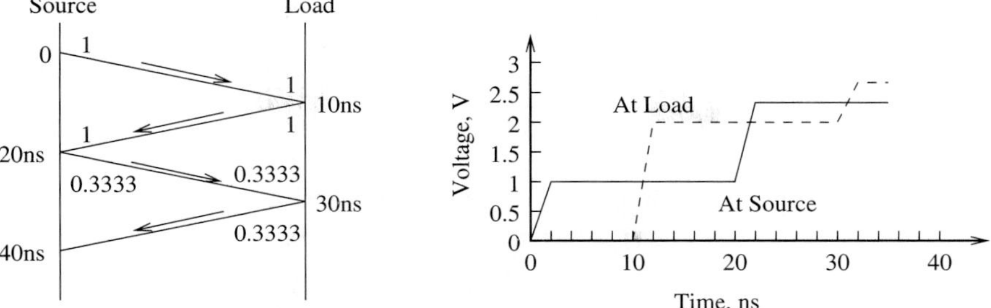

13. Assuming the data is launched on the rising edge, the DDR clock must be shifted 90° or doubled in frequency to enable clocking in the data in the center of the eye. The use of either a DLL or a PLL introduces some jitter into the timing. The clock and data lose additional synchronization due to delays introduced by crosstalk and differences in propagation delay. The timing diagram is

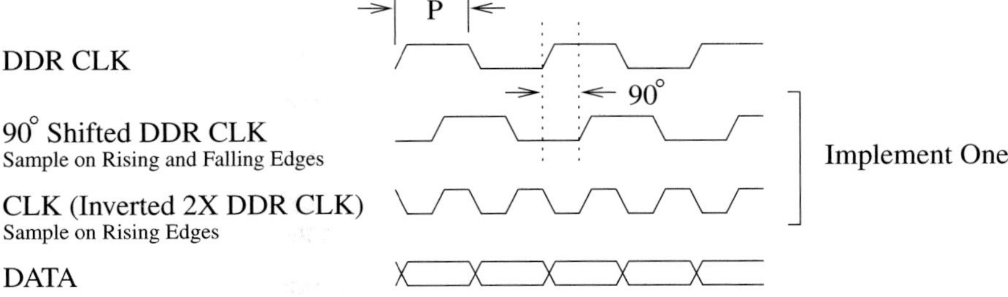

The requirement on the setup time for a 50% duty cycle clock is

$$P/2 > t_S + \Delta t_{\mathrm{TOF}} + \Delta t_{jitter} + \Delta t_{margin}.$$

The hold time constraint is symmetrical to the setup time constraint, so

$$P/2 > t_H + \Delta t_{\mathrm{TOF}} + \Delta t_{jitter} + \Delta t_{margin}.$$

14. Inserting the far-end crosstalk from (2.61) into the approximation for the skew in (2.65) causes the derivative to approximately cancel. The skew is then approximately independent of the edge rate.

15. (a) Set $Z_L = 0$ in (2.67). A plotted result is

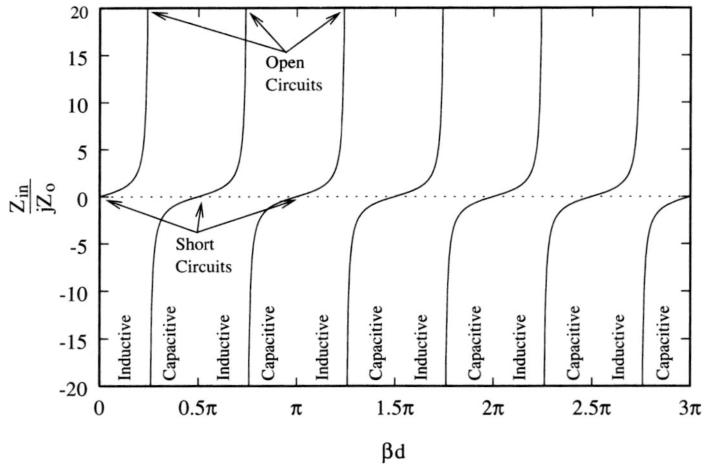

16. Because of the limit operation as the size of the section is reduced to zero, the order of the elements in the lumped model is not significant. The results for the transmission line equations are unchanged.

18. The voltages and currents at the junction are related by

$$v_n^+ + v_n^- = v(t) + v_{n+1}^+ + v_{n+1}^-,$$

where $v(t)$ is the applied voltage at the junction. For the current,

$$\frac{v_n^+ - v_n^-}{Z_{o,n}} = \frac{v_{n+1}^+ - v_{n+1}^-}{Z_{o,n+1}}.$$

To obtain reflection and transmission coefficients, assume that no wave is incident from the $n+1$ section, so $v_{n+1}^- = 0$. Solving for the reflection coefficient yields

$$\Gamma = \frac{\frac{Z_{o,n+1}}{Z_{o,n}} - 1 + \frac{v(t)}{v_n^+}}{\frac{Z_{o,n+1}}{Z_{o,n}} + 1}.$$

For no reflection, $\Gamma = 0$ requires that

$$v(t) = v_n^+ \left(1 - \frac{Z_{o,n+1}}{Z_{o,n}}\right).$$

19. Using the "O" SPICE element for both lossy and lossless simulations yields

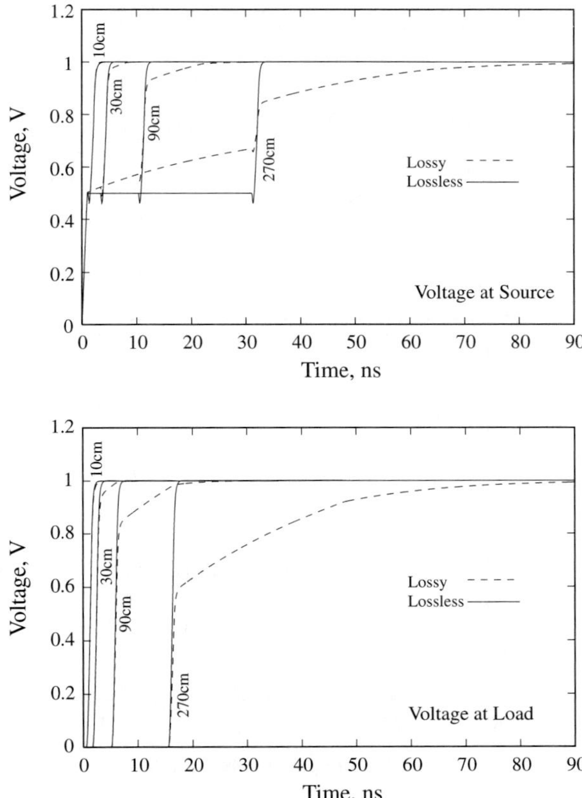

Up to the 90cm line length, the waveforms are close enough for most applications. For longer lines, the discrepancy approaches the trigger point of receivers and could significantly affect timing.

From (2.37), the maximum length of line for lossless analysis is estimated as

$$d \ll \frac{2}{20}\sqrt{\frac{289 \times 10^{-9}}{115 \times 10^{-12}}} = 5\text{m},$$

so take $d < 50$cm, which is in good agreement with the SPICE simulation.

Chapter 3

4. Voltage measurements across partial inductances are not unique because partial inductance itself is not unique. For this problem, there is no significant bypass capacitance on-chip, so the rail collapse is due either to ground bounce from high-to-low transitions or to power droop from low-to-high transitions. Therefore, the on-chip rail noise provides a good measure of SSN while avoiding measurements across partial inductances.

   To obtain three different edge rates, the netlist in Appendix C is run with the same 18Ω driver in parallel. The netlist is augmented with a 50Ω resistor in series with the voltage source vin to alleviate convergence difficulties induced by the parallel drivers. For one, two, and three drivers in parallel, then

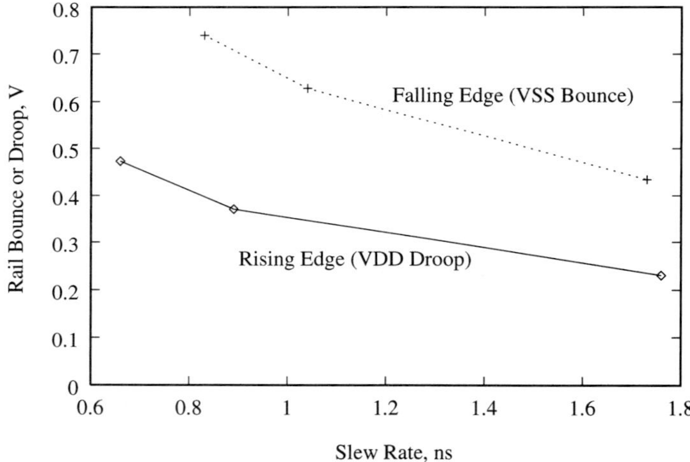

   From the plot, faster edge rates cause higher SSN.

5. The three driver setups are simulated with no power supply or signal inductance; the driver simply drives a 50pF load. The edge rates for the three setups are 1.97ns, 1.03ns, and 0.73ns. The driver impedance is computed by assuming that $t_r = 2.2RC$, where $C = 50\text{pF}$. The driver impedances are then 17.9Ω, 9.4Ω, and 6.6Ω. For a high-to-low transition, the effective induc-

tance is 3.47nH, while for the low-to-high transition, it is 2.15nH. Using the impedances, edge rates, effective inductances, and $V_T = 0.85$V, then (3.4) is applied to obtain the following results:

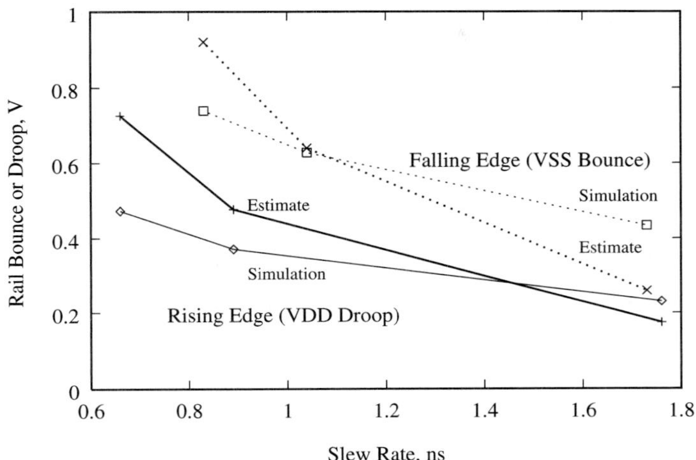

The estimates are not very good, but the trends are correct. Analytic estimates of SSN are good for spotting trends and for comparison studies. In this example, note how lower SSN results for the low-to-high transition due to the smaller effective inductance.

Chapter 4

1. From (4.13), $\overline{\overline{S}} = (\overline{\overline{Z}}\,\overline{\overline{k}}^{-1} + \overline{\overline{k}})^{-1}(\overline{\overline{Z}}\,\overline{\overline{k}}^{-1} - \overline{\overline{k}})$. Noting that $(\overline{\overline{Z}}\,\overline{\overline{k}}^{-1} \pm \overline{\overline{k}})^T = (\overline{\overline{Z}}\,\overline{\overline{k}}^{-1} \pm \overline{\overline{k}})$ because $\overline{\overline{Z}}$ is symmetric and $\overline{\overline{k}}$ is diagonal (so it is symmetric, and multiplication is commutative), then

$$\begin{aligned}\overline{\overline{S}} &= [(\overline{\overline{Z}}\,\overline{\overline{k}}^{-1} - \overline{\overline{k}})^T((\overline{\overline{Z}}\,\overline{\overline{k}}^{-1} + \overline{\overline{k}})^{-1})^T]^T \\ &= [(\overline{\overline{Z}}\,\overline{\overline{k}}^{-1} - \overline{\overline{k}})(\overline{\overline{Z}}\,\overline{\overline{k}}^{-1} + \overline{\overline{k}})^{-1}]^T,\end{aligned}$$

where the relationship $(\overline{\overline{A}}\,\overline{\overline{B}})^T = \overline{\overline{B}}^T\overline{\overline{A}}^T$ is used. For symmetry, $\overline{\overline{S}} = \overline{\overline{S}}^T$, so $(\overline{\overline{Z}}\,\overline{\overline{k}}^{-1} - \overline{\overline{k}})(\overline{\overline{Z}}\,\overline{\overline{k}}^{-1} + \overline{\overline{k}})^{-1} = (\overline{\overline{Z}}\,\overline{\overline{k}}^{-1} + \overline{\overline{k}})^{-1}(\overline{\overline{Z}}\,\overline{\overline{k}}^{-1} - \overline{\overline{k}})$ is required. Multiplying through to eliminate the matrix inverses and expanding, then

$$
\begin{aligned}
(\overline{\overline{Z}}\,\overline{\overline{k}}^{-1} + \overline{\overline{k}})(\overline{\overline{Z}}\,\overline{\overline{k}}^{-1} - \overline{\overline{k}}) &= (\overline{\overline{Z}}\,\overline{\overline{k}}^{-1} - \overline{\overline{k}})(\overline{\overline{Z}}\,\overline{\overline{k}}^{-1} + \overline{\overline{k}}) \\
(\overline{\overline{Z}}\,\overline{\overline{k}}^{-1})^2 + \overline{\overline{k}}\,\overline{\overline{Z}}\,\overline{\overline{k}}^{-1} - \overline{\overline{Z}} - \overline{\overline{k}}^2 &= (\overline{\overline{Z}}\,\overline{\overline{k}}^{-1})^2 - \overline{\overline{k}}\,\overline{\overline{Z}}\,\overline{\overline{k}}^{-1} + \overline{\overline{Z}} - \overline{\overline{k}}^2 \\
\overline{\overline{k}}\,\overline{\overline{Z}}\,\overline{\overline{k}}^{-1} &= \overline{\overline{Z}} \\
\overline{\overline{k}}\,\overline{\overline{Z}} &= \overline{\overline{Z}}\,\overline{\overline{k}},
\end{aligned}
$$

and since $\overline{\overline{k}}\,\overline{\overline{Z}} = \overline{\overline{Z}}\,\overline{\overline{k}}$ due to $\overline{\overline{k}}$ being diagonal, then $\overline{\overline{S}} = \overline{\overline{S}}^T$ and $\overline{\overline{S}}$ is symmetric.

2. From (4.4), the total incident power is $P_i = \frac{1}{2}\sum |a_k|^2$, while the reflected power is $P_r = \frac{1}{2}\sum |b_k|^2$. The total dissipated power is then the difference. Noting that $\sum |c_k|^2 = \overline{c}^T \cdot \overline{c}^*$, then the dissipated power is $P = P_i - P_r = \frac{1}{2}\overline{a}^T \cdot \overline{a}^* - \frac{1}{2}\overline{b}^T \cdot \overline{b}^*$. Since $\overline{b} = \overline{\overline{S}}\,\overline{a}$, then

$$
\begin{aligned}
P &= \frac{1}{2}\overline{a}^T \cdot \overline{a}^* - \frac{1}{2}(\overline{\overline{S}}\,\overline{a})^T \cdot (\overline{\overline{S}}\,\overline{a})^* \\
&= \frac{1}{2}\overline{a}^T \cdot \overline{a}^* - \frac{1}{2}\overline{a}^T \overline{\overline{S}}^T \cdot \overline{\overline{S}}^* \overline{a}^* \\
&= \frac{1}{2}\overline{a}^T(\overline{\overline{I}} - \overline{\overline{S}}^T \overline{\overline{S}}^*)\overline{a}^*.
\end{aligned}
$$

Note that the dissipated power depends on how the network is driven.

3. The average power delivered to a port is $P_{j,\text{avg}} = \frac{1}{2}\text{Re}[V_j I_j^*]$. Using the fact that $\text{Re}[z] = \frac{z+z^*}{2}$, then $P_{j,\text{avg}} = \frac{1}{4}(V_j I_j^* + V_j^* I_j)$. Summing across all ports, then $P_{\text{total}} = \frac{1}{4}\sum_j (V_j I_j^* + V_j^* I_j) = \frac{1}{4}(\overline{V}^T \overline{I}^* + \overline{I}^T \overline{V}^*) = \frac{1}{4}(\overline{I}^T \overline{\overline{Z}}^T \overline{I}^* + \overline{I}^T \overline{\overline{Z}}^* \overline{I}^*) = \frac{1}{4}\overline{I}^T(\overline{\overline{Z}}^T + \overline{\overline{Z}}^*)\overline{I}^*$.

4. The schematic for this problem is

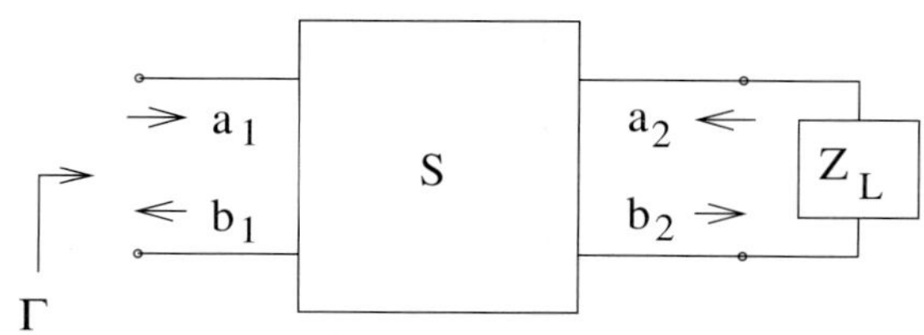

Note that $\Gamma_L = \frac{a_2}{b_2}$. The S-parameters relate the waves as

$$\begin{aligned} b_1 &= S_{11}a_1 + S_{12}a_2 = S_{11}a_1 + S_{12}b_2\Gamma_L \\ b_2 &= S_{21}a_1 + S_{22}a_2 = S_{21}a_1 + S_{22}b_2\Gamma_L. \end{aligned}$$

From the second equation, $b_2 = \frac{S_{21}a_1}{1-S_{22}\Gamma_L}$, then the first equation yields $b_1 = S_{11}a_1 + S_{12}\Gamma_L\frac{S_{21}a_1}{1-S_{22}\Gamma_L}$, or $\Gamma = S_{11} + \frac{S_{12}S_{21}\Gamma_L}{1-S_{22}\Gamma_L}$.

7. Introducing traveling waves, with $\overline{a}_1$ and $\overline{b}_1$ at ports 1 to N on $\overline{\overline{S}}_1$, $\overline{a}_2$ and $\overline{b}_2$ at ports $N+1$ to $2N$ on $\overline{\overline{S}}_1$, and $\overline{a}_3$ and $\overline{b}_3$ at ports $N+1$ to $2N$ on $\overline{\overline{S}}_2$, then the original S-parameter matrices relate the traveling waves according to

$$\begin{bmatrix} \overline{b}_1 \\ \overline{b}_2 \end{bmatrix} = \begin{bmatrix} \overline{\overline{S}}_{1A} & \overline{\overline{S}}_{1B} \\ \overline{\overline{S}}_{1C} & \overline{\overline{S}}_{1D} \end{bmatrix} \begin{bmatrix} \overline{a}_1 \\ \overline{a}_2 \end{bmatrix}$$

and

$$\begin{bmatrix} \overline{a}_2 \\ \overline{b}_3 \end{bmatrix} = \begin{bmatrix} \overline{\overline{S}}_{2A} & \overline{\overline{S}}_{2B} \\ \overline{\overline{S}}_{2C} & \overline{\overline{S}}_{2D} \end{bmatrix} \begin{bmatrix} \overline{b}_2 \\ \overline{a}_3 \end{bmatrix}.$$

Expanding, these provide four equations in four unknowns. Eliminate first $\overline{b}_2$ and then $\overline{a}_2$ to get

$$\begin{aligned} \overline{b}_1 &= (\overline{\overline{S}}_{1A} + \overline{\overline{S}}_{1B}(\overline{\overline{I}} - \overline{\overline{S}}_{2A}\overline{\overline{S}}_{1D})^{-1}\overline{\overline{S}}_{2A}\overline{\overline{S}}_{1C})\overline{a}_1 \\ &+ (\overline{\overline{S}}_{1B}(\overline{\overline{I}} - \overline{\overline{S}}_{2A}\overline{\overline{S}}_{1D})^{-1}\overline{\overline{S}}_{2B})\overline{a}_3 \\ \overline{b}_3 &= (\overline{\overline{S}}_{2C}\overline{\overline{S}}_{1C} + \overline{\overline{S}}_{2C}\overline{\overline{S}}_{1D}(\overline{\overline{I}} - \overline{\overline{S}}_{2A}\overline{\overline{S}}_{1D})^{-1}\overline{\overline{S}}_{2A}\overline{\overline{S}}_{1C})\overline{a}_1 \\ &+ (\overline{\overline{S}}_{2D} + \overline{\overline{S}}_{2C}\overline{\overline{S}}_{1D}(\overline{\overline{I}} - \overline{\overline{S}}_{2A}\overline{\overline{S}}_{1D})^{-1}\overline{\overline{S}}_{2B})\overline{a}_3. \end{aligned}$$

Similarly, eliminate $\overline{a}_2$ then $\overline{b}_2$ to obtain

$$\begin{aligned} \overline{b}_1 &= (\overline{\overline{S}}_{1A} + \overline{\overline{S}}_{1B}\overline{\overline{S}}_{2A}(\overline{\overline{I}} - \overline{\overline{S}}_{1D}\overline{\overline{S}}_{2A})^{-1}\overline{\overline{S}}_{1C})a_1 \\ &+ (\overline{\overline{S}}_{1B}\overline{\overline{S}}_{2A}(\overline{\overline{I}} - \overline{\overline{S}}_{1D}\overline{\overline{S}}_{2A})^{-1}\overline{\overline{S}}_{1D}\overline{\overline{S}}_{2B} + \overline{\overline{S}}_{1B}\overline{\overline{S}}_{2B})a_3 \\ \overline{b}_3 &= (\overline{\overline{S}}_{2C}(\overline{\overline{I}} - \overline{\overline{S}}_{1D}\overline{\overline{S}}_{2A})^{-1}\overline{\overline{S}}_{1C})a_1 \\ &+ (\overline{\overline{S}}_{2C}(\overline{\overline{I}} - \overline{\overline{S}}_{1D}\overline{\overline{S}}_{2A})^{-1}\overline{\overline{S}}_{1D}\overline{\overline{S}}_{2B} + \overline{\overline{S}}_{2D})a_3. \end{aligned}$$

Both results are equivalent, with slightly different forms. The final result comes from pulling $\overline{\overline{S}}_A$, $\overline{\overline{S}}_B$, and $\overline{\overline{S}}_D$ from the first result and $\overline{\overline{S}}_C$ from the second.

If the networks are reciprocal, then $\overline{\overline{S}}_C$ should equal $\overline{\overline{S}}_B^T$. Noting that $\overline{\overline{S}}_{1C} = \overline{\overline{S}}_{1B}^T$, $\overline{\overline{S}}_{2C} = \overline{\overline{S}}_{2B}^T$, $\overline{\overline{S}}_{2A} = \overline{\overline{S}}_{2A}^T$, and $\overline{\overline{S}}_{1D} = \overline{\overline{S}}_{1D}^T$, then it can be easily shown that $\overline{\overline{S}}_C = \overline{\overline{S}}_B^T$.

8. Let the voltages and currents be assigned as shown below:

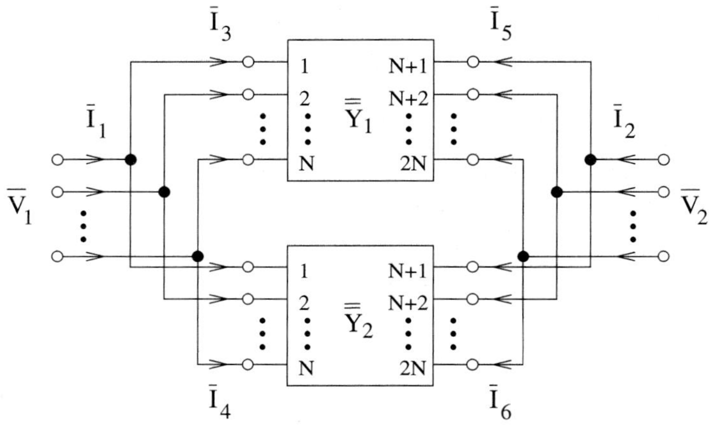

Then the Y-parameter matrices can be expanded as

$$\begin{bmatrix} \overline{I}_3 \\ \overline{I}_5 \end{bmatrix} = \begin{bmatrix} \overline{\overline{Y}}_{1A} & \overline{\overline{Y}}_{1B} \\ \overline{\overline{Y}}_{1C} & \overline{\overline{Y}}_{1D} \end{bmatrix} \begin{bmatrix} \overline{V}_1 \\ \overline{V}_2 \end{bmatrix}$$

and

$$\begin{bmatrix} \overline{I}_4 \\ \overline{I}_6 \end{bmatrix} = \begin{bmatrix} \overline{\overline{Y}}_{2A} & \overline{\overline{Y}}_{2B} \\ \overline{\overline{Y}}_{2C} & \overline{\overline{Y}}_{2D} \end{bmatrix} \begin{bmatrix} \overline{V}_1 \\ \overline{V}_2 \end{bmatrix}.$$

Then Kirchhoff's current law requires

$$\begin{aligned} \overline{I}_1 &= \overline{I}_3 + \overline{I}_4 \\ &= \overline{\overline{Y}}_{1A}\overline{V}_1 + \overline{\overline{Y}}_{1B}\overline{V}_2 + \overline{\overline{Y}}_{2A}\overline{V}_1 + \overline{\overline{Y}}_{2B}\overline{V}_2 \\ \overline{I}_2 &= \overline{I}_5 + \overline{I}_6 \\ &= \overline{\overline{Y}}_{1C}\overline{V}_1 + \overline{\overline{Y}}_{1D}\overline{V}_2 + \overline{\overline{Y}}_{2C}\overline{V}_1 + \overline{\overline{Y}}_{2D}\overline{V}_2. \end{aligned}$$

Collecting terms and assembling into an array yields the desired result.

9. Set up the six equations describing the problem as

$$\begin{aligned}
V_1 &= Z_{11}I_1 + Z_{12}I_2 \\
V_2 &= Z_{21}I_1 + Z_{22}I_2 \\
V_3 &= Z_{33}I_3 + Z_{34}I_4 \\
V_4 &= Z_{43}I_3 + Z_{44}I_4 \\
V_2 &= V_3 \\
I_2 &= -I_3
\end{aligned}$$

and eliminate V_2, V_3, I_2 and I_3.

10. Assign the traveling waves as shown below,

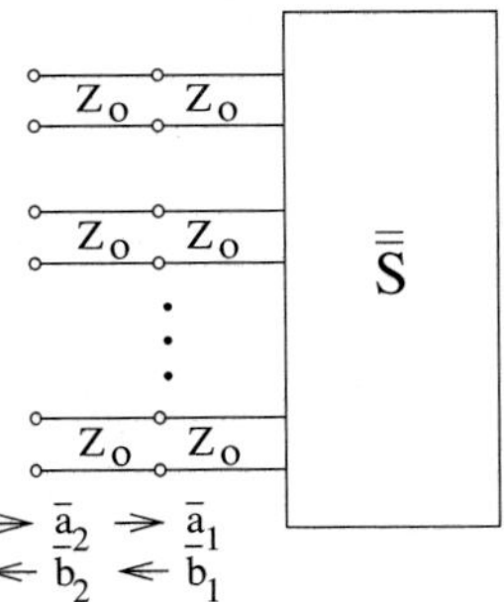

and since the characteristic impedances of the lines are matched to the port impedances, then

$$\bar{a}_1 = \begin{bmatrix} e^{-j\theta_1} & 0 & \cdots & 0 \\ 0 & e^{-j\theta_2} & \cdots & 0 \\ \vdots & \vdots & \ddots & \vdots \\ 0 & 0 & \cdots & e^{-j\theta_N} \end{bmatrix} \bar{a}_2$$

$$\bar{b}_2 = \begin{bmatrix} e^{-j\theta_1} & 0 & \cdots & 0 \\ 0 & e^{-j\theta_2} & \cdots & 0 \\ \vdots & \vdots & \ddots & \vdots \\ 0 & 0 & \cdots & e^{-j\theta_N} \end{bmatrix} \bar{b}_1.$$

Since $\overline{b}_1 = \overline{\overline{S}}\,\overline{a}_1$, then

$$\begin{bmatrix} e^{j\theta_1} & 0 & \cdots & 0 \\ 0 & e^{j\theta_2} & \cdots & 0 \\ \vdots & \vdots & \ddots & \vdots \\ 0 & 0 & \cdots & e^{j\theta_N} \end{bmatrix} \overline{b}_2 = \overline{\overline{S}} \begin{bmatrix} e^{-j\theta_1} & 0 & \cdots & 0 \\ 0 & e^{-j\theta_2} & \cdots & 0 \\ \vdots & \vdots & \ddots & \vdots \\ 0 & 0 & \cdots & e^{-j\theta_N} \end{bmatrix} \overline{a}_2.$$

Inverting the diagonal matrix yields the desired result.

11. (a) The voltage and current on a lossless transmission line are

$$V(z) = V^+e^{-j\beta z} + V^-e^{j\beta z}$$
$$I(z) = y_o(V^+e^{-j\beta z} - V^-e^{j\beta z}).$$

Let port 1 be located at $z = -\ell$ with voltage and current

$$V_1 = V^+e^{j\beta\ell} + V^-e^{-j\beta\ell}$$

and

$$I_1 = y_o(V^+e^{j\beta\ell} - V^-e^{-j\beta\ell}).$$

Let port 2 be located at $z = 0$ with voltage and current

$$V_2 = V^+ + V^-$$

and

$$I_2 = y_o(-V^+ + V^-),$$

where the current is reversed to take into account that positive current enters the DUT for an admittance matrix. As in problem 1 in chapter 2, solving these two for V^+ and V^- yields

$$V^+ = \frac{V_2 - z_oI_2}{2}$$

and

$$V^- = \frac{V_2 + z_oI_2}{2}.$$

Eliminating V^+ and V^- from the equations for port 1 and solving for the currents yields the desired result.

(b) Substitute $\beta = -\jmath\gamma$ in the prior result and use the identities $\cos(\jmath x) = \cosh(x)$ and $\sin(\jmath x) = \jmath\sinh(x)$.

12. Use the given matrix in (4.8) with $Z_o = 50$ and invert to obtain

$$\overline{\overline{Y}} = \begin{bmatrix} -\jmath 0.00101307 & +\jmath 0.0143216 \\ +\jmath 0.0143216 & -\jmath 0.00101307 \end{bmatrix}.$$

From problem 11,

$$\overline{\overline{Y}} = \jmath 0.0143216 \begin{bmatrix} -0.070737 & 1 \\ 1 & -0.070737 \end{bmatrix},$$

so $Y_o \sin(\beta d) = 0.0143216$ and $\cos(\beta d) = 0.070737$, leading to $\beta d = 1.5$ and $Y_o = 0.014358$ or $Z_o = 69.65\Omega$. From (2.24), $\beta = \omega\sqrt{\epsilon_{\text{eff}}}/c_o$, so solving for the effective dielectric constant yields $\epsilon_{\text{eff}} = 4$.

13. The short-circuit admittances at port 1 can be found as

$$\begin{aligned} y_{11} &= \left.\frac{I_1}{V_1}\right|_{V_2=0} \\ y_{21} &= \left.\frac{I_2}{V_1}\right|_{V_2=0}. \end{aligned}$$

The short at port 2 shorts out y_3, leaving y_1 in parallel to y_2. y_{11} is just the input admittance, and since parallel admittances add, then y_{11} can be directly written down as

$$y_{11} = y_1 + y_2.$$

The current at port 2 is found by the current divider as

$$I_2 = -I_1 \frac{y_2}{y_1 + y_2},$$

where the negative sign is needed since positive current flows into the ports. Divide through by V_1 and substitute for y_{11} and y_{21}. The remainder is filled in by observation due to symmetry.

15. Since the characteristic impedance is the same in both transmission lines, the voltage traveling waves can be represented directly by a and b. Match port 2, then $a_2 = 0$ and S_{11} and S_{21} can be found. The voltage is continuous across the capacitor, so

$$a_1 + b_1 = b_2.$$

The current through the capacitor is

$$\jmath\omega C b_2 = \frac{1}{Z_o}a_1 - \frac{1}{Z_o}b_1 - \frac{1}{Z_o}b_2,$$

where b_2 is the voltage across the capacitor. Eliminate b_2 to obtain $S_{11} = b_1/a_1$. Eliminate b_1 to find $S_{21} = b_2/a_1$. Symmetry then provides S_{12} and S_{22}.

Chapter 5

2. The partial mutual inductance is defined in (5.47), where the numerator of the integrand is $\overline{d\ell}' \cdot \overline{d\ell}$. If these two vectors are orthogonal, then $\overline{d\ell}' \cdot \overline{d\ell} = 0$, so $M_p = 0$.

3. Define the various vectors as

$$\begin{aligned}
\overline{r} &= x\hat{x} \\
\overline{r}' &= r'(\cos\theta\hat{x} + \sin\theta\hat{y}) \\
|\overline{r}' - \overline{r}| &= \sqrt{(r'\cos\theta - x)^2 + (r'\sin\theta)^2} \\
\overline{d\ell} &= dx\hat{x} \\
\overline{d\ell}' &= dr'(\cos\theta\hat{x} + \sin\theta\hat{y}) \\
\overline{d\ell}' \cdot \overline{d\ell} &= \cos\theta dx dr',
\end{aligned}$$

then the partial mutual inductance from (5.47) is given by

$$M_p = \frac{\mu}{4\pi}\int_0^a\int_0^a \frac{\cos\theta dx dr'}{\sqrt{(r'\cos\theta - x)^2 + (r'\sin\theta)^2}}.$$

Solution with a symbolic manipulator yields the result.

4. The partial self-inductances from three legs must be added to the partial mutual inductances between the legs. Using (5.53) and the results above with $\cos 60° = \frac{1}{2}$, then

$$\begin{aligned} L &= 3\frac{\mu b}{2\pi}\left[\ln\left(\frac{2b}{a}\right) - \frac{3}{4}\right] \\ &- 6\frac{\mu b}{2\pi}\frac{1}{2}\ln\left[1 + \sqrt{\frac{2}{1-\frac{1}{2}}}\right]. \end{aligned}$$

Simplification yields the result.

5. The partial self-inductances from four legs must be added to the partial mutual inductance between the legs. Noting that the mutual inductance between adjacent sides is zero, then from (5.49) and (5.53)

$$\begin{aligned} L &= 4\frac{\mu b}{2\pi}\left[\ln\left(\frac{2b}{a}\right) - \frac{3}{4}\right] \\ &- 4\frac{\mu b}{2\pi}\left[\ln\left(\frac{b}{b} + \sqrt{1 + \left(\frac{b}{b}\right)^2}\right) - \sqrt{1 + \left(\frac{b}{b}\right)^2} + \frac{b}{b}\right]. \end{aligned}$$

Simplification yields the required result.

6. From (5.47), the partial mutual inductance between a closed loop and a contour is

$$M_p = \frac{\mu}{4\pi}\oint_{square}\int_{wire}\frac{\overline{d\ell}' \cdot \overline{d\ell}}{|\overline{r} - \overline{r}'|}.$$

The closed integral over the square can be broken into four path integrals as

$$\begin{aligned} M_p &= \frac{\mu}{4\pi}\int_{bottom}\int_{wire}\frac{\overline{d\ell}' \cdot \overline{d\ell}}{|\overline{r} - \overline{r}'|} \\ &+ \frac{\mu}{4\pi}\int_{left}\int_{wire}\frac{\overline{d\ell}' \cdot \overline{d\ell}}{|\overline{r} - \overline{r}'|} \\ &+ \frac{\mu}{4\pi}\int_{top}\int_{wire}\frac{\overline{d\ell}' \cdot \overline{d\ell}}{|\overline{r} - \overline{r}'|} \\ &+ \frac{\mu}{4\pi}\int_{right}\int_{wire}\frac{\overline{d\ell}' \cdot \overline{d\ell}}{|\overline{r} - \overline{r}'|}. \end{aligned}$$

The integrals with the top and bottom vanish due to orthogonality. The remaining two integrals calculate the mutual inductance between two parallel straight thin wires with the solution from (5.49). Assuming that this formula is written as $f(\ell, s)$, where ℓ is the length of the two wires and s is the spacing, then

$$M_p = f(a, a+b) - f(a, b).$$

7. In the gap, Ampere's law requires

$$\begin{aligned}\oint_C \overline{H} \cdot \overline{d\ell} &= \iint_S \overline{J} \cdot \overline{ds} \\ 2\pi r H_\phi &= i \\ H_\phi &= \frac{i}{2\pi r}.\end{aligned}$$

Since $\overline{B} = \mu \overline{H}$, then $B_\phi = \frac{\mu i}{2\pi r}$. From (5.15) then

$$L = \frac{1}{i^2} \ell \int_a^b 2\pi r \frac{i}{2\pi r} \frac{\mu i}{2\pi r} dr,$$

where for external inductance, the integration takes place only over the space external to the conductors (the field is zero outside the outer conductor). Completing the integration yields the desired result.

8. From Ampere's law, the magnetic field strength produced external to the inner conductor is

$$H_\phi = \frac{i}{2\pi r}.$$

The magnetic field strength produced inside the outer conductor is

$$H_\phi = -\frac{1}{2\pi r} \frac{r^2 - b^2}{(b+t)^2 - b^2} i,$$

where the area fraction of the current enclosed by the contour modifies the magnitude. Note that the sign is reversed because the current direction is reversed from the direction in the center. Summing the two produces the total field inside the outer conductor:

$$H_\phi = \frac{i}{2\pi r}\left(1 - \frac{r^2 - b^2}{(b+t)^2 - b^2}\right).$$

Plugging into (5.15) yields

$$\frac{L}{\ell} = \frac{\mu}{2\pi}\int_b^{b+t} \frac{1}{r}\left(\frac{(b+t)^2 - r^2}{(b+t)^2 - b^2}\right)^2 dr,$$

where the integration is only over the outer conductor. Completing the integration yields the desired result.

9. Inside the conductor, for uniform current the magnetic field strength is

$$H_\phi = \frac{1}{2\pi r}\frac{r^2 - a^2}{(a+t)^2 - a^2} i.$$

Plugging into (5.15) yields

$$\frac{L}{\ell} = \frac{\mu}{2\pi}\int_a^{a+t} \frac{1}{r}\left(\frac{r^2 - a^2}{(a+t)^2 - a^2}\right)^2 dr.$$

Completing the integration yields the desired result.

12. The trick is to add a third conductor to the problem to act as the return path for the two conductors forming the annular coax. The third conductor is a conductive ring with radius c and thickness t. The value of c is arbitrary as long as $c > b + t$.

    The partial self-inductance for the inner conductor is just the inductance for the annular coax formed from the inner conductor and the added return path. All of this inductance is assigned to the inner conductor, thereby transforming it into a partial inductance. The same is done for the outer conductor. The partial mutual inductance is then defined by the requirement that the loop inductance using the partial inductances must add up to the correct inductance value:

$$L(a,b,t) = L(a,c,t) + L(b,c,t) - 2M.$$

Since the value for c is arbitrary, there are an infinite number of valid partial inductance values for any given geometry, and they all give the correct inductance.

14. The fields above the plane are identical to the fields of the geometry where the plane is replaced by an image wire a distance d below the location of the plane. This problem is just the two-wire transmission line with $a_1 = a_2 = a$ and $S = 2d$ in (5.55). Take half of the result since the fields actually exist only above the plane.

15. Simulating with SPICE yields:

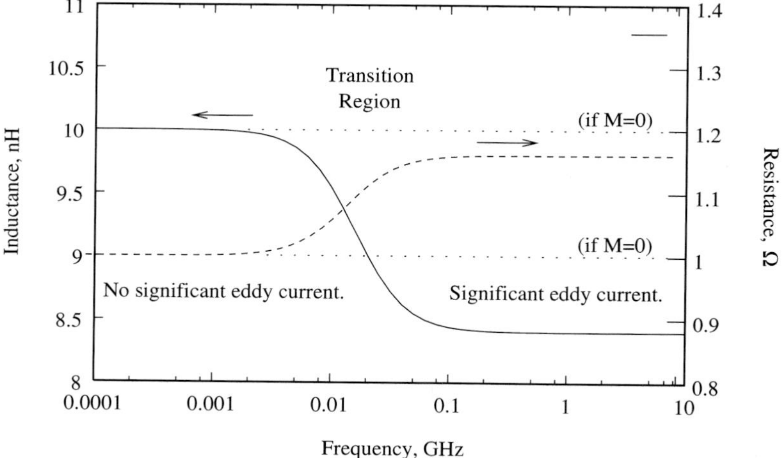

If there is no coupling, then there is no eddy current and R and L are constant.

16. For the mesh with partial inductances, Kirchhoff's voltage law requires

$$-v + L_1 \frac{di}{dt} - M\frac{di}{dt} - M\frac{di}{dt} + L_2\frac{di}{dt} = 0.$$

For the mesh with inductance,

$$-v + L\frac{di}{dt} = 0,$$

so collecting terms and equating the two expressions provides the required formula.

18. First, find $\overline{\overline{T}}^T \overline{\overline{T}}$ as

$$
\begin{aligned}
\overline{\overline{T}}^T \overline{\overline{T}} &= \begin{bmatrix} \bar{i}_1^T \\ \bar{i}_2^T \\ \vdots \\ \bar{i}_N^T \end{bmatrix} \begin{bmatrix} \bar{i}_1 & \bar{i}_2 & \cdots & \bar{i}_N \end{bmatrix} \\
&= \begin{bmatrix} \bar{i}_1 \cdot \bar{i}_1 & \bar{i}_1 \cdot \bar{i}_2 & \cdots & \bar{i}_1 \cdot \bar{i}_N \\ \bar{i}_2 \cdot \bar{i}_1 & \bar{i}_2 \cdot \bar{i}_2 & \cdots & \bar{i}_2 \cdot \bar{i}_N \\ \vdots & \vdots & \ddots & \vdots \\ \bar{i}_N \cdot \bar{i}_1 & \bar{i}_N \cdot \bar{i}_2 & \cdots & \bar{i}_N \cdot \bar{i}_N \end{bmatrix} \\
&= \begin{bmatrix} 1 & 0 & \cdots & 0 \\ 0 & 1 & \cdots & 0 \\ \vdots & \vdots & \ddots & \vdots \\ 0 & 0 & \cdots & 1 \end{bmatrix} \\
&= \overline{\overline{U}},
\end{aligned}
$$

where $(\bar{i}_m \cdot \bar{i}_n = 0, \quad m \neq n)$ is used. Then find $\overline{\overline{T}}\,\overline{\overline{T}}^T$ by noting that

$$
\begin{aligned}
\overline{\overline{T}}^T \overline{\overline{T}} &= \overline{\overline{U}} \\
\overline{\overline{T}}^T \overline{\overline{T}}\,\overline{\overline{T}}^T &= \overline{\overline{T}}^T \\
\overline{\overline{T}}\,\overline{\overline{T}}^T &= \left(\overline{\overline{T}}^T\right)^{-1} \overline{\overline{T}}^T \\
\overline{\overline{T}}\,\overline{\overline{T}}^T &= \left(\overline{\overline{T}}^{-1}\right)^T \overline{\overline{T}}^T \\
\overline{\overline{T}}\,\overline{\overline{T}}^T &= \left(\overline{\overline{T}}\,\overline{\overline{T}}^{-1}\right)^T \\
\overline{\overline{T}}\,\overline{\overline{T}}^T &= \left(\overline{\overline{U}}\right)^T \\
\overline{\overline{T}}\,\overline{\overline{T}}^T &= \overline{\overline{U}}
\end{aligned}
$$

19.

$$\overline{\overline{T}}^T\,\overline{\overline{L}}\,\overline{\overline{T}} = \begin{bmatrix} \bar{i}_1^T \\ \bar{i}_2^T \\ \vdots \\ \bar{i}_N^T \end{bmatrix} \overline{\overline{L}} \begin{bmatrix} \bar{i}_1 & \bar{i}_2 & \cdots & \bar{i}_N \end{bmatrix}$$

$$= \begin{bmatrix} \bar{i}_1^T \\ \bar{i}_2^T \\ \vdots \\ \bar{i}_N^T \end{bmatrix} \begin{bmatrix} \lambda_1\bar{i}_1 & \lambda_2\bar{i}_2 & \cdots & \lambda_N\bar{i}_N \end{bmatrix}$$

$$= \begin{bmatrix} \lambda_1\bar{i}_1\cdot\bar{i}_1 & \lambda_2\bar{i}_1\cdot\bar{i}_2 & \cdots & \lambda_N\bar{i}_1\cdot\bar{i}_N \\ \lambda_1\bar{i}_2\cdot\bar{i}_1 & \lambda_2\bar{i}_2\cdot\bar{i}_2 & \cdots & \lambda_N\bar{i}_2\cdot\bar{i}_N \\ \vdots & \vdots & \ddots & \vdots \\ \lambda_1\bar{i}_N\cdot\bar{i}_1 & \lambda_2\bar{i}_N\cdot\bar{i}_2 & \cdots & \lambda_N\bar{i}_N\cdot\bar{i}_N \end{bmatrix}$$

$$= \begin{bmatrix} \lambda_1 & 0 & \cdots & 0 \\ 0 & \lambda_2 & \cdots & 0 \\ \vdots & \vdots & \ddots & \vdots \\ 0 & 0 & \cdots & \lambda_N \end{bmatrix}$$

$$= \overline{\overline{\lambda}}$$

20. (a) Working with a symbolic manipulator, the following are found:

$$\overline{\overline{T}} = \begin{bmatrix} -1/\sqrt{2} & 1/\sqrt{5} & \sqrt{3/10} \\ 0 & -\sqrt{3/5} & 2/\sqrt{10} \\ 1/\sqrt{2} & 1/\sqrt{5} & \sqrt{3/10} \end{bmatrix}$$

and

$$\overline{\overline{\lambda}} = \begin{bmatrix} L - M/\sqrt{3} & 0 & 0 \\ 0 & L - 2M/\sqrt{3} & 0 \\ 0 & 0 & L + \sqrt{3}M \end{bmatrix}.$$

(b) The circuit must extract modal voltages from total voltages using $\overline{v}_m = \overline{\overline{T}}^T \overline{v}$ and total currents from modal currents using $\overline{i} = \overline{\overline{T}}\overline{i}_m$. The resulting schematic follows.

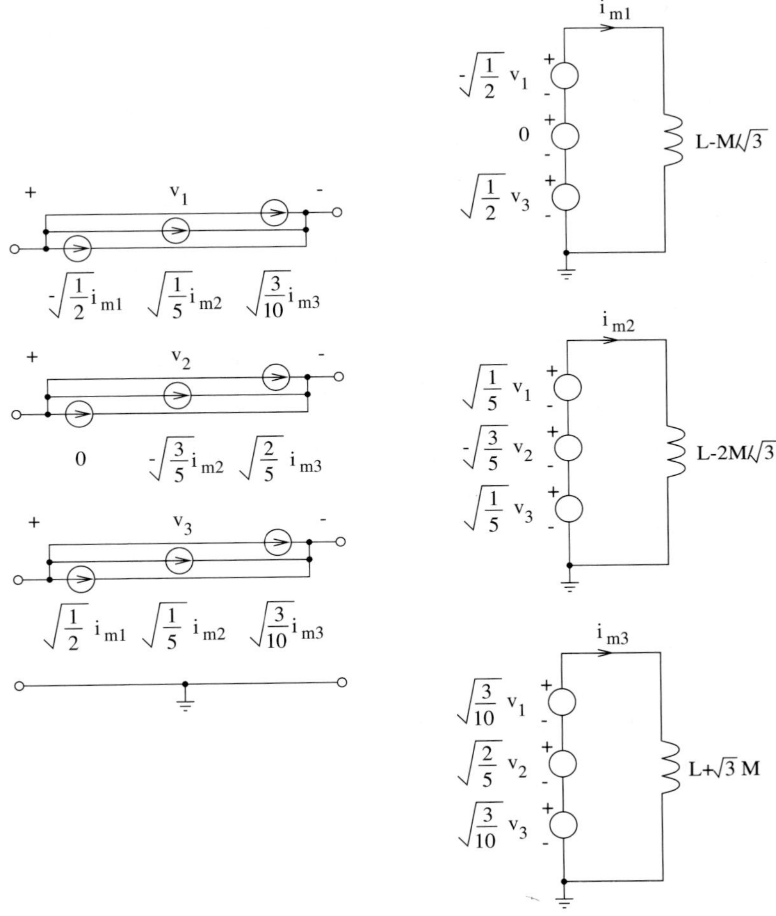

(c) All modal inductances must be positive. For $M < 0$, $L > -\sqrt{3}M$, while for $M > 0$, $L > 2M/\sqrt{3}$.

24. $\overline{\overline{R}}$ is already diagonal, so no substitution is made. Substitute $\overline{\overline{L}} = \overline{\overline{T}}\,\overline{\overline{\lambda}}\,\overline{\overline{T}}^T$ to

get

$$\overline{v} = \overline{\overline{R}}\,\overline{i} + \overline{\overline{T}}\,\overline{\overline{\lambda}}\,\overline{\overline{T}}^T \frac{d\overline{i}}{dt}.$$

Rearrange to get

$$\overline{\overline{T}}^T \overline{v} = \left(\overline{\overline{T}}^T \overline{\overline{R}}\,\overline{\overline{T}}\right) \overline{\overline{T}}^T \overline{i} + \overline{\overline{\lambda}} \frac{d\overline{\overline{T}}^T \overline{i}}{dt}.$$

Noting the modal voltage and current, then

$$\overline{v}_m = \left(\overline{\overline{T}}^T \overline{\overline{R}}\,\overline{\overline{T}}\right) \overline{i}_m + \overline{\overline{\lambda}} \frac{d\overline{i}_m}{dt}.$$

The term in parenthesis is not diagonal since $\overline{\overline{R}}$ is diagonal. Therefore, the equation does not decouple and a modal decomposition is not available. Another way to look at it is that with no inductance, resistance is diagonal and the equation is decoupled. With no resistance, a modal decomposition on inductance can decouple the equation. However, when both are present, decoupling is not possible.

25. The voltage drop is

$$\overline{v} = \overline{\overline{R}}\,\overline{i},$$

and since $\overline{\overline{R}}$ is diagonal, these equations are decoupled and can be solved one at a time. The instantaneous power dissipated in the kth resistor is

$$p_k(t) = R_k i_k^2(t).$$

The total power dissipated up to time t is

$$E_k(t) = R_k \int_{-\infty}^{t} i_k^2(\tau) d\tau.$$

Due to the square, the integral is always positive, so E_k is positive when R_k is positive.

Chapter 6

1. The provided conditions are that $v_1 = 0$, $v_3 = 1$, and $Q_2 = 0$ since line 2 is floating. The relation $\overline{Q} = \overline{\overline{C}}\,\overline{v}$ for the 3×3 case is

$$\begin{bmatrix} Q_1 \\ 0 \\ Q_3 \end{bmatrix} = \begin{bmatrix} C_{11} & C_{12} & C_{13} \\ C_{12} & C_{22} & C_{23} \\ C_{13} & C_{23} & C_{33} \end{bmatrix} \begin{bmatrix} 0 \\ v_2 \\ 1 \end{bmatrix}.$$

The second equation requires that $0 = C_{22}v_2 + C_{23}$, or

$$v_2 = -\frac{C_{23}}{C_{22}},$$

which is positive because $C_{23} < 0$ while $C_{22} > 0$ and less than 1 since $C_{22} > -C_{23}$. The charges are found by multiplying out the equation, so

$$Q_1 = C_{13} - \frac{C_{12}C_{23}}{C_{22}}$$

and

$$Q_3 = C_{33} - \frac{C_{23}^2}{C_{22}}.$$

4. From (6.3), the electric field is constant between the plates, so $v = Ed$, where d is the separation between the plates. With $D = \epsilon E$, (6.8) provides

$$C = \frac{1}{v^2} \int_0^d \int_0^W \int_0^L \epsilon \left(\frac{v}{d}\right)^2 dx\, dy\, dz,$$

where W and L are the width and length of the capacitor, respectively. Integrating and setting $A = WL$ yields the desired result.

5. (a) Working with a symbolic manipulator, the following are found:

$$\overline{\overline{T}} = \begin{bmatrix} -1/\sqrt{2} & 1/\sqrt{5} & \sqrt{3/10} \\ 0 & -\sqrt{3/5} & 2/\sqrt{10} \\ 1/\sqrt{2} & 1/\sqrt{5} & \sqrt{3/10} \end{bmatrix}$$

and

$$\overline{\overline{\lambda}} = \begin{bmatrix} C + C_m/\sqrt{3} & 0 & 0 \\ 0 & C + 2C_m/\sqrt{3} & 0 \\ 0 & 0 & C - \sqrt{3}C_m \end{bmatrix}.$$

(b) The circuit must extract modal currents from total currents using $\overline{i}_m = \overline{\overline{T}}^T \overline{i}$ and total voltages from modal voltages using $\overline{v} = \overline{\overline{T}}\overline{v}_m$. The resulting schematic follows.

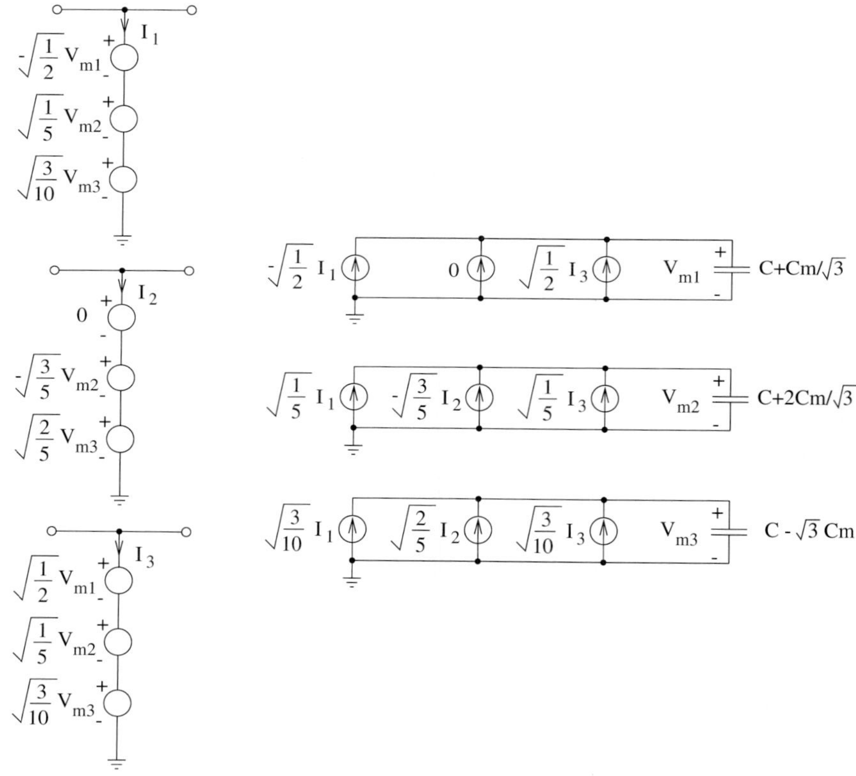

(c) All modal capacitances must be positive. Since $C > 0$ and $C_m > 0$, the first two modal capacitances are guaranteed to be positive. Summing the second row of the capacitance matrix yields the capacitance to ground for the second lead as $C_{20} = C - 2C_m$. Rearranging and subtracting $\sqrt{3}C_m$, then $C - \sqrt{3}C_m = C_{20} + 2C_m - \sqrt{3}C_m > 0$ since $C_{20} >$ and $2C_m > \sqrt{3}C_m$, so the third modal capacitance is guaranteed to be positive as well.

6. (a) Delete row and column 2 to obtain

$$\overline{\overline{C}} = \begin{bmatrix} 10 & -1 \\ -1 & 5 \end{bmatrix}.$$

(b) Invert to get

$$\overline{\overline{C}}^{-1} = \frac{1}{148} \begin{bmatrix} 21 & 17 & 11 \\ 17 & 49 & 23 \\ 11 & 23 & 41 \end{bmatrix}.$$

Delete row and column 2, then invert the result to get

$$\overline{\overline{C}}' = \frac{1}{5} \begin{bmatrix} 41 & -11 \\ -11 & 21 \end{bmatrix}.$$

Chapter 7

1. The impedance matrix to be reduced is given by

$$\overline{\overline{Z}}_p = \begin{bmatrix} R + j\omega L & j\omega M_1 & j\omega M_2 \\ j\omega M_1 & R + j\omega L & j\omega M_1 \\ j\omega M_2 & j\omega M_1 & R + j\omega L \end{bmatrix}.$$

Grounding one end of the three branches leaves just one node, so the incidence matrix is a 3×1 array given by

$$\overline{\overline{A}} = \begin{bmatrix} 1 \\ 1 \\ 1 \end{bmatrix}.$$

Because there are no internal nodes, $\overline{\overline{B}} = 1$. Then the equivalent matrix can be computed as

$$\overline{\overline{Z}}_{eq} = \left(\begin{bmatrix} 1 & 1 & 1 \end{bmatrix} \begin{bmatrix} R + j\omega L & j\omega M_1 & j\omega M_2 \\ j\omega M_1 & R + j\omega L & j\omega M_1 \\ j\omega M_2 & j\omega M_1 & R + j\omega L \end{bmatrix}^{-1} \begin{bmatrix} 1 \\ 1 \\ 1 \end{bmatrix} \right)^{-1}.$$

2. The impedance matrix to be reduced is given by

$$\overline{\overline{Z}}_p = \begin{bmatrix} R + j\omega L & j\omega M_1 & j\omega M_3 \\ j\omega M_1 & R + j\omega L & j\omega M_2 \\ j\omega M_3 & j\omega M_2 & R + j\omega L \end{bmatrix}.$$

Taking the node voltages as V_1 and V leads to the 3×2 incidence matrix

$$\overline{\overline{A}} = \begin{bmatrix} 1 & -1 \\ 0 & -1 \\ 0 & -1 \end{bmatrix}.$$

Node V can be eliminated using

$$\overline{\overline{B}} = \begin{bmatrix} 1 \\ 0 \end{bmatrix}.$$

The equivalent impedance is then given by

$$\overline{\overline{Z}}_{eq} = \begin{bmatrix} 1 & 0 \end{bmatrix} \left(\begin{bmatrix} 1 & 0 & 0 \\ -1 & -1 & -1 \end{bmatrix} \begin{bmatrix} R + j\omega L & j\omega M_1 & j\omega M_3 \\ j\omega M_1 & R + j\omega L & j\omega M_2 \\ j\omega M_3 & j\omega M_2 & R + j\omega L \end{bmatrix}^{-1} \begin{bmatrix} 1 & -1 \\ 0 & -1 \\ 0 & -1 \end{bmatrix} \right)^{-1} \begin{bmatrix} 1 \\ 0 \end{bmatrix}.$$

5. At low frequencies, Ampere's law in integral form is

$$\oint_C \overline{H} \cdot d\ell = \iint_S \overline{J} \cdot ds.$$

Assume that the x-direction is horizontal along the conductor's width, the y-direction is vertical, and the z-direction is the direction of current flow. For a wide conductor, $\overline{H}$ is horizontal and constant with respect to x. If the origin is in the center of the slab, then $\overline{H}$ is in the $-x$ direction above the slab and $+x$ direction below. Applying Ampere's law, there is contribution to the line integral from above and below, yielding

$$2WH_x(y) = i,$$

where W is the width of the slab, and i is the current in the slab between $\pm y$. Faraday's law in differential form inside a good conductor is

$$\nabla \times \overline{J} = -\sigma\mu\frac{\partial \overline{H}}{\partial t}.$$

With $\overline{J}$ and $\overline{H}$ only in the z and x directions, respectively, then

$$\frac{\partial J_z}{\partial y} = \sigma\mu\frac{\partial H_x}{\partial t}.$$

Substituting the result from Ampere's law, then

$$\frac{\partial J_z}{\partial y} = \frac{\sigma\mu}{2W}\frac{\partial i}{\partial t}.$$

Let i_k represent the sum of the currents from all slabs out to a distance $\pm y_k$. With the slabs numbered from the outside in, then in the kth slab

$$\frac{\partial J_z}{\partial y} \approx \frac{(i_{k-1} - i_k)/A_{k-1} - (i_k - i_{k+1})/A_k}{y_{k-1} - y_k} = \frac{\sigma\mu}{2W}\frac{\partial i_k}{\partial t},$$

where $A_k = 2Wt_k$ is the area of the two slabs k at $\pm y_k$, and t_k are the slabs' thicknesses. Rearranging yields (7.16) with

$$R_k = \frac{1}{\sigma t_k}$$

and

$$L_k = \frac{\mu}{2}(y_{k-1} - y_k).$$

So, the ladder network is exactly the same, but with different resistor and inductor values.

Chapter 8

1. The Y_{ii} can be directly measured at port i with the IA, with all other ports shorted to ground. For $Y_{ij} = Y_{ji}$, short all ports to ground except for ports i and j, then

$$\begin{bmatrix} I_1 \\ I_2 \end{bmatrix} = \begin{bmatrix} Y_{ii} & Y_{ij} \\ Y_{ij} & Y_{jj} \end{bmatrix} \begin{bmatrix} V_1 \\ V_2 \end{bmatrix}.$$

Invert the matrix to get

$$\frac{1}{\Delta} \begin{bmatrix} Y_{jj} & -Y_{ij} \\ -Y_{ij} & Y_{ii} \end{bmatrix} \begin{bmatrix} I_1 \\ I_2 \end{bmatrix} = \begin{bmatrix} V_1 \\ V_2 \end{bmatrix},$$

where $\Delta = Y_{ii}Y_{jj} - Y_{ij}^2$. Apply the fact that $I_1 = -I_2 = I_{\text{meas}}$ to get

$$\frac{I_{\text{meas}}}{\Delta} \begin{bmatrix} Y_{jj} + Y_{ij} \\ -Y_{ij} - Y_{ii} \end{bmatrix} = \begin{bmatrix} V_1 \\ V_2 \end{bmatrix}.$$

Apply the fact that $V_{\text{meas}} = V_1 - V_2$ to get

$$V_{\text{meas}} = \frac{I_{\text{meas}}}{\Delta} \left(Y_{ii} + 2Y_{ij} + Y_{jj} \right).$$

Then the measured admittance is

$$Y_{\text{meas}} = \frac{\Delta}{Y_{ii} + 2Y_{ij} + Y_{jj}}$$

or,

$$Y_{\text{meas}} = \frac{Y_{ii}Y_{jj} - Y_{ij}^2}{Y_{ii} + 2Y_{ij} + Y_{jj}}.$$

Since Y_{ii} and Y_{jj} are known from prior one-port measurements, Y_{ij} can be found.

2. Measurement at ports 1 and 2 produce

$$Z_{11} = \frac{1}{\jmath\omega \left(C_{10} + C_{12}C_{20}/(C_{12} + C_{20}) \right)}$$

and

$$Z_{22} = \frac{1}{\jmath\omega\left(C_{20} + C_{12}C_{10}/(C_{12} + C_{10})\right)}.$$

The bridged measurement yields

$$Z_{\text{meas}} = \frac{1}{\jmath\omega\left(C_{12} + C_{10}C_{20}/(C_{10} + C_{20})\right)}.$$

Solve for Z_{12}, build the 2×2 impedance matrix, then invert to get

$$\overline{\overline{C}} = \begin{bmatrix} C_{10} + C_{12} & -C_{12} \\ -C_{12} & C_{20} + C_{12} \end{bmatrix}.$$

3. The perfect VNA measures a_1 and b_1, while a_2 and b_2 exist between the interconnect and the DUT. The traveling waves are related to the S-parameters of the interconnect by

$$\begin{bmatrix} b_1 \\ b_2 \end{bmatrix} = \begin{bmatrix} S_{11} & S_{12} \\ S_{21} & S_{22} \end{bmatrix} \begin{bmatrix} a_1 \\ a_2 \end{bmatrix}.$$

Divide through by the a's, noting that $S_{11M} = b_1/a_1$ is the measured reflection coefficient and $S_{11L} = a_2/b_2$ is the actual reflection coefficient, then eliminating a_1/a_2 yields

$$\frac{1}{S_{11L}} = \frac{S_{12}S_{21}}{S_{11M} - S_{11}} + S_{22}.$$

This formula gives the correction from the measured reflection coefficient (S_{11M}) to the actual reflection coefficient (S_{11L}) if three parameters are known for the interconnect: S_{11}, S_{22}, and $S_{12}S_{21}$. Note that it is not necessary to actually find S_{12} and S_{21} separately.

The three unknowns can be found by measuring three known standards. Rearranging the relationship between S_{11M} and S_{11L} yields

$$S_{11M} = S_{11} + \frac{S_{11L}S_{12}S_{21}}{1 - S_{11L}S_{22}}.$$

For a short-circuit load, $S_{11L} = -1$ and

$$S_{11MS} = S_{11} - \frac{S_{12}S_{21}}{1 + S_{22}}.$$

For an open-circuit load, $S_{11L} = +1$ and

$$S_{11MO} = S_{11} + \frac{S_{12}S_{21}}{1 - S_{22}}.$$

For a matched load, $S_{11L} = 0$ and

$$S_{11ML} = S_{11}.$$

Solving these three equations yields formulas for S_{11}, S_{22}, and $S_{12}S_{21}$ in terms of the measured quantities S_{11MS}, S_{11MO}, and S_{11ML}.

4. (a) For the inductance matrix, short all terminals on one side to ground. On the other side, a two-port TDR can be applied at terminals i and j, with all others open-circuited. The voltages at terminals i and j are

$$v_i = L_{ii}\frac{di_i}{dt} + L_{ij}\frac{di_j}{dt}$$

and

$$v_j = L_{ji}\frac{di_i}{dt} + L_{jj}\frac{di_j}{dt}.$$

The inductances can be found by pulsing terminal i then j, and using (8.12) and (8.13) to extract the relevant matrix entries. The measurement is repeated for all matrix elements.

5. The differential equation describing the problem is

$$v_r = Ri + L\frac{di}{dt}.$$

Integrating yields

$$\int_{-\infty}^{t} v_r(\tau)d\tau = \int_{-\infty}^{t} Ri(\tau)d\tau + Li(t).$$

For large t, $i(t) = i_{DC} = v_o/(Z_o + R)$. Solving for L yields

$$L = \frac{Z_o + R}{v_o} \int_{-\infty}^{t} (v_r(\tau) - Ri(\tau))d\tau.$$

Noting that $i(t) = (v_s(t) - v_r(t))/Z_o$, then

$$L = \frac{Z_o + R}{v_o} \left[\int_{-\infty}^{t} v_r(\tau)\left(1 + \frac{R}{Z_o}\right) d\tau - \frac{R}{Z_o} \int_{-\infty}^{t} v_s(\tau)d\tau \right].$$

Therefore, the time-dependent profile of the source is required in addition to the DC value. An open-circuit measurement can yield $v_s(t)$. As before, a short-circuit measurement can be used to de-embed the fixturing.

7. (a) From chapter 2, equation (2.11), the inductance is $L = 60 * 0.1 \times 10^{-9} =$ 6nH.

   (b) $S_{11} = (Z_L - Z_o)/(Z_L + Z_o) = (\jmath\omega 6 \times 10^{-9} - 50)/(\jmath\omega 6 \times 10^{-9} + 50)$. Plug in $\omega = 2\pi 100 \times 10^6$ and simplify.

9. For a general 2×2 matrix,

$$\begin{bmatrix} a_{11} & a_{12} \\ a_{21} & a_{22} \end{bmatrix}^{-1} = \frac{1}{a_{11}a_{22} - a_{12}a_{21}} \begin{bmatrix} a_{22} & -a_{12} \\ -a_{21} & a_{11} \end{bmatrix}.$$

From chapter 4, equation (4.8),

$$\begin{aligned}
\begin{bmatrix} z_{11} & z_{12} \\ z_{21} & z_{22} \end{bmatrix} &= z_o \left(\begin{bmatrix} 1 & 0 \\ 0 & 1 \end{bmatrix} + \begin{bmatrix} s_{11} & s_{12} \\ s_{21} & s_{22} \end{bmatrix} \right) \left(\begin{bmatrix} 1 & 0 \\ 0 & 1 \end{bmatrix} - \begin{bmatrix} s_{11} & s_{12} \\ s_{21} & s_{22} \end{bmatrix} \right)^{-1} \\
&= z_o \begin{bmatrix} 1 + s_{11} & s_{12} \\ s_{21} & 1 + s_{22} \end{bmatrix} \begin{bmatrix} 1 - s_{11} & -s_{12} \\ -s_{21} & 1 - s_{22} \end{bmatrix}^{-1} \\
&= \frac{z_o}{\Delta} \begin{bmatrix} 1 + s_{11} & s_{12} \\ s_{21} & 1 + s_{22} \end{bmatrix} \begin{bmatrix} 1 - s_{22} & s_{12} \\ s_{21} & 1 - s_{11} \end{bmatrix} \\
&= \frac{z_o}{\Delta} \begin{bmatrix} (1 + s_{11})(1 - s_{22}) + s_{12}s_{21} & 2s_{12} \\ 2s_{21} & (1 - s_{11})(1 + s_{22}) + s_{12}s_{21} \end{bmatrix},
\end{aligned}$$

where $\Delta = (1 - S_{11})(1 - S_{22}) - S_{12}S_{21}$.

Chapter 9

1. Plug the S-parameters into chapter 8, equation (8.5) or into chapter 4, equation (4.8) to obtain

$$\overline{\overline{Z}} = \begin{bmatrix} 0.1 + \jmath\omega 5 \times 10^{-9} & \jmath\omega 2 \times 10^{-9} \\ \jmath\omega 2 \times 10^{-9} & 0.15 + \jmath\omega 6 \times 10^{-9} \end{bmatrix}.$$

By inspection, the model for these two leads is

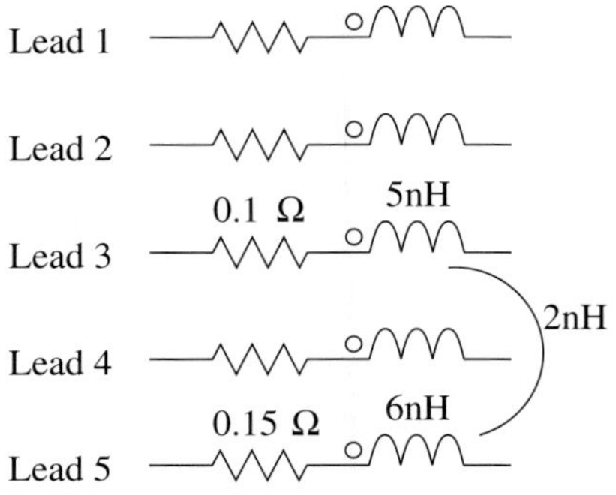

Models for the remaining leads will require more measurements.

2. (a)

$$V_1^L - V_1^R = R_1 I_1^L + \jmath\omega L_1 I_1^L + \jmath\omega M_{12} I_2^L + \jmath\omega M_{13} I_3^L$$

$$V_2^L - V_2^R = \jmath\omega M_{12} I_1^L + R_2 I_2^L + \jmath\omega L_2 I_2^L + \jmath\omega M_{23} I_3^L$$

$$V_3^L - V_3^R = \jmath\omega M_{13} I_1^L + \jmath\omega M_{23} I_2^L + R_3 I_3^L + \jmath\omega L_3 I_3^L$$

(b) Combining into vectors and matrices, then

$$\begin{bmatrix} V_1^L \\ V_2^L \\ V_3^L \end{bmatrix} - \begin{bmatrix} V_1^R \\ V_2^R \\ V_3^R \end{bmatrix} = \begin{bmatrix} R_1 & 0 & 0 \\ 0 & R_2 & 0 \\ 0 & 0 & R_3 \end{bmatrix} \begin{bmatrix} I_1^L \\ I_2^L \\ I_3^L \end{bmatrix} + \jmath\omega \begin{bmatrix} L_1 & M_{12} & M_{13} \\ M_{12} & L_2 & M_{23} \\ M_{13} & M_{23} & L_3 \end{bmatrix} \begin{bmatrix} I_1^L \\ I_2^L \\ I_3^L \end{bmatrix}.$$

Replacing the vectors and matrices with generalized notation yields (9.2).

3. (a)

$$I_1^L - I_1^R = \jmath\omega C_{12}(V_1 - V_2) + \jmath\omega C_{13}(V_1 - V_3) + \jmath\omega C_{10}V_1$$

$$I_2^L - I_2^R = \jmath\omega C_{12}(V_2 - V_1) + \jmath\omega C_{23}(V_2 - V_3) + \jmath\omega C_{20}V_2$$

$$I_3^L - I_3^R = \jmath\omega C_{13}(V_3 - V_1) + \jmath\omega C_{23}(V_3 - V_2) + \jmath\omega C_{30}V_3$$

(b) Collecting terms, then

$$I_1^L - I_1^R = \jmath\omega(C_{10} + C_{12} + C_{13})V_1 - \jmath\omega C_{12}V_2 - \jmath\omega C_{13}V_3$$

$$I_2^L - I_2^R = -\jmath\omega C_{12}V_1 + \jmath\omega(C_{20} + C_{12} + C_{23})V_2 - \jmath\omega C_{23}V_3$$

$$I_3^L - I_3^R = -\jmath\omega C_{13}V_1 - \jmath\omega C_{23}V_2 + \jmath\omega(C_{30} + C_{13} + C_{23})V_3.$$

Combining into vectors and matrices, then

$$\begin{bmatrix} I_1^L \\ I_2^L \\ I_3^L \end{bmatrix} - \begin{bmatrix} I_1^R \\ I_2^R \\ I_3^R \end{bmatrix} = \jmath\omega \begin{bmatrix} C_{10}+C_{12}+C_{13} & -C_{12} & -C_{13} \\ -C_{12} & C_{20}+C_{12}+C_{23} & -C_{23} \\ -C_{13} & -C_{23} & C_{30}+C_{13}+C_{23} \end{bmatrix} \begin{bmatrix} V_1 \\ V_2 \\ V_3 \end{bmatrix}.$$

Replacing the vectors and matrices with generalized notation yields (9.3).

4. From chapter 2, equations (2.11) and (2.12), $L = 15\text{nH}$ and $C = 0.667\text{pF}$. For a π-network, split the capacitance into two 0.333pF pieces. Combining (9.20) and (2.10) yields the bandwidth limit

$$f < \frac{0.62}{2\pi\text{TOF}} = \frac{0.62}{2\pi 0.1 \times 10^{-9}} = 987\text{MHz}.$$

Using $t_r = 1/BW$ implies that the minimum edge rate supported by the model is 1ns.

5. Using $t_r = 1/BW$, the bandwidth requirement is 500MHz. Combining (9.20) and (2.10) yields the maximum TOF as

$$\text{TOF} < \frac{0.62}{2\pi f} = 0.197\text{ns}.$$

The length of the line is then given by distance=rate×time as

$$d = \frac{3 \times 10^8}{\sqrt{4}} 0.197 \times 10^{-9} = 2.96\text{cm}.$$

6. By trial and error on a calculator, $\beta d = 0.385$ to achieve 5% error. Since $\beta d = \omega\sqrt{\ell c}d = \omega\sqrt{LC}$, then the frequency limit is $\omega < 0.385/\sqrt{LC}$, which is a little more than half the value in (9.20). The iterative impedance concept is rooted more directly in circuit behavior and leads to a less restrictive bandwidth limitation.

7. To find the iterative impedance, set $Z_{\text{in}} = Z_L = Z_i$ to obtain

$$Z_i = Z_o \frac{Z_i + Z_o \tanh(\gamma d)}{Z_o + Z_i \tanh(\gamma d)}.$$

Multiply through to get

$$Z_o Z_i + Z_i^2 \tanh(\gamma d) = Z_o Z_i + Z_o^2 \tanh(\gamma d),$$

then $Z_i = Z_o$. This is a restatement of the well known transmission line property that the input impedance of a transmission line of any length is Z_o when loaded by Z_o.

8. Any of the three ports can be grounded for the shorted sample. If port 1 is grounded and current is applied only to port 2, the voltages at ports 2 and 3 at sufficiently low frequencies are

$$V_2 = \jmath\omega L_2 I_2 + \jmath\omega L_1 I_2$$

and

$$V_3 = \jmath\omega M_{23} I_2 + \jmath\omega L_1 I_2.$$

Similarly, the current can be applied only to port 3 to obtain

$$V_2 = \jmath\omega M_{23} I_3 + \jmath\omega L_1 I_3$$

and

$$V_3 = \jmath\omega L_3 I_3 + \jmath\omega L_1 I_3.$$

Collecting these results yields the impedance matrix looking into ports 2 and 3 as

$$\overline{\overline{Z}} = \jmath\omega \begin{bmatrix} L_1 + L_2 & L_1 + M_{23} \\ L_1 + M_{23} & L_1 + L_3 \end{bmatrix}.$$

For the open sample, open port 1 and measure the capacitance, C, at either of ports 2 or 3 at a sufficiently low frequency. The circuit model is shown below.

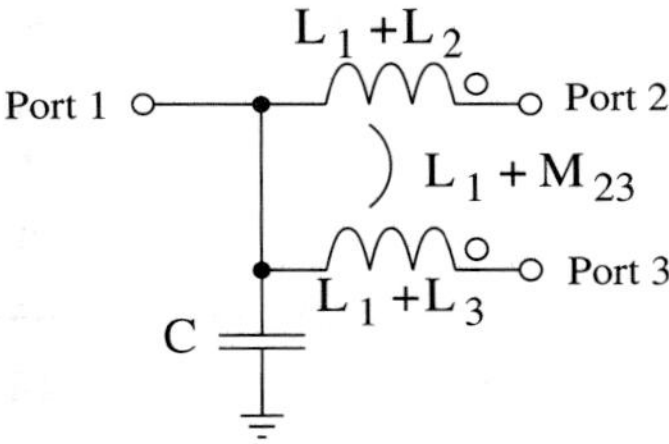

The original and model topologies are equivalent at low frequencies.

Chapter 10

1.

$$\begin{aligned} \frac{\sum_{i=k}^{N} a_i s^i}{\sum_{i=k}^{N} b_i s^i} &= \frac{a_k}{b_k} - \frac{a_k}{b_k} + \frac{\sum_{i=k}^{N} a_i s^i}{\sum_{i=k}^{N} b_i s^i} \\ &= \frac{a_k}{b_k} + \frac{\sum_{i=k}^{N} (a_i - \frac{a_k}{b_k} b_i) s^i}{\sum_{i=k}^{N} b_i s^i} \\ &= \frac{a_k}{b_k} + \frac{\sum_{i=k+1}^{N} (a_i - \frac{a_k}{b_k} b_i) s^i}{\sum_{i=k}^{N} b_i s^i} \end{aligned}$$

The numerator is reduced from the bottom.

$$\begin{aligned}
\frac{\sum_{i=k}^{N} a_i s^i}{\sum_{i=k+1}^{N} b_i s^i} &= \frac{a_k}{b_{k+1}}\frac{1}{s} - \frac{a_k}{b_{k+1}}\frac{1}{s} + \frac{\sum_{i=k}^{N} a_i s^i}{\sum_{i=k+1}^{N} b_i s^i} \\
&= \frac{a_k}{b_{k+1}}\frac{1}{s} + \frac{\sum_{i=k}^{N} a_i s^i - \sum_{i=k+1}^{N} \frac{a_k}{b_{k+1}} b_i s^{i-1}}{\sum_{i=k+1}^{N} b_i s^i} \\
&= \frac{a_k}{b_{k+1}}\frac{1}{s} + \frac{\sum_{i=k+1}^{N} a_i s^i - \sum_{i=k+2}^{N} \frac{a_k}{b_{k+1}} b_i s^{i-1}}{\sum_{i=k+1}^{N} b_i s^i} \\
&= \frac{a_k}{b_{k+1}}\frac{1}{s} + \frac{\sum_{i=k+1}^{N} a_i s^i - \sum_{i=k+1}^{N-1} \frac{a_k}{b_{k+1}} b_{i+1} s^i}{\sum_{i=k+1}^{N} b_i s^i} \\
&= \frac{a_k}{b_{k+1}}\frac{1}{s} + \frac{\sum_{i=k+1}^{N-1} (a_i - \frac{a_k}{b_{k+1}} b_{i+1}) s^i + a_N s^N}{\sum_{i=k+1}^{N} b_i s^i}
\end{aligned}$$

The numerator is again reduced from the bottom.

2. Answer:

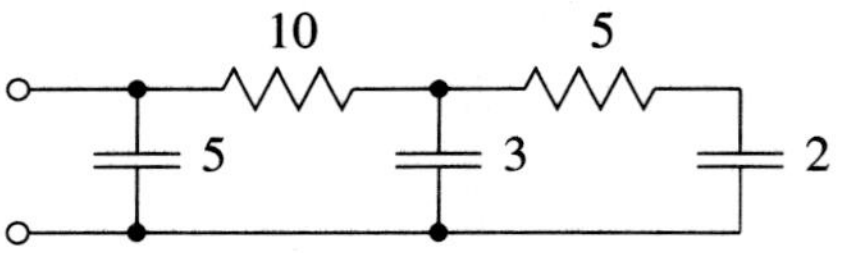

3. Let the rational function be

$$Y(s) = \frac{\sum_{i=0}^{M} a_i s^i}{\sum_{i=0}^{N} b_i s^i},$$

then

$$Y(s \to 0) = \frac{a_0}{b_0} = Y_{DC},$$

where Y_{DC} is assumed to be known. Incorporating this information into the numerator yields

$$\begin{aligned}
Y(s) &= \frac{a_0 + \sum_{i=1}^{M} a_i s^i}{\sum_{i=0}^{N} b_i s^i} \\
&= \frac{Y_{DC} b_0 + \sum_{i=1}^{M} a_i s^i}{\sum_{i=0}^{N} b_i s^i}.
\end{aligned}$$

Evaluating this equation at P frequencies yields the matrix equation

$$\left(\overline{\overline{B}} - \overline{\overline{C}}\right)\overline{b} = \overline{\overline{A}}^{'}\overline{a}^{'},$$

where $\overline{\overline{B}}$ and $\overline{b}$ are defined as before, and

$$\overline{\overline{A}}^{'} = \begin{bmatrix} s_1 & s_1^2 & \cdots & s_1^M \\ s_2 & s_2^2 & \cdots & s_2^M \\ s_3 & s_3^2 & \cdots & s_3^M \\ \vdots & \vdots & \ddots & \vdots \\ s_P & s_P^2 & \cdots & s_P^M \end{bmatrix},$$

$$\overline{a}^{'} = [\; a_1 \quad \cdots \quad a_M \;]^T,$$

and

$$\overline{\overline{C}} = \begin{bmatrix} Y_{DC} & 0 & \cdots & 0 \\ Y_{DC} & 0 & \cdots & 0 \\ Y_{DC} & 0 & \cdots & 0 \\ \vdots & \vdots & \ddots & \vdots \\ Y_{DC} & 0 & \cdots & 0 \end{bmatrix}$$

is a $P \times (N+1)$ matrix. The matrix equation can be solved with the same technique as used for (10.40).

4. Let the rational function be

$$Y(s) = \frac{\sum_{i=0}^{N} a_i s^i}{\sum_{i=0}^{N} b_i s^i},$$

then

$$Y(s \to \infty) = \frac{a_N}{b_N} = Y_\infty,$$

where Y_∞ is assumed to be known. Incorporating this information into the numerator yields

$$Y(s) = \frac{\sum_{i=0}^{N-1} a_i s^i + a_N s^N}{\sum_{i=0}^{N} b_i s^i} = \frac{\sum_{i=0}^{N-1} a_i s^i + b_N Y_\infty s^N}{\sum_{i=0}^{N} b_i s^i}.$$

Evaluating this equation at P frequencies yields the matrix equation

$$\left(\overline{\overline{B}} - \overline{\overline{C}}\right)\overline{b} = \overline{\overline{A}}' \overline{a}',$$

where $\overline{\overline{B}}$ and $\overline{b}$ are defined as before, and

$$\overline{\overline{A}}' = \begin{bmatrix} 1 & s_1 & s_1^2 & \cdots & s_1^{N-1} \\ 1 & s_2 & s_2^2 & \cdots & s_2^{N-1} \\ 1 & s_3 & s_3^2 & \cdots & s_3^{N-1} \\ \vdots & \vdots & \vdots & \ddots & \vdots \\ 1 & s_P & s_P^2 & \cdots & s_P^{N-1} \end{bmatrix},$$

$$\overline{a}' = [\ a_0 \quad a_1 \quad \cdots \quad a_{N-1}\ ]^T,$$

and

$$\overline{\overline{C}} = \begin{bmatrix} 0 & 0 & \cdots & Y_\infty s_1^N \\ 0 & 0 & \cdots & Y_\infty s_2^N \\ 0 & 0 & \cdots & Y_\infty s_3^N \\ \vdots & \vdots & \ddots & \vdots \\ 0 & 0 & \cdots & Y_\infty s_P^N \end{bmatrix}.$$

The matrix equation can be solved with the same technique as used for (10.40).

5. (a) Multiply top and bottom by $\cos(\beta d)$ to get

$$Z_{\text{in}} = Z_o \frac{Z_L \cos(\beta d) + \jmath Z_o \sin(\beta d)}{Z_o \cos(\beta d) + \jmath Z_L \sin(\beta d)},$$

then using the facts that

$$\sin(x) = \sum_{n=1}^{\infty} (-1)^{n+1} \frac{x^{2n-1}}{(2n-1)!}$$

and

$$\cos(x) = 1 + \sum_{n=1}^{\infty}(-1)^n \frac{x^{2n}}{(2n)!},$$

then Z_{in} becomes

$$\begin{aligned}
Z_{\text{in}} &= Z_o \frac{Z_L(1+\sum_{n=1}^{\infty}(-1)^n \frac{(\beta d)^{2n}}{(2n)!}) + \jmath Z_o \sum_{n=1}^{\infty}(-1)^{n+1}\frac{(\beta d)^{2n-1}}{(2n-1)!}}{Z_o(1+\sum_{n=1}^{\infty}(-1)^n \frac{(\beta d)^{2n}}{(2n)!}) + \jmath Z_L \sum_{n=1}^{\infty}(-1)^{n+1}\frac{(\beta d)^{2n-1}}{(2n-1)!}} \\
&= Z_o \frac{Z_L + \sum_{n=1}^{\infty}\left(Z_L(-1)^n \frac{(\beta d)^{2n}}{(2n)!} + \jmath Z_o(-1)^{n+1}\frac{(\beta d)^{2n-1}}{(2n-1)!}\right)}{Z_o + \sum_{n=1}^{\infty}\left(Z_o(-1)^n \frac{(\beta d)^{2n}}{(2n)!} + \jmath Z_L(-1)^{n+1}\frac{(\beta d)^{2n-1}}{(2n-1)!}\right)}.
\end{aligned}$$

This is a rational function in the form of

$$Z_{\text{in}} = \frac{\sum_{i=0}^{\infty} a_i(\beta d)^i}{\sum_{i=0}^{\infty} b_i(\beta d)^i}.$$

(b) Because the series goes to ∞ in the numerator, there are infinitely many zeros. Similarly for the denominator, there are infinitely many poles. The implication is that a ladder network would require infinitely many components to exactly represent a transmission line. In other words, transmission lines can only be approximately modeled using resistors, inductors, and capacitors.

(c) Keeping just two terms in the numerator and denominator means keeping terms through βd, so

$$Z_{\text{in}} \approx Z_o \frac{Z_L + \jmath Z_o \beta d}{Z_o + \jmath Z_L \beta d}.$$

Deriving the continued fraction by applying (10.51) and (10.52) yields

$$\begin{aligned} Z_{\text{in}} &= \frac{Z_o^2}{Z_L} - \frac{Z_o^2}{Z_L} + Z_o \frac{Z_L + \jmath Z_o \beta d}{Z_o + \jmath Z_L \beta d} \\ &= \frac{Z_o^2}{Z_L} + Z_o \frac{1 - \left(\frac{Z_o}{Z_L}\right)^2}{\frac{Z_o}{Z_L} + \jmath \beta d} \\ &= \frac{Z_o^2}{Z_L} + \frac{1}{\frac{1}{Z_o} \frac{\frac{Z_o}{Z_L} + \jmath \beta d}{1 - \left(\frac{Z_o}{Z_L}\right)^2}} \\ &= \frac{Z_o^2}{Z_L} + \frac{1}{\frac{1}{Z_o} \frac{\frac{Z_o}{Z_L}}{1 - \left(\frac{Z_o}{Z_L}\right)^2} + \jmath \frac{1}{Z_o} \frac{\beta d}{1 - \left(\frac{Z_o}{Z_L}\right)^2}}. \end{aligned}$$

(d) The circuit is

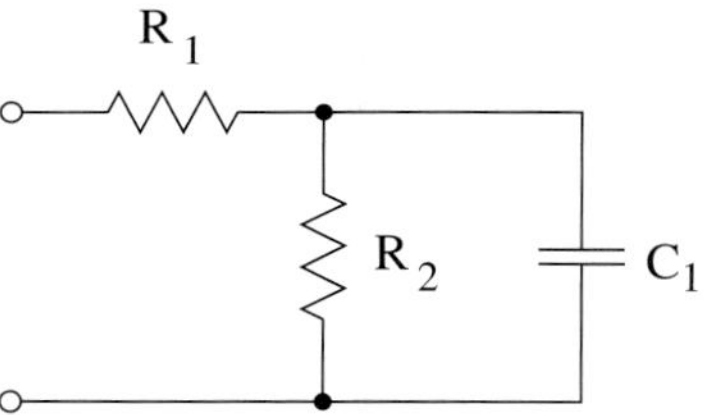

where $R_1 = Z_o^2/Z_L$, $R_2 = Z_L\left[1 - (Z_o/Z_L)^2\right]$, and $C_1 = (\beta d)/\left(Z_o\left[1 - (Z_o/Z_L)^2\right]\right)$. It is interesting to note that the circuit is passive when $Z_o < Z_L$ and active when $Z_o > Z_L$. When $Z_L = Z_o$, the input impedance to the circuit is just Z_o, as it should be.

6. The minimum rise time is 0.5ns, so initially set the delay of each lumped section at 0.05ns. The delay of the line is $0.025/(3 \times 10^8/\sqrt{9.8}) = 0.26$ns, so the required number of sections is $0.26/0.05 = 5.2 \approx 5$. For five sections, the delay each 0.5cm section is then 0.052ns. From chapter 2, equations (2.11) and (2.12), the lumped inductance and capacitance are 3.9nH and 0.693pF. The model is given in the following schematic.

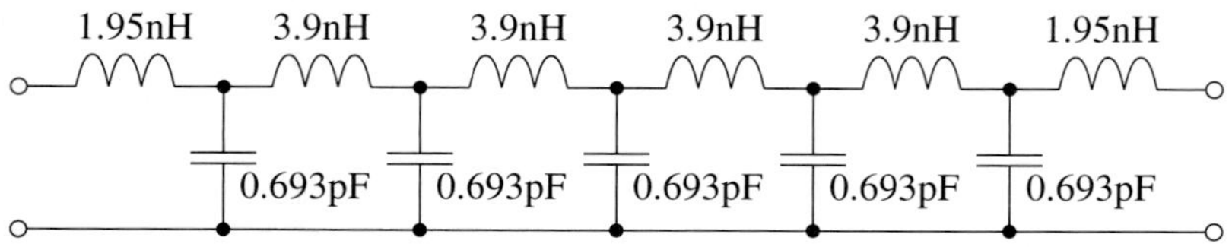

7. With $\overline{\overline{\ell}} = \overline{\overline{g}} = 0$, (10.5) and (10.6) become

$$\frac{\partial \overline{v}}{\partial z} = -\overline{\overline{r}}\,\overline{i}$$

and

$$\frac{\partial \overline{i}}{\partial z} = -\overline{\overline{c}}\frac{\partial \overline{v}}{\partial t}.$$

Eliminate i to get

$$\frac{\partial^2 \overline{v}}{\partial z^2} = \overline{\overline{r}}\,\overline{\overline{c}}\frac{\partial \overline{v}}{\partial t}.$$

Noting that $\overline{\overline{r}}\,\overline{\overline{c}} = \overline{\overline{T}}_V \overline{\overline{\lambda}}\, \overline{\overline{T}}_V^{-1}$, where $\overline{\overline{\lambda}}$ is diagonal, then

$$\frac{\partial^2 \overline{\overline{T}}_V^{-1} \overline{v}}{\partial z^2} = \overline{\overline{\lambda}}\frac{\partial \overline{\overline{T}}_V^{-1} \overline{v}}{\partial t}.$$

Defining the modal voltage as before, then

$$\frac{\partial^2 \overline{v}_m}{\partial z^2} = \overline{\overline{\lambda}}\frac{\partial \overline{v}_m}{\partial t}$$

represents a set of decoupled equations.

Eliminating v yields a similar equation in i with $\overline{\overline{c}}\,\overline{\overline{r}} = \overline{\overline{T}}_I \overline{\overline{\lambda}}\, \overline{\overline{T}}_I^{-1}$. The modal currents are also defined as before.

The first-order equations can be rewritten as

$$\frac{\partial \overline{v}_m}{\partial z} = -\overline{\overline{r}}_m \overline{i}_m$$

and

$$\frac{\partial \overline{i}_m}{\partial z} = -\overline{\overline{c}}_m \frac{\partial \overline{v}_m}{\partial t},$$

where $\overline{\overline{r}}_m = \overline{\overline{T}}_V^{-1} \overline{\overline{r}}\, \overline{\overline{T}}_I$ and $\overline{\overline{c}}_m = \overline{\overline{T}}_I^{-1} \overline{\overline{c}}\, \overline{\overline{T}}_V$.

Since no special properties are used in the prior derivation regarding $\overline{\overline{\ell}}$ and $\overline{\overline{c}}$, the conclusions also apply to $\overline{\overline{r}}$ and $\overline{\overline{c}}$ in this derivation. Namely, $\overline{\overline{r}}_m$ and $\overline{\overline{c}}_m$ are diagonal.

8. Equating the even- and odd-mode phase velocities from (10.24) and (10.25) yields the required result. Far-end noise is eliminated for symmetric lines in a homogeneous medium. Stripline and buried microstrip are typical examples of transmission lines that meet this criteria.

9. From (10.24) and the fact that $m/\ell = c_m/c$ in a homogeneous medium, the even-mode phase velocity is

$$\begin{aligned} v_e &= \frac{1}{\sqrt{(\ell + m)(c - c_m)}} \\ &= \frac{1}{\sqrt{\ell c + mc - \ell c_m - mc_m}} \\ &= \frac{1}{\sqrt{\ell c - mc_m}}. \end{aligned}$$

A similar derivation shows the same result for the odd mode. Then

$$\frac{1}{v_e^2} = \frac{1}{v_o^2} = \ell c - mc_m = \frac{1}{v_o^2},$$

and, again incorporating $m/\ell = c_m/c$,

$$\begin{aligned} \overline{\overline{\ell}}\,\overline{\overline{c}} &= \begin{bmatrix} \ell & m \\ m & \ell \end{bmatrix} \begin{bmatrix} c & -c_m \\ -c_m & c \end{bmatrix} \\ &= \begin{bmatrix} \ell c - mc_m & 0 \\ 0 & \ell c - mc_m \end{bmatrix} \\ &= \begin{bmatrix} \frac{1}{v_o^2} & 0 \\ 0 & \frac{1}{v_o^2} \end{bmatrix} \\ &= \frac{1}{v_o^2}. \end{aligned}$$

10. (a) Working with a symbolic manipulator, it is found that

$$\overline{\overline{T}}_V = \overline{\overline{T}}_I = \begin{bmatrix} 1/\sqrt{3} & -1/\sqrt{2} & -1/\sqrt{6} \\ 1/\sqrt{3} & 0 & \sqrt{2/3} \\ 1/\sqrt{3} & 1/\sqrt{2} & -1/\sqrt{6} \end{bmatrix}.$$

While not needed, it is interesting to note that

$$\overline{\overline{\lambda}} = \begin{bmatrix} (c-2c_m)(\ell+2m) & 0 & 0 \\ 0 & (c+c_m)(\ell-m) & 0 \\ 0 & 0 & (c+c_m)(\ell-m) \end{bmatrix},$$

which has two identical eigenvalues, reflecting the fact that the problem has a high degree of symmetry.

The modal inductances are then found with (10.16) to be

$$\overline{\overline{\ell}}_m = \begin{bmatrix} \ell+2m & 0 & 0 \\ 0 & \ell-m & 0 \\ 0 & 0 & \ell-m \end{bmatrix},$$

while from (10.17) the modal capacitances are

$$\overline{\overline{c}}_m = \begin{bmatrix} c-2c_m & 0 & 0 \\ 0 & c+c_m & 0 \\ 0 & 0 & c+c_m \end{bmatrix}.$$

The modal characteristic impedances are

$$Z_{o1} = \sqrt{\frac{\ell+2m}{c-2c_m}}$$

and

$$Z_{o2} = Z_{o3} = \sqrt{\frac{\ell-m}{c+c_m}},$$

while the modal phase velocities are

$$v_{p1} = \frac{1}{\sqrt{(\ell+2m)(c-2c_m)}}$$

and

$$v_{p2} = v_{p3} = \frac{1}{\sqrt{(\ell-m)(c+c_m)}}.$$

(b) Using

$$\overline{\overline{T}}_V^{-1} = \overline{\overline{T}}_I^{-1} = \begin{bmatrix} 1/\sqrt{3} & 1/\sqrt{3} & 1/\sqrt{3} \\ -1/\sqrt{2} & 0 & 1/\sqrt{2} \\ -1/\sqrt{6} & \sqrt{2/3} & -1/\sqrt{6} \end{bmatrix}$$

along with $\overline{\overline{T}}_V$ and $\overline{\overline{T}}_I$ to convert between total and modal quantities, the schematic implementing the modal conversion is given below.

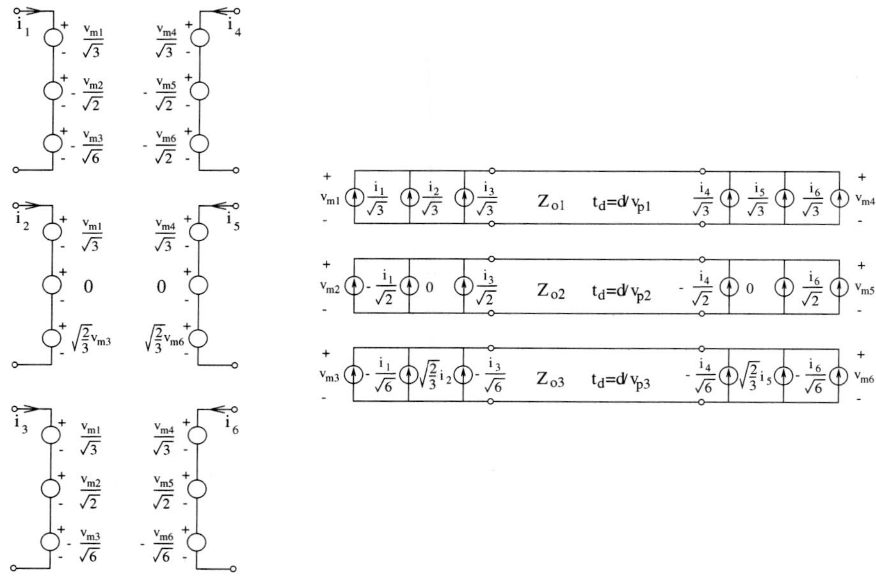

11. (a) From (10.28), for small Z_2 the third term in the denominator dominates when $Z_2 \ll 2Z_{oe}Z_{oo}/(Z_{oe}+Z_{oo})$, then

$$K_{ne} \approx \frac{Z_{oe} - Z_{oo}}{2\frac{Z_{oe}Z_{oo}}{Z_2}}.$$

Rearrange to obtain the given form.

(b) Substituting the expressions for the weak forms for Z_{oe} and Z_{oo} from (10.29) and (10.30) yields

$$
\begin{aligned}
K_{ne} &\approx \frac{\sqrt{\frac{\ell}{c}\frac{1+\frac{1}{2}m/\ell}{1-\frac{1}{2}c_m/c}} - \sqrt{\frac{\ell}{c}\frac{1-\frac{1}{2}m/\ell}{1+\frac{1}{2}c_m/c}}}{2\frac{\sqrt{\frac{\ell}{c}\frac{1+\frac{1}{2}m/\ell}{1-\frac{1}{2}c_m/c}}\sqrt{\frac{\ell}{c}\frac{1-\frac{1}{2}m/\ell}{1+\frac{1}{2}c_m/c}}}{Z_2}} \\
&= \frac{1}{2}Z_2\sqrt{\frac{c}{\ell}}\frac{(1+\frac{1}{2}m/\ell)(1+\frac{1}{2}c_m/c)-(1-\frac{1}{2}m/\ell)(1-\frac{1}{2}c_m/c)}{(1+\frac{1}{2}m/\ell)(1-\frac{1}{2}m/\ell)} \\
&\approx \frac{1}{2}Z_2\sqrt{\frac{c}{\ell}}\left(\frac{m}{\ell}+\frac{c_m}{c}\right),
\end{aligned}
$$

neglecting the second-order term $1/4(m/\ell)^2$.

12. Multiply out the numerator and denominator to get

$$
Z(s) = \frac{a_1 s + a_2 s^3 + \cdots + a_N s^{2N-1}}{b_0 + b_1 s^2 + b_2 s^4 + \cdots + b_N s^{2N}}.
$$

Multiply through to get

$$
Z(s)b_0 + Z(s)b_1 s^2 + Z(s)b_2 s^4 + \cdots + Z(s)b_N s^{2N} = a_1 s + a_2 s^3 + \cdots + a_N s^{2N-1}.
$$

Evaluate at P frequencies, then in matrix form

$$
\overline{\overline{B}}\,\overline{b} = \overline{\overline{A}}\,\overline{a},
$$

which is equivalent to (10.40), where

$$
\overline{a} = \begin{bmatrix} a_1 & a_2 & \cdots & a_N \end{bmatrix}^T,
$$

$$
\overline{b} = \begin{bmatrix} b_0 & b_1 & \cdots & b_N \end{bmatrix}^T,
$$

$$
\overline{\overline{A}} = \begin{bmatrix} s_1 & s_1^3 & \cdots & s_1^{2N-1} \\ s_2 & s_2^3 & \cdots & s_2^{2N-1} \\ s_3 & s_3^3 & \cdots & s_3^{2N-1} \\ \vdots & \vdots & \ddots & \vdots \\ s_P & s_P^3 & \cdots & s_P^{2N-1} \end{bmatrix},
$$

and

$$\overline{\overline{B}} = \begin{bmatrix} Z(s_1) & Z(s_1)s_1^2 & Z(s_1)s_1^4 & \cdots & Z(s_1)s_1^{2N} \\ Z(s_2) & Z(s_2)s_2^2 & Z(s_2)s_2^4 & \cdots & Z(s_2)s_2^{2N} \\ Z(s_3) & Z(s_3)s_3^2 & Z(s_3)s_3^4 & \cdots & Z(s_3)s_3^{2N} \\ \vdots & \vdots & \vdots & \ddots & \vdots \\ Z(s_P) & Z(s_P)s_P^2 & Z(s_P)s_P^4 & \cdots & Z(s_P)s_P^{2N} \end{bmatrix}.$$

Chapter 11

1. For all cases, the driver holds one output high at V_{DD} with driver impedance R, and it holds the other low to ground with impedance R. For the optimal termination, V_{DD} has a DC path to ground through the resistance

$$\begin{aligned} R_{\text{eq}} &= R + Z_{oe} \| (2Z_{oe}Z_{oo}/(Z_{oe} - Z_{oo}) + R \| Z_{oe}) \\ &= \frac{2(R + Z_{oe})(R + Z_{oo})}{2R + Z_{oe} + Z_{oo}}. \end{aligned}$$

The static power dissipation is then

$$P_{\text{optimal}} = \frac{V_{DD}^2(2R + Z_{oe} + Z_{oo})}{2(R + Z_{oe})(R + Z_{oo})}.$$

For bridged termination, the equivalent resistance is $R_{\text{eq}} = 2(R + Z_{oo})$ for a static power dissipation of

$$P_{\text{bridged}} = \frac{V_{DD}^2}{2(R + Z_{oo})}.$$

For single-ended termination, the power dissipation is determined by the output holding high, so

$$P_{\text{single-ended}} = \frac{V_{DD}^2}{R + Z_{oo}}.$$

Looking at the ratios

$$\frac{P_{\text{single-ended}}}{P_{\text{bridged}}} = 2$$

and

$$\frac{P_{\text{optimal}}}{P_{\text{bridged}}} = \frac{2R + Z_{oe} + Z_{oo}}{R + Z_{oe}}$$
$$= 1 + \frac{R + Z_{oo}}{R + Z_{oe}},$$

both are greater than one, so bridged termination dissipates the least static power.

2. Plugging the results from the solution of chapter 10, problem 10 into (11.4) yields

$$\overline{\overline{Y}} = \begin{bmatrix} 1/\sqrt{3} & -1/\sqrt{2} & -1/\sqrt{6} \\ 1/\sqrt{3} & 0 & \sqrt{2/3} \\ 1/\sqrt{3} & 1/\sqrt{2} & -1/\sqrt{6} \end{bmatrix} \begin{bmatrix} \sqrt{\frac{c-2c_m}{\ell+2m}} & 0 & 0 \\ 0 & \sqrt{\frac{c+c_m}{\ell-m}} & 0 \\ 0 & 0 & \sqrt{\frac{c+c_m}{\ell-m}} \end{bmatrix}$$
$$\begin{bmatrix} 1/\sqrt{3} & 1/\sqrt{3} & 1/\sqrt{3} \\ -1/\sqrt{2} & 0 & 1/\sqrt{2} \\ -1/\sqrt{6} & \sqrt{2/3} & -1/\sqrt{6} \end{bmatrix}$$
$$= \frac{1}{3} \begin{bmatrix} Y_{o1} + 2Y_{o2} & Y_{o1} - Y_{o2} & Y_{o1} - Y_{o2} \\ Y_{o1} - Y_{o2} & Y_{o1} + 2Y_{o2} & Y_{o1} - Y_{o2} \\ Y_{o1} - Y_{o2} & Y_{o1} - Y_{o2} & Y_{o1} + 2Y_{o2} \end{bmatrix},$$

where

$$Y_{o1} = \sqrt{\frac{c - 2c_m}{\ell + 2m}}$$

and

$$Y_{o2} = \sqrt{\frac{c + c_m}{\ell - m}}.$$

Using the method in chapter 9, section 9.9 for implementing admittance matrices, the following circuit is the optimal termination network.

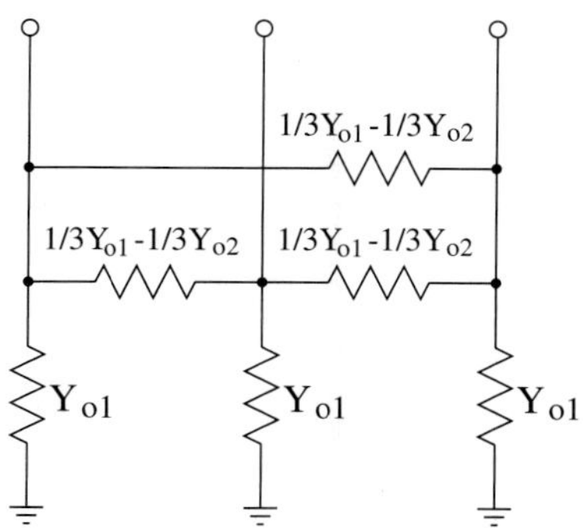

3. The quick answer is that $V_{\text{detected}} = \pm 0.0025 \times 100 = \pm 250\text{mV}$ since the receiver rejects common-mode noise. The point of the exercise is to practice manipulations with modal decompositions.

   Call the current on the top wire i_1 and the current on the bottom wire i_2. For continuity, $i_1(t) = i(t)$ and $i_2(t) = -i(t)$. From chapter 10, equations (10.11) and (10.20)

$$\bar{i}_m = \frac{1}{\sqrt{2}} \begin{bmatrix} 1 & 1 \\ -1 & 1 \end{bmatrix} \begin{bmatrix} i(t) \\ -i(t) \end{bmatrix} = \begin{bmatrix} 0 \\ -\sqrt{2}\, i(t) \end{bmatrix}.$$

   The odd mode propagates to the far end of the line at the odd-mode phase velocity, while the even mode picks up noise. At the far end, the modal currents are

$$\bar{i}_m = \begin{bmatrix} i_{\text{noise}} \\ -\sqrt{2}\, i(t - d/v_o) \end{bmatrix}.$$

   Using (10.19), the total currents at the far end are

$$\begin{bmatrix} i_1 \\ i_2 \end{bmatrix} = \frac{1}{\sqrt{2}} \begin{bmatrix} 1 & -1 \\ 1 & 1 \end{bmatrix} \begin{bmatrix} i_{\text{noise}} \\ -\sqrt{2} i(t - d/v_o) \end{bmatrix} = \begin{bmatrix} \frac{1}{\sqrt{2}} i_{\text{noise}} + i(t - d/v_o) \\ \frac{1}{\sqrt{2}} i_{\text{noise}} - i(t - d/v_o) \end{bmatrix}.$$

Using the results from chapter 4, problem 14, the voltage at the load for a symmetric T-network termination is

$$\begin{aligned}
\begin{bmatrix} v_1 \\ v_2 \end{bmatrix} &= \begin{bmatrix} 50+z_2 & z_2 \\ z_2 & 50+z_2 \end{bmatrix} \begin{bmatrix} \frac{1}{\sqrt{2}} i_{\text{noise}} + i(t-d/v_o) \\ \frac{1}{\sqrt{2}} i_{\text{noise}} - i(t-d/v_o) \end{bmatrix} \\
&= \begin{bmatrix} \frac{1}{\sqrt{2}}(50+2z_2) i_{\text{noise}} + 50 i(t-d/v_o) \\ \frac{1}{\sqrt{2}}(50+2z_2) i_{\text{noise}} - 50 i(t-d/v_o) \end{bmatrix}.
\end{aligned}$$

The receiver detects the differential voltage across the resistor, so

$$v_{\text{detected}} = v_1 - v_2 = 100 i(t - d/v_o),$$

which matches the quick result above. Note how z_2 is left in the termination network to avoid mathematical difficulties, but its effect drops out in the end.

4. (a) Using (11.5) with $y_1 = 0$ and $y_2 = 1/(2Z_o)$ yields the required result.

   (b) The odd-mode reflection coefficient is not zero.

   (c) The odd-mode impedance is

$$\begin{aligned}
Z_{oo} &= \sqrt{\frac{\ell - m}{c + c_m}} \\
&= \sqrt{\frac{\ell}{c}} \sqrt{\frac{1 - m/\ell}{1 + c_m/c}} \\
&= \sqrt{\frac{\ell}{c}} \sqrt{\frac{1 - 0.1}{1 + 0.1}} \\
&= 0.905 \sqrt{\frac{\ell}{c}}.
\end{aligned}$$

   Since $Z_o = \sqrt{\ell/c}$, the odd-mode reflection coefficient is

$$\begin{aligned}
\Gamma_o &= \frac{1 - 0.905}{1 + 0.905} \\
&= 0.05.
\end{aligned}$$

   Since just 5% of the incident odd-mode wave is reflected, a reflection coefficient of 0.05 is good.

5. The modal reflection coefficient from (11.2) using (10.19) is

$$\overline{\Gamma}_m = \left(\begin{bmatrix} 1 & -1 \\ 1 & 1 \end{bmatrix} \begin{bmatrix} Y_{o1} & 0 \\ 0 & Y_{o2} \end{bmatrix} + \begin{bmatrix} y_1 & 0 \\ 0 & y_3 \end{bmatrix} \begin{bmatrix} 1 & -1 \\ 1 & 1 \end{bmatrix} \right)^{-1}$$
$$\left(\begin{bmatrix} 1 & -1 \\ 1 & 1 \end{bmatrix} \begin{bmatrix} Y_{o1} & 0 \\ 0 & Y_{o2} \end{bmatrix} - \begin{bmatrix} y_1 & 0 \\ 0 & y_3 \end{bmatrix} \begin{bmatrix} 1 & -1 \\ 1 & 1 \end{bmatrix} \right).$$

Multiplying this out yields

$$\Gamma_{m11} = \frac{(y_1 + y_3)(Y_{o1} - Y_{o2}) - 2(y_1 y_3 - Y_{o1} Y_{o2})}{(y_1 + y_3)(Y_{o1} + Y_{o2}) + 2(y_1 y_3 + Y_{o1} Y_{o2})},$$

$$\Gamma_{m12} = \frac{2Y_{o2}(y_1 - y_3)}{(y_1 + y_3)(Y_{o1} + Y_{o2}) + 2(y_1 y_3 + Y_{o1} Y_{o2})},$$

$$\Gamma_{m21} = \frac{2Y_{o1}(y_1 - y_3)}{(y_1 + y_3)(Y_{o1} + Y_{o2}) + 2(y_1 y_3 + Y_{o1} Y_{o2})},$$

and

$$\Gamma_{m22} = \frac{(y_1 + y_3)(-Y_{o1} + Y_{o2}) - 2(y_1 y_3 - Y_{o1} Y_{o2})}{(y_1 + y_3)(Y_{o1} + Y_{o2}) + 2(y_1 y_3 + Y_{o1} Y_{o2})}.$$

Mode conversion from the incident even mode to the reflected odd mode is given by Γ_{21}. For no mode conversion, a balanced load is required with $y_1 = y_3$. Otherwise, the reflected odd-mode voltage due to the incidence even-mode waveform is

$$v_o^- = \Gamma_{21} v_e^+,$$

which is the voltage detected by a perfect differential receiver.

6. From (11.3) with (10.19) and (10.20),

$$\overline{\overline{Z}} = \frac{1}{2} \begin{bmatrix} 1 & -1 \\ 1 & 1 \end{bmatrix} \begin{bmatrix} Z_{oe} & 0 \\ 0 & Z_{oo} \end{bmatrix} \begin{bmatrix} 1 & 1 \\ -1 & 1 \end{bmatrix}$$
$$= \frac{1}{2} \begin{bmatrix} Z_{oe} + Z_{oo} & Z_{oe} - Z_{oo} \\ Z_{oe} - Z_{oo} & Z_{oe} + Z_{oo} \end{bmatrix}.$$

Matching up terms with the impedance matrix from chapter 4, problem 14 yields

$$\begin{aligned} z_2 &= \frac{Z_{oe} - Z_{oo}}{2} \\ z_3 &= z_1 \\ z_1 &= Z_{oo}. \end{aligned}$$

7. (a) The cavity is resonant when the input impedance has no reactance. From the given equation, this occurs when $\cot(\beta d) = 0$, or $\beta d = \pi/2 + n\pi$. Since $\beta = \omega/v_p$, then $\beta d = 0.1 \times 2\pi f/(3 \times 10^8/\sqrt{4})$, so the resonant frequencies are 375MHz, 1.125GHz, 1.875GHz, and so on.

   (b) The wavelength at 3GHz is 5cm. For lumped modeling, the maximum length of each section is 0.5cm, so a 20×20 sectioning of the PCB is required. Assuming parallel-plate formulas with DC current distribution, the resistance of each section is

$$R = 2\frac{1}{5.7 \times 10^7}\frac{0.005}{0.005 \times 0.18 \times 10^{-6}} = 0.195\Omega,$$

   the characteristic impedance from (11.6) is

$$Z_o = \sqrt{\frac{4\pi \times 10^{-7}}{4 \times 8.854 \times 10^{-12}}}\frac{0.0005}{0.005} = 18.84\Omega,$$

   and the phase velocity is $v_p = 1.5 \times 10^8$m/s, so the TOF, or t_d, is 33.33ps (666.7ps/20). From chapter 2, equations (2.11) and (2.12), the inductance and capacitance are 0.628nH and 1.77pF. Building and running a SPICE model produces the following result:

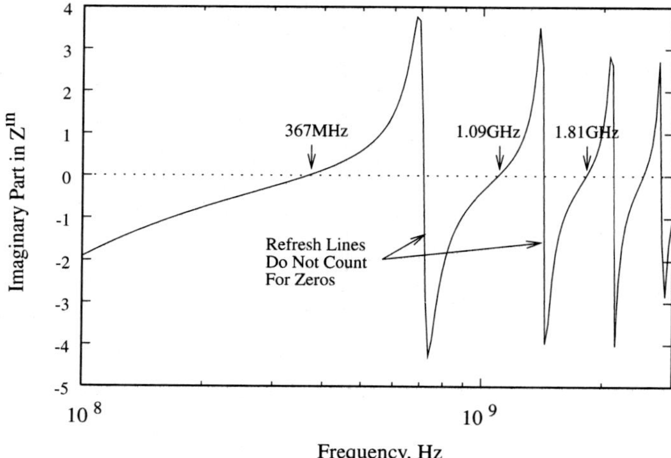

(c) There are slight differences due to the losses included in the SPICE simulation.

(d) The total capacitance is $C = 400 \times 1.77$pf=708pF. At resonance, only the resistance is seen, so the ESR is $0.195 \times 20/20 = 0.195\Omega$. The ESL can be computed from the resonant frequency of 375MHz using the lumped formula $\omega = 1/\sqrt{LC}$ to obtain ESL=0.254nH. Note that the ESL is considerably less than 0.628nH because the lumped approximation does not apply at 375MHz. For a small component like a ceramic capacitor, the ESL will be close to the inductance.

(e) The currents must spread to fill the PCB, so resistance and inductance increase. The higher inductance leads to significantly lower resonant frequencies.

Appendix B

COAXIAL PEEC CALCULATOR

```
#include <stdlib.h>
#include <string.h>
#include <stdio.h>
#include <zmatrix.h>
#include <matrix2.h>
#include <math.h>

/*
Requires the Meschach library of mathematical subroutines
available from the netlib2.cs.utk.edu archive.

Computes the resistance and inductance of solid coax as a function of
frequency.  The outer conductor thickness is equal to the radius of the
center conductor less 1%.  Uses the PEEC method with the inner and
outer conductors broken down into concentric current rings of equal
thickness.

The program is hard-wired limited to 50 rings for the inner and outer
conductors.  Change the array declarations and the count check to
increase.

Once the current has crowded to the inside rings of the inner and outer
conductors, the resistance and inductance values stop changing with
```

```
frequency.  If enough rings are included, then the inductance can
be a close approximation of the asymptotic high-frequency value.  However,
the resistance will be completely in error because resistance continues
increasing with a square root dependence (i.e. there is no asymptotic value).
*/

char current_file[8]="i.dat";  /* branch currents */
char voltage_file[8]="v.dat";  /* branch voltages */

double rho=1.77e-8;            /* resistivity of copper */
//double rho=5.7e-8;            /* resistivity of molybdenum */

double ring_resistance (double , double );
double coax (double , double , double );
double annular_coax (double , double , double );
int ring_partials (double , double , double , double , double *,
   double *, double *, double *, double *);
complex equivalent_Z (ZMAT **, ZMAT **);

void main(int argc, char *argv[])
{
  int number_of_rings,number_of_frequencies;
  double inner_radius,outer_radius,fmin,fmax;

  VEC *ring_radii,*ring_R;
  MAT *ring_L;
  ZMAT *Zp,*incidence;
  char syscommand[100];
  int i,j,k;
  double t,a,c,outer_thickness,R1,R2,fmult,frequency,pi;
  complex Zeq;

  /* get the order of the numerator */
  if (argc != 7)
  {
    printf ("Usage: coax N Ri Ro fmin fmax Nf\n\n");
    printf ("where N is the number of rings per conductor,\n");
```

```
  printf ("Ri and Ro are the inner and outer radii in m,\n");
  printf ("fmin is the lowest frequency to compute in Hz,\n");
  printf ("fmax is the highest frequency to compute in Hz,\n");
  printf ("Nf is the number of frequency points to use.\n");
  exit (1);
}
sscanf (argv[1],"%d",&number_of_rings);
sscanf (argv[2],"%lf",&inner_radius);
sscanf (argv[3],"%lf",&outer_radius);
sscanf (argv[4],"%lf",&fmin);
sscanf (argv[5],"%lf",&fmax);
sscanf (argv[6],"%d",&number_of_frequencies);

pi=4.*atan(1.);

fmult=pow(fmax/fmin,1./(float)(number_of_frequencies-1));

/* remove the output files */
sprintf (syscommand,"rm -f %s %s",current_file,voltage_file);
system (syscommand);

/* allocate space for the ring information */
ring_radii=v_get(2*number_of_rings);
ring_R=v_get(2*number_of_rings);
ring_L=m_get(2*number_of_rings,2*number_of_rings);
Zp=zm_get(2*number_of_rings,2*number_of_rings);
incidence=zm_get(2,2*number_of_rings);

/* compute ring geometries */

t=(inner_radius-inner_radius/100)/number_of_rings;

   /* inner conductor */

a=inner_radius/100;
i=0;
while (i < number_of_rings)
```

```
{
  ring_radii->ve[i]=a;
  a+=t;
  i++;
}

  /* outer conductor */

a=outer_radius;
i=0;
while (i < number_of_rings)
{
  ring_radii->ve[i+number_of_rings]=a;
  a+=t;
  i++;
}
outer_thickness=a-outer_radius;

/* set the reference distance - any value with c>outer_radius+outer_thickness
   is fine as long as it does not cause numerical instability */
c=a+outer_thickness;

/* print out a summary */
printf ("# inner radius = %12.4e\n",inner_radius);
printf ("# outer radius = %12.4e\n",outer_radius);
printf ("# outer conductor thickness = %12.4e\n",outer_thickness);
printf ("# reference radius = %12.4e\n",c);
printf ("# number of rings per conductor = %d\n",number_of_rings);
printf ("# DC inductance = %e H/m\n",coax(inner_radius,outer_radius,outer_thickness));
printf ("# High frequency inductance = %e H/m\n",2.e-7*log(outer_radius/inner_radius));
printf ("\n# frequency   resistance   inductance \n");

/* compute the ring resistances */
i=0;
while (i < 2*number_of_rings)
{
  ring_R->ve[i]=ring_resistance(ring_radii->ve[i],t);
```

```
  i++;
}

/* compute the partial inductance matrix */
i=0;
while (i < 2*number_of_rings-1)
{
  j=i+1;
  while (j < 2*number_of_rings)
  {
    k=ring_partials (ring_radii->ve[i],ring_radii->ve[j],c,t,
      &ring_L->me[i][i],&ring_L->me[j][j],&ring_L->me[i][j],&R1,&R2);
    if (k != 0) exit (1);
    ring_L->me[j][i]=ring_L->me[i][j];
    j++;
  }
  i++;
}

/* build the incidence matrix */
i=0;
while (i < number_of_rings)
{
  incidence->me[0][i]=zmake(1,0);
  incidence->me[0][i+number_of_rings]=zmake(0,0);
  incidence->me[1][i]=zmake(-1,0);
  incidence->me[1][i+number_of_rings]=zmake(-1,0);
  i++;
}

/* compute impedance at each frequency */
k=0; frequency=fmin;
while (k < number_of_frequencies)
{

  /* build the partial impedance matrix */
  i=0;
```

```
    while (i < 2*number_of_rings)
    {
      Zp->me[i][i]=zmake(ring_R->ve[i],2*pi*frequency*ring_L->me[i][i]);
      j=0;
      while (j < 2*number_of_rings)
      {
        if (i != j) Zp->me[i][j]=zmake(0.,2*pi*frequency*ring_L->me[i][j]);
        j++;
      }
      i++;
    }

    /* compute the equivalent impedance */
    Zeq=equivalent_Z (&incidence,&Zp);

    /* print out the results */
    printf ("%12.4e %12.4e %12.4e\n",frequency,Zeq.re,Zeq.im/(2*pi*frequency));

    frequency*=fmult;
    k++;
  }

}
/***************************************************************************/
/*          -1 T -1                                                        */
/* Compute (AZ  A )                                                        */
/*                                                                         */
/***************************************************************************/
complex equivalent_Z (ZMAT **A, ZMAT **Z)
{
   ZMAT *Zi,*At,*t1,*t2,*t3;
   complex Zeq;
   int i,j;

   /* find the inverse of Z */
   Zi=zm_copy(*Z,ZMNULL);
   zm_inverse(*Z,Zi);
```

```
  /* form the transpose of A */
  At=zm_get((*A)->n,(*A)->m);
  i=0;
  while (i < At->m)
  {
    j=0;
    while (j < At->n)
    {
      At->me[i][j]=(*A)->me[j][i];
      j++;
    }
    i++;
  }

  t1=zm_get(Zi->m,At->n);
  zm_mlt(Zi,At,t1);

  t2=zm_get((*A)->m,t1->n);
  zm_mlt(*A,t1,t2);

  t3=zm_get(t2->m,t2->n);
  zm_inverse(t2,t3);

  Zeq=t3->me[0][0];

  zm_free(Zi); zm_free(At); zm_free(t1); zm_free(t2); zm_free(t3);

  return Zeq;
}
/*************************************************************************/
/* Computes a set of partial inductances for annular coax based on       */
/* a return path formed by a ring of radius c>b+t.  Note that c should be */
/* consistent for all calls to this subroutine.                           */
/* Also computes the ring resistances.                                    */
/*   a   -  inner radius                                                  */
/*   b   -  outer radius                                                  */
```

```
/*   c   -  return radius                                                */
/*   L1  -  partial self-inductance for the inner ring                   */
/*   L2  -  partial self-inductance for the outer ring                   */
/*   M   -  partial mutual inductance between the two rings              */
/*   rho -  resistivity of the conductor in m/S                          */
/*   R1  -  DC resistance for the inner ring                             */
/*   R2  -  DC resistance for the outer ring                             */
/*************************************************************************/
int ring_partials (double a, double b, double c, double t, double *L1,
   double *L2, double *M, double *R1, double *R2)
{
  double L;

  if (c < b+t) return 1;

  *L1=annular_coax(a,c,t);
  *L2=annular_coax(b,c,t);
  L=annular_coax(a,b,t);

  *M=(*L1+*L2-L)/2;

  *R1=ring_resistance(a,t);
  *R2=ring_resistance(b,t);

  return 0;
}
/*************************************************************************/
/* Compute the inductance per-unit-length for annular coax with          */
/* DC current distribution.                                              */
/*   a - inner radius                                                    */
/*   b - outer radius                                                    */
/*   t - shell thickness                                                 */
/*   returns inductance in H/m                                           */
/*************************************************************************/
double annular_coax (double a, double b, double t)
{
  double mu,pi,L;
```

```
  pi=4.*atan(1.);
  mu=4.e-7*pi;  /* permeability of free space (and nonmagnetic metals) */

  L=mu/(2.*pi)*(log(b/(a+t))+3./8.-a/(4.*t)-t/(8.*(2.*a+t))+
    log((a+t)/a)/(pow(t/a,2)*pow(2.+t/a,2))+
    pow(b+t,4)/(pow(t,2)*pow(2.*b+t,2))*
    log((b+t)/b)-5./8.-b/(4.*t)-t/(8.*(2.*b+t)));

  return L;
}
/************************************************************************/
/* Compute the inductance per-unit-length for a coax with              */
/* DC current distribution.                                             */
/*   a - inner radius                                                   */
/*   b - outer radius                                                   */
/*   t - shell thickness                                                */
/*   returns inductance in H/m                                          */
/************************************************************************/
double coax (double a, double b, double t)
{
  double mu,pi,L;
  pi=4.*atan(1.);
  mu=4.e-7*pi;  /* permeability of free space (and nonmagnetic metals) */

  L=mu/(2.*pi)*(log(b/a)+1./4.+pow(b+t,4)/(pow(t,2)*pow(2.*b+t,2))*
    log((b+t)/b)-5./8.-b/(4.*t)-t/(8.*(2.*b+t)));

  return L;
}
/************************************************************************/
double ring_resistance (double radius, double thickness)
{
   double R,pi;
   pi=4.*atan(1.);
   R=rho/(pi*pow(radius+thickness,2)-pi*pow(radius,2));
   return R;
}
```

Appendix C

SAMPLE SPICE SSN SIMULATIONS

```
* SSN demonstration circuit
* Run on unmodified SPICE3f4

* i1 - on-chip vdd
* i5 - on-chip vss

* o1 - off-chip vdd
* o5 - off-chip ground

* i2 - active driver
* i3 - quiet driver - holding high
* i4 - quiet driver - holding low

xint i1 o1 i2 o2 i3 o3 i4 o4 i5 o5 interconnect
xdrv1 in1 i2 i1 i5 driver
xdrv2 i1 i3 i1 i5 driver
xdrv3 i5 i4 i1 i5 driver
xdrv4 in in1 i1 i5 inv

vvdd t1 0 3.3
Lsupply t1 o1 5n
vgnd o5 0 0
```

```
c1n o2 o5 25p
c2n o3 o5 25p
c3n o4 o5 25p

c1p o2 o1 25p
c2p o3 o1 25p
c3p o4 o1 25p

* off-chip bypass capacitance
cbypass1 o1 o5 10u

* on-chip bypass capacitance
*cbypass2 i1 i5 10n

vin in i5 pulse 0 3.3 1n .5n .5n 9n 20n

.control
tran 1n 50n
set width=100
set nobreak
print v(i1)-v(i5)+2,v(o2)+8,v(o3)+2,v(o4)+.6 > results.dat
quit
.endc

.end

.subckt driver in out ovdd ognd

* optimized to driver a 50pF load in 2ns over
* a 3.3 volt swing

* last stage
mp1 out in1 ovdd ovdd pmos_book l=1u w=1500u
mn1 out in1 ognd ognd nmos_book l=1u w=600u

* 2nd to last stage
mp2 in1 in2 ovdd ovdd pmos_book l=1u w=750u
```

```
mn2 in1 in2 ognd ognd nmos_book l=1u w=300u

* 3rd to last stage
mp3 in2 in3 ovdd ovdd pmos_book l=1u w=375u
mn3 in2 in3 ognd ognd nmos_book l=1u w=150u

* 4th to last stage
mp4 in3 in4 ovdd ovdd pmos_book l=1u w=188u
mn4 in3 in4 ognd ognd nmos_book l=1u w=75u

* 5th to last stage
mp5 in4 in5 ovdd ovdd pmos_book l=1u w=94u
mn5 in4 in5 ognd ognd nmos_book l=1u w=37u

* 6th to last stage
mp6 in5 in ovdd ovdd pmos_book l=1u w=47u
mn6 in5 in ognd ognd nmos_book l=1u w=19u


.ends driver

.subckt inv in out vdd gnd

mp1 out in vdd vdd pmos_book l=1u w=23u
mn1 out in gnd gnd nmos_book l=1u w=9u

.ends inv

* 5-wire linear interconnect

*  O   O   O   O   O
*  |<->|<->|<->|<->|
*   0.3 0.3 0.3 0.3  mm

* wires are 3 mm long
* spaced at 0.3 mm
* radius of 0.05 mm
```

```
* Use the round wire formula for the partial self inductances.
* Use the current filament approximation, centered on the wires,
* for the partial mutual inductances.

.subckt interconnect i1 o1 i2 o2 i3 o3 i4 o4 i5 o5
L1 i1 o1 2.272n
K12 L1 L2 0.5268
K13 L1 L3 0.3440
K14 L1 L4 0.2369
K15 L1 L5 0.1609
L2 i2 o2 2.272n
K23 L2 L3 0.5268
K24 L2 L4 0.3440
K25 L2 L5 0.2369
L3 i3 o3 2.272n
K34 L3 L4 0.5268
K35 L3 L5 0.3440
L4 i4 o4 2.272n
K45 L4 L5 0.5268
L5 i5 o5 2.272n
.ends interconnect

* NMOS example model for book simulations
* From Table 4.4 Bakoglu
* with LD=0
*
.model nmos_book nmos
+ LEVEL=1
+ VTO=0.85
+ KP=53U
+ GAMMA=0.175
+ PHI=0.56
+ LAMBDA=0.03
+ PB=0.89
+ CGSO=2.4e-10
+ CGDO=2.4e-10
```

```
+ CGBO=2.e-10
+ CJ=13e-5
+ CJSW=6.5e-10
+ NSUB=7e14
+ TOX=400e-10
*+ XJ=0.35e-6
+ LD=0.
+ UO=650

* PMOS example model for book simulations
* From Table 4.4 Bakoglu
* with LD=0
*
.model pmos_book pmos
+ LEVEL=1
+ VTO=-0.85
+ KP=21U
+ GAMMA=0.715
+ PHI=0.7
+ LAMBDA=0.066
+ PB=1.0
+ CGSO=3.4e-10
+ CGDO=3.4e-10
+ CGBO=2.e-10
+ CJ=23e-5
+ CJSW=15e-10
+ NSUB=1.1e16
+ TOX=400e-10
*+ XJ=0.5e-6
+ LD=0.
+ UO=240
```

Appendix D

SAMPLE MODAL DECOMPOSITION CODE

```
#include <stdio.h>
#include <stdlib.h>
#include <string.h>
#include <matrix2.h>
#include <math.h>

/*
   Requires the Meschach library of mathematical subroutines
   available from the netlib2.cs.utk.edu archive.
*/

void eigen (MAT*, MAT*, MAT*, VEC*, VEC*);
void get_mn (FILE*, int*, int*);
int eigen_real (MAT* ,MAT* ,VEC* );
void subckt_line (FILE *,char *,int ,int,char *,char *);
void call_line (FILE *,char *,int ,int ,int ,int ,int );

double Length=.2;       /* length of multiconductor transmission line in m */
double tr_min=1e-9;     /* minimum edge rate the lumped model will support */
int wrap=5;             /* number of node pairs per line in subcircuit     */
#define MAXNORM 1e-14   /* maximum allowed error in matrix checks          */

main ()
```

```
{
   FILE *fp;
   MAT *C,*L,*LC,*CL,*Tv,*Ti,*iTv,*iTi,*T1,*T2,*Lm,*Cm;
   VEC *lv,*li;
   char node[10];
   double Z0,TD,vp,vpmax,sub_Length,Cgnd;
   int i,j,k,m,n,p,q,n_sub;

   /* get the inductance matrix */
   fp = fopen("L.matrix","r");
   get_mn (fp,&m,&n);
   L=m_get(m,n);
   m_finput(fp,L);
   fclose (fp);
   if (m != n) { puts ("m not equal to n in L.\n"); exit(1); }
   i=m;

   /* get the capacitance matrix */
   fp = fopen("C.matrix","r");
   get_mn (fp,&m,&n);
   C=m_get(m,n);
   m_finput(fp,C);
   fclose (fp);
   if (m != n) { puts ("m not equal to n in C.\n"); exit(1); }
   if (m != i) { puts ("L and C are not the same size.\n"); exit(1); }

   LC=m_get(m,n); CL=m_get(m,n); Tv=m_get(m,n); Ti=m_get(m,n);
   lv=v_get(m); li=v_get(m);

   m_mlt (L,C,LC);  m_mlt (C,L,CL);

   if(eigen_real(LC,Tv,lv) != 0) exit (1);
   if(eigen_real(CL,Ti,li) != 0) exit (1);

   /* compute the inverse transformation */
   iTv=m_get(m,n);  iTi=m_get(m,n);
```

```
m_inverse (Tv,iTv);  m_inverse (Ti,iTi);

/* find the modal inductances and capacitances */
Lm=m_get(m,n);  Cm=m_get(m,n);
T1=m_get(m,n);  T2=m_get(m,n);

m_mlt(iTv,L,T2);  m_mlt(T2,Ti,Lm);

i=0;
do
{
   Lm->me[i][i]=fabs(Lm->me[i][i]);  /* eliminate negative inductances */
   i++;
} while (i<m);

/* recompute Tv to eliminate sign ambiguity */
m_inverse (Lm,T2);  m_mlt(L,Ti,T1);  m_mlt(T1,T2,Tv);

m_inverse (Tv,iTv);

/* recompute Cm with new Tv */
m_mlt(iTi,C,T2);  m_mlt(T2,Tv,Cm);

/****************************************************/
/* write the spice netlist for modal decomposition */
/****************************************************/

fp = fopen("mctl.ckt","w");
subckt_line (fp,"mctl",m,1,"i","o");

/* put in the voltage sources on the inputs and outputs */
i=0;
do
{
   j=0; p=0; q=0;
   do
   {
```

```
      fprintf (fp,"EL%d_%d TL%d_%d TL%d_%d %d 0 %e\n",
          i,j,i,p,i,p+1,j+1,Tv->me[i][j]);
      p++; q++; j++;
    } while (j<m);
    fprintf (fp,"VL%d_1 TL%d_0 i%d 0\n",i,i,i+1);
    fprintf (fp,"VL%d_2 TL%d_%d 0 0\n",i,i,m);
    i++;
  } while (i<m);
  fputs ("\n",fp);

  i=0;
  do
  {
    j=0; p=0; q=0;
    do
    {
      fprintf (fp,"ER%d_%d TR%d_%d TR%d_%d %d 0 %e\n",
          i,j,i,p,i,p+1,j+1+m,Tv->me[i][j]);
      p++; q++; j++;
    } while (j<m);
    fprintf (fp,"VR%d_1 TR%d_0 o%d 0\n",i,i,i+1);
    fprintf (fp,"VR%d_2 TR%d_%d 0 0\n",i,i,m);
    i++;
  } while (i<m);
  fputs ("\n",fp);

  /* put in the decoupled tranmsission lines */

  i=0;
  do
  {

    Z0=sqrt(Lm->me[i][i]/Cm->me[i][i]);
    TD=Length*sqrt(Lm->me[i][i]*Cm->me[i][i]);
    fprintf (fp,"T%d %d 0 %d 0 Z0=%f TD=%e\n",i,i+1,i+1+m,Z0,TD);
    i++;
  } while (i<m);
```

```
fputs ("\n",fp);

/* put in the current sources */
i=0;
do
{
  j=0;
  do
  {
      fprintf (fp,"FL%d_%d 0 %d EL%d_0 %e\n",i,j,i+1,j,iTi->me[i][j]);
    j++;
  } while (j<m);
  fputs ("\n",fp);
  i++;
} while (i<m);

i=0;
do
{
  j=0;
  do
  {
      fprintf (fp,"FR%d_%d 0 %d ER%d_0 %e\n",i,j,i+1+m,j,iTi->me[i][j]);
    j++;
  } while (j<m);
  fputs ("\n",fp);
  i++;
} while (i<m);

fputs (".ends mctl\n",fp);
fclose (fp);

/**********************************************/
/* write the spice netlist for a lumped model */
/**********************************************/

/* find the max modal phase velocity */
```

```
i=0;
vpmax=0;
do
{
   vp=1/sqrt(Lm->me[i][i]*Cm->me[i][i]);
   if (vp > vpmax) vpmax=vp;
   i++;
} while (i<m);

/* use the fastest edge rate to find the length of the subsections */
sub_Length=0.1*vpmax*tr_min;
n_sub=Length/sub_Length;
if (n_sub == 0) n_sub=1;
sub_Length=Length/n_sub;
printf ("Number of subsections = %d\n",n_sub);

fp=fopen ("lumped.ckt","w");

/* plan on a T-circuit, so output the inductance matrix *.5 */
subckt_line (fp,"Lmatrix_by_2",m,1,"i","o");
i=0;
do
{
  fprintf (fp,"L%d i%d o%d %e\n",i+1,i+1,i+1,L->me[i][i]*sub_Length/2);
  j=i+1;
  while (j<m)
  {
    fprintf (fp,"K%d_%d L%d L%d %e\n",i+1,j+1,i+1,j+1,
        L->me[i][j]/sqrt(L->me[i][i]*L->me[j][j]));
    j++;
  }
  i++;
} while (i<m);
fputs (".ends Lmatrix_by_2\n\n",fp);

/* inductance matrix */
subckt_line (fp,"Lmatrix",m,1,"i","o");
```

```
i=0;
do
{
  fprintf (fp,"L%d i%d o%d %e\n",i+1,i+1,i+1,L->me[i][i]*sub_Length);
  j=i+1;
  while (j<m)
  {
    fprintf (fp,"K%d_%d L%d L%d %e\n",i+1,j+1,i+1,j+1,
        L->me[i][j]/sqrt(L->me[i][i]*L->me[j][j]));
    j++;
  }
  i++;
} while (i<m);
fputs (".ends Lmatrix\n\n",fp);

/* the capacitance matrix */
subckt_line (fp,"Cmatrix",m,2,"i","o");
i=0;
do
{
  j=0;
  Cgnd=0;
  do
  {
      Cgnd=Cgnd+C->me[i][j];
      j++;
  } while (j<m);
  fprintf (fp,"C%d_0 i%d 0 %e\n",i+1,i+1,Cgnd*sub_Length);
  j=i+1;
  while (j<m)
  {
    fprintf (fp,"C%d_%d i%d i%d %e\n",i+1,j+1,i+1,j+1,
        -1*C->me[i][j]*sub_Length);
    j++;
  }
  i++;
} while (i<m);
```

```
   fputs (".ends Cmatrix\n\n",fp);

   /* cascade to create the the transmission line */
   sprintf (node,"%d_",n_sub+1);
   subckt_line (fp,"mctl",m,1,"1_",node);
   call_line (fp,"Lmatrix_by_2",1,m,1,1,2);

   i=1;
   while (i<n_sub-1)
   {
      i++;
      call_line (fp,"Cmatrix",i,m,2,i,i);
      call_line (fp,"Lmatrix",i,m,1,i,i+1);
   }
   i++;
   call_line (fp,"Cmatrix",i,m,2,i,i);
   call_line (fp,"Lmatrix_by_2",i,m,1,i,i+1);

   fputs (".ends mctl\n\n",fp);
   fclose (fp);
}

/* output the subcircuit call line */
void subckt_line (FILE *fp,char name[],int m,int option,char n1[],char n2[])
{
   int i,j;

   fprintf (fp,".subckt %s",name);
   j=0; i=0;
   do
   {
      if (option==1) fprintf (fp," %s%d %s%d",n1,i+1,n2,i+1);
      if (option==2) fprintf (fp," %s%d",n1,i+1);
      if (j++ == wrap-1 && i+1<m) { fputs("\n+    ",fp); j=0; }
      i++;
   } while (i<m);
   fputs ("\n",fp);
```

```
}

/* output a line of subcircuit call */
void call_line (FILE *fp, char subckt[], int instance,
                 int m, int option, int n1, int n2)
{
   int i,j;

   fprintf (fp,"X%s%d",subckt,instance);
   i=0; j=0;
   do
   {
      if (option==1) fprintf (fp," %d_%d %d_%d",n1,i+1,n2,i+1);
      if (option==2) fprintf (fp," %d_%d",n1,i+1);
      if (j++==wrap-1 && i+1<m) {fputs("\n+   ",fp); j=0;}
      i++;
   } while (i<m);
   fprintf (fp," %s\n",subckt);
}

/* scan the standard matrix format for the array size */
void get_mn (FILE *fp, int *m, int *n)
{
  char line[100],word[100];

  fseek (fp,0,SEEK_SET);
  while (fgets(line,100,fp) != NULL)
  {
     sscanf (line,"%s",word);
     if (strcmp(word,"Matrix:") == 0)
     {
         sscanf (line,"%s%d%s%d",word,m,word,n);
         fseek (fp,0,SEEK_SET);
         return;
     }
  }
}
```

```
/* solve the eigenvalue problem and do some checks */
int eigen_real (MAT *A,MAT *A_re,VEC *ei_re)
{
   MAT *A_im,*diag,*T1,*T2;
   VEC *ei_im;
   int i,m,n;

   m=A->m; n=A->n;

   A_im=m_get(m,n);  ei_im=v_get(m);
   eigen (A,A_re,A_im,ei_re,ei_im);   /* solve eigenvalue problem */

   /* check for imaginary parts */
   if (m_norm_inf(A_im) > 1.e-14) return 1;
   if (v_norm_inf(ei_im) > 1.e-14) return 2;

   /* put eigenvalues into a diagonal matrix */
   diag=m_get(m,n);
   i=0;
   do { diag->me[i][i]=ei_re->ve[i];  i++; } while (i < m);

   /* check for orthonormal properties */
   T1=m_get(m,n);
   T2=m_get(m,n);
   m_inverse(A_re,T1);
   m_mlt(A_re,diag,T2);
   m_mlt(T2,T1,diag);

   m_inverse(diag,T1);
   m_mlt(A,T1,T2);
   i=0;
   do { T2->me[i][i]-=1; i++; } while (i<m);
   if (m_norm_inf(T2) > MAXNORM) return 3;
   return 0;
}
```

```
void eigen (MAT *A, MAT *X_re, MAT *X_im, VEC *evals_re, VEC *evals_im)
{
  MAT *T,*Q;

/* A is the matrix whose eigenvalues and eigenvectors are sought */
/* k'th eigenvector is k'th column of (X_re + i*X_im) */

  Q = m_get(A->m,A->n);
  T = m_copy(A,MNULL);

  schur(T,Q);                       /* compute Schur form: A = Q*T*Q^T */
  schur_evals(T,evals_re,evals_im); /* extract eigenvalues */
  schur_vecs(T,Q,X_re,X_im);        /* Q not needed for eiegenvalues */
}

* spice subcircuit for modal decomposition
.subckt mctl i1 o1 i2 o2 i3 o3 i4 o4 i5 o5
EL0_0 TL0_0 TL0_1 1 0 3.651740e-01
EL0_1 TL0_1 TL0_2 2 0 5.499348e-01
EL0_2 TL0_2 TL0_3 3 0 5.570483e-01
EL0_3 TL0_3 TL0_4 4 0 4.259384e-01
EL0_4 TL0_4 TL0_5 5 0 -2.300951e-01
VL0_1 TL0_0 i1 0
VL0_2 TL0_5 0 0
EL1_0 TL1_0 TL1_1 1 0 4.795472e-01
EL1_1 TL1_1 TL1_2 2 0 -1.064700e-01
EL1_2 TL1_2 TL1_3 3 0 4.361856e-01
EL1_3 TL1_3 TL1_4 4 0 -5.627495e-01
EL1_4 TL1_4 TL1_5 5 0 5.158394e-01
VL1_1 TL1_0 i2 0
VL1_2 TL1_5 0 0
EL2_0 TL2_0 TL2_1 1 0 5.198346e-01
EL2_1 TL2_1 TL2_2 2 0 -6.073649e-01
EL2_2 TL2_2 TL2_3 3 0 2.583636e-03
EL2_3 TL2_3 TL2_4 4 0 -8.657632e-03
EL2_4 TL2_4 TL2_5 5 0 -6.117483e-01
VL2_1 TL2_0 i3 0
```

```
VL2_2 TL2_5 0 0
EL3_0 TL3_0 TL3_1 1 0 4.811427e-01
EL3_1 TL3_1 TL3_2 2 0 -1.160280e-01
EL3_2 TL3_2 TL3_3 3 0 -4.328936e-01
EL3_3 TL3_3 TL3_4 4 0 5.692112e-01
EL3_4 TL3_4 TL3_5 5 0 5.072893e-01
VL3_1 TL3_0 i4 0
VL3_2 TL3_5 0 0
EL4_0 TL4_0 TL4_1 1 0 3.673636e-01
EL4_1 TL4_1 TL4_2 2 0 5.512542e-01
EL4_2 TL4_2 TL4_3 3 0 -5.586016e-01
EL4_3 TL4_3 TL4_4 4 0 -4.216790e-01
EL4_4 TL4_4 TL4_5 5 0 -2.222332e-01
VL4_1 TL4_0 i5 0
VL4_2 TL4_5 0 0

ER0_0 TR0_0 TR0_1 6 0 3.651740e-01
ER0_1 TR0_1 TR0_2 7 0 5.499348e-01
ER0_2 TR0_2 TR0_3 8 0 5.570483e-01
ER0_3 TR0_3 TR0_4 9 0 4.259384e-01
ER0_4 TR0_4 TR0_5 10 0 -2.300951e-01
VR0_1 TR0_0 o1 0
VR0_2 TR0_5 0 0
ER1_0 TR1_0 TR1_1 6 0 4.795472e-01
ER1_1 TR1_1 TR1_2 7 0 -1.064700e-01
ER1_2 TR1_2 TR1_3 8 0 4.361856e-01
ER1_3 TR1_3 TR1_4 9 0 -5.627495e-01
ER1_4 TR1_4 TR1_5 10 0 5.158394e-01
VR1_1 TR1_0 o2 0
VR1_2 TR1_5 0 0
ER2_0 TR2_0 TR2_1 6 0 5.198346e-01
ER2_1 TR2_1 TR2_2 7 0 -6.073649e-01
ER2_2 TR2_2 TR2_3 8 0 2.583636e-03
ER2_3 TR2_3 TR2_4 9 0 -8.657632e-03
ER2_4 TR2_4 TR2_5 10 0 -6.117483e-01
VR2_1 TR2_0 o3 0
VR2_2 TR2_5 0 0
```

```
ER3_0 TR3_0 TR3_1 6 0 4.811427e-01
ER3_1 TR3_1 TR3_2 7 0 -1.160280e-01
ER3_2 TR3_2 TR3_3 8 0 -4.328936e-01
ER3_3 TR3_3 TR3_4 9 0 5.692112e-01
ER3_4 TR3_4 TR3_5 10 0 5.072893e-01
VR3_1 TR3_0 o4 0
VR3_2 TR3_5 0 0
ER4_0 TR4_0 TR4_1 6 0 3.673636e-01
ER4_1 TR4_1 TR4_2 7 0 5.512542e-01
ER4_2 TR4_2 TR4_3 8 0 -5.586016e-01
ER4_3 TR4_3 TR4_4 9 0 -4.216790e-01
ER4_4 TR4_4 TR4_5 10 0 -2.222332e-01
VR4_1 TR4_0 o5 0
VR4_2 TR4_5 0 0

T0 1 0 6 0 Z0=91.448571 TD=1.211573e-09
T1 2 0 7 0 Z0=62.786004 TD=1.067471e-09
T2 3 0 8 0 Z0=75.211097 TD=1.126285e-09
T3 4 0 9 0 Z0=54.415470 TD=1.031954e-09
T4 5 0 10 0 Z0=49.699831 TD=1.013872e-09

FL0_0 0 1 EL0_0 3.652535e-01
FL0_1 0 1 EL1_0 4.796516e-01
FL0_2 0 1 EL2_0 5.199477e-01
FL0_3 0 1 EL3_0 4.812474e-01
FL0_4 0 1 EL4_0 3.674436e-01

FL1_0 0 2 EL0_0 5.500591e-01
FL1_1 0 2 EL1_0 -1.064941e-01
FL1_2 0 2 EL2_0 -6.075022e-01
FL1_3 0 2 EL3_0 -1.160542e-01
FL1_4 0 2 EL4_0 5.513788e-01

FL2_0 0 3 EL0_0 5.571499e-01
FL2_1 0 3 EL1_0 4.362651e-01
FL2_2 0 3 EL2_0 2.584107e-03
FL2_3 0 3 EL3_0 -4.329725e-01
```

```
FL2_4 0 3 EL4_0 -5.587034e-01

FL3_0 0 4 EL0_0 4.260161e-01
FL3_1 0 4 EL1_0 -5.628523e-01
FL3_2 0 4 EL2_0 -8.659213e-03
FL3_3 0 4 EL3_0 5.693151e-01
FL3_4 0 4 EL4_0 -4.217560e-01

FL4_0 0 5 EL0_0 -2.301111e-01
FL4_1 0 5 EL1_0 5.158755e-01
FL4_2 0 5 EL2_0 -6.117910e-01
FL4_3 0 5 EL3_0 5.073247e-01
FL4_4 0 5 EL4_0 -2.222487e-01

FR0_0 0 6 ER0_0 3.652535e-01
FR0_1 0 6 ER1_0 4.796516e-01
FR0_2 0 6 ER2_0 5.199477e-01
FR0_3 0 6 ER3_0 4.812474e-01
FR0_4 0 6 ER4_0 3.674436e-01

FR1_0 0 7 ER0_0 5.500591e-01
FR1_1 0 7 ER1_0 -1.064941e-01
FR1_2 0 7 ER2_0 -6.075022e-01
FR1_3 0 7 ER3_0 -1.160542e-01
FR1_4 0 7 ER4_0 5.513788e-01

FR2_0 0 8 ER0_0 5.571499e-01
FR2_1 0 8 ER1_0 4.362651e-01
FR2_2 0 8 ER2_0 2.584107e-03
FR2_3 0 8 ER3_0 -4.329725e-01
FR2_4 0 8 ER4_0 -5.587034e-01

FR3_0 0 9 ER0_0 4.260161e-01
FR3_1 0 9 ER1_0 -5.628523e-01
FR3_2 0 9 ER2_0 -8.659213e-03
FR3_3 0 9 ER3_0 5.693151e-01
FR3_4 0 9 ER4_0 -4.217560e-01
```

```
FR4_0 0 10 ER0_0 -2.301111e-01
FR4_1 0 10 ER1_0 5.158755e-01
FR4_2 0 10 ER2_0 -6.117910e-01
FR4_3 0 10 ER3_0 5.073247e-01
FR4_4 0 10 ER4_0 -2.222487e-01

.ends mctl

* inductance and capacitance matrices for lumped approximation
.subckt Lmatrix_by_2 i1 o1 i2 o2 i3 o3 i4 o4 i5 o5
L1 i1 o1 3.710000e-09
K1_2 L1 L2 2.250156e-01
K1_3 L1 L3 7.654063e-02
K1_4 L1 L4 3.380271e-02
K1_5 L1 L5 1.771760e-02
L2 i2 o2 3.678000e-09
K2_3 L2 L3 2.233583e-01
K2_4 L2 L4 7.605328e-02
K2_5 L2 L5 3.377573e-02
L3 i3 o3 3.677000e-09
K3_4 L3 L4 2.233519e-01
K3_5 L3 L5 7.649294e-02
L4 i4 o4 3.680000e-09
K4_5 L4 L5 2.249209e-01
L5 i5 o5 3.712000e-09
.ends Lmatrix_by_2

.subckt Lmatrix i1 o1 i2 o2 i3 o3 i4 o4 i5 o5
L1 i1 o1 7.420000e-09
K1_2 L1 L2 2.250156e-01
K1_3 L1 L3 7.654063e-02
K1_4 L1 L4 3.380271e-02
K1_5 L1 L5 1.771760e-02
L2 i2 o2 7.356000e-09
K2_3 L2 L3 2.233583e-01
K2_4 L2 L4 7.605328e-02
```

```
K2_5 L2 L5 3.377573e-02
L3 i3 o3 7.354000e-09
K3_4 L3 L4 2.233519e-01
K3_5 L3 L5 7.649294e-02
L4 i4 o4 7.360000e-09
K4_5 L4 L5 2.249209e-01
L5 i5 o5 7.424000e-09
.ends Lmatrix

.subckt Cmatrix i1 i2 i3 i4 i5
C1_0 i1 0 1.441976e-12
C1_2 i1 i2 2.066000e-13
C1_3 i1 i3 1.465600e-14
C1_4 i1 i4 5.744000e-15
C1_5 i1 i5 3.224000e-15
C2_0 i2 0 1.274436e-12
C2_3 i2 i3 2.042000e-13
C2_4 i2 i4 1.383000e-14
C2_5 i2 i5 5.734000e-15
C3_0 i3 0 1.267930e-12
C3_4 i3 i4 2.040000e-13
C3_5 i3 i5 1.461400e-14
C4_0 i4 0 1.274226e-12
C4_5 i4 i5 2.068000e-13
C5_0 i5 0 1.443028e-12
.ends Cmatrix

.subckt mctl 1_1 11_1 1_2 11_2 1_3 11_3 1_4 11_4 1_5 11_5
XLmatrix_by_21 1_1 2_1 1_2 2_2 1_3 2_3 1_4 2_4 1_5 2_5 Lmatrix_by_2
XCmatrix2 2_1 2_2 2_3 2_4 2_5 Cmatrix
XLmatrix2 2_1 3_1 2_2 3_2 2_3 3_3 2_4 3_4 2_5 3_5 Lmatrix
XCmatrix3 3_1 3_2 3_3 3_4 3_5 Cmatrix
XLmatrix3 3_1 4_1 3_2 4_2 3_3 4_3 3_4 4_4 3_5 4_5 Lmatrix
XCmatrix4 4_1 4_2 4_3 4_4 4_5 Cmatrix
XLmatrix4 4_1 5_1 4_2 5_2 4_3 5_3 4_4 5_4 4_5 5_5 Lmatrix
XCmatrix5 5_1 5_2 5_3 5_4 5_5 Cmatrix
XLmatrix5 5_1 6_1 5_2 6_2 5_3 6_3 5_4 6_4 5_5 6_5 Lmatrix
```

```
XCmatrix6 6_1 6_2 6_3 6_4 6_5 Cmatrix
XLmatrix6 6_1 7_1 6_2 7_2 6_3 7_3 6_4 7_4 6_5 7_5 Lmatrix
XCmatrix7 7_1 7_2 7_3 7_4 7_5 Cmatrix
XLmatrix7 7_1 8_1 7_2 8_2 7_3 8_3 7_4 8_4 7_5 8_5 Lmatrix
XCmatrix8 8_1 8_2 8_3 8_4 8_5 Cmatrix
XLmatrix8 8_1 9_1 8_2 9_2 8_3 9_3 8_4 9_4 8_5 9_5 Lmatrix
XCmatrix9 9_1 9_2 9_3 9_4 9_5 Cmatrix
XLmatrix9 9_1 10_1 9_2 10_2 9_3 10_3 9_4 10_4 9_5 10_5 Lmatrix
XCmatrix10 10_1 10_2 10_3 10_4 10_5 Cmatrix
XLmatrix_by_210 10_1 11_1 10_2 11_2 10_3 11_3 10_4 11_4 10_5 11_5 Lmatrix_by_2
.ends mctl

* Modal decomposition demonstration circuit

* choose the interconnect model
.include lumped.ckt
*.include mctl.ckt

xint i1 o1 i2 o2 i3 o3 i4 o4 i5 o5 mctl
v1 n1 0 pulse 0 3.3 1n 2n 2n 8n 20n
r1 n1 i1 50
v2 n2 0 pulse 0 3.3 1n 2n 2n 8n 20n
r2 n2 i2 50
r3 i3 0 50
v4 n4 0 pulse 0 3.3 1n 2n 2n 8n 20n
r4 n4 i4 50
v5 n5 0 pulse 0 3.3 1n 2n 2n 8n 20n
r5 n5 i5 50

* capacitive loads
c1 o1 0 6p
c2 o2 0 6p
c3 o3 0 6p
c4 o4 0 6p
c5 o5 0 6p

.control
```

```
tran 1n 61n
set width=150
set nobreak
print i1,i2,i3,i4,i5 > results_i.dat
print o1,o2,o3,o4,o5 > results_o.dat
quit
.endc

.end
```

Appendix E

SAMPLE LAYER PEELING CODE

```
#include <stdlib.h>
#include <string.h>
#include <stdio.h>
#include <math.h>

/* Sample layer peeling program for removing multiple reflections
   from TDR traces. */

int MAXPOINTS=1024;  /* Maximum number of data points */
int MAXLINE=100;     /* Maximum number of characters per line
                        in the data files */
double sysZo=50;     /* characteristic impedance of the TDR */

int main(int argc, char *argv[])
{
  FILE *fp;
  char *input,*match,*output,line[100];
  int i,j,maxdata,istart;
  double time[MAXPOINTS],vpL[MAXPOINTS],vnL[MAXPOINTS],
         vpR[MAXPOINTS],vnR[MAXPOINTS],
         Zo[MAXPOINTS],Zraw[MAXPOINTS],vr[MAXPOINTS],vm[MAXPOINTS];
```

```
if (argc != 4)
{
   printf ("usage: peel inputfile matchedfile outputfile\n\n");
   exit (0);
}

input=argv[1];
match=argv[2];
output=argv[3];

/* load the input data */
fp=fopen(input,"r");
i=0;
while (fgets(line,MAXLINE,fp) != NULL)
{
   /* assumes that the line contains: integer time step number,
      double time, double voltage */
   /* no blank lines are allowed */
   if (i < MAXPOINTS) sscanf (line,"%d%lf%lf",&j,&time[i],&vr[i]);
   i++;
}
fclose (fp);
if (i >= MAXPOINTS) printf ("Warning: too many data points.");
maxdata=i;

/* load the matched data */
fp=fopen(match,"r");
i=0;
while (fgets(line,MAXLINE,fp) != NULL)
{
   /* assumes that the line contains: integer time step number,
      double time, double voltage */
   if (i < MAXPOINTS) sscanf (line,"%d%lf%lf",&j,&time[i],&vm[i]);
   i++;
}
fclose (fp);
if (i >= MAXPOINTS) printf ("Warning: too many data points.");
```

```
/* find the incident and reflected waves */
i=0;
while (i < maxdata)
{
  vpL[i]=vm[i];
  vnL[i]=vr[i]-vm[i];
  i++;
}

/* assume that the first set of data is already on the right side */
i=0;
while (i < maxdata) {vpR[i]=vpL[i]; vnR[i]=vnL[i]; i++;}

/* find the first non-zero data */
i=0;
while (vpR[i] == 0) i++;
istart=i;

/* peeling loop for the remaining sections */
i=istart; Zo[istart]=sysZo;
while (i < maxdata-1)
{
  /* Step 1: propagate across the section (skip if first set of data) */
  if (i != istart)
  {
    j=0;
    while (j < maxdata-1)
    {
      vpR[j+1]=vpL[j];
      if (j > 0) vnR[j-1]=vnL[j];
      j++;
    }
  }

  /* Step 2: compute next characteristic impedance */
  Zo[i+1]=Zo[i]*(vpR[i]+vnR[i])/(vpR[i]-vnR[i]);
```

```
    if (i == 0) Zo[i+1]=sysZo;

    /* Step 3: cross the boundary */
    j=i;
    while (j < maxdata)
    {
      vpL[j]=(1+Zo[i+1]/Zo[i])/2*vpR[j]+(1-Zo[i+1]/Zo[i])/2*vnR[j];
      vnL[j]=(1-Zo[i+1]/Zo[i])/2*vpR[j]+(1+Zo[i+1]/Zo[i])/2*vnR[j];
      j++;
    }

    i++;
  }

  /* compute the basic TDR profile */
  i=istart;
  while (i < maxdata)
  {
     Zraw[i]=sysZo*vr[i]/(2*vm[i]-vr[i]);
     i++;
  }

  /* save the basic TDR impedance profile and the peeled profile */
  fp=fopen(output,"w");
  fprintf (fp,"# Time   Zo(raw)  Zo(Peeled)\n");
  i=istart;
  while (i < maxdata/2)
  {
     fprintf (fp,"%e %e %e\n",time[i],Zraw[2*i],Zo[i]);
     i++;
  }
  fclose (fp);

}
```

INDEX

ABOUT THE AUTHOR

Brian Young is a Member of the Technical Staff at the Somerset Design Center, Semiconductor Product Sector, Motorola, working on packaging, interconnects, and I/O design for PowerPC^{TM} microprocessors and the RapidIO^{TM} Interconnect Architecture. For over seven years he has specialized in simulation, modeling, measurement, and performance studies for high-speed signaling with microprocessors, fast static RAMs, and DSPs. He has served as an Assistant Professor in the Department of Electrical Engineering at Texas A&M University, College Station, and as an adjunct professor in the Department of Electrical Engineering at the University of Texas, Austin. Dr. Young holds a Ph.D. from the University of Texas, Austin, and holds six patents related to packaging. He has published numerous articles in conferences and journals.